DATE DUE

DEMCO, INC. 38-2931

AN INTRODUCTION TO FUZZY SETS

AN INTRODUCTION TO FUZZY SETS
Analysis and Design

Witold Pedrycz and Fernando Gomide

A Bradford Book
The MIT Press
Cambridge, Massachusetts
London, England

This book was set in Times New Roman on the Monotype "Prism Plus" PostScript Image-setter by Asco Trade Typesetting Ltd., Hong Kong and was printed and bound in the United States of America.

Library of Congress Cataloging-in-Publication Data

Pedrycz, Witold, 1953–
     An introduction to fuzzy sets : analysis and design / Witold
Pedrycz and Fernando Gomide.
        p.   cm. — (Complex adaptive systems)
     "A Bradford book."
     Includes bibliographical references and index.
     ISBN 0-262-16171-0 (hc)
     1. Fuzzy sets. I. Gomide, Fernando. II. Title. III. Series.
QA248.5.P38   1998
006.3'01'51132—dc21   97-34598
                                                              CIP

To Ewa, Thais, Adam, Tiago, Barbara, Flavia, and Arthur

Contents

Foreword

As an introduction to the theory of fuzzy sets and fuzzy logic, *An Introduction to Fuzzy Sets: Analysis and Design*, by Professors Witold Pedrycz and Fernando Gomide, leaves nothing to be desired. Written by two of the leading contributors to fuzzy logic and its applications, the book combines a very lucid and authoritative exposition of the fundamentals of fuzzy sets and fuzzy logic with an insightful discussion of their applications, illustrated by numerous examples and case studies.

Venturing beyond fuzzy sets and fuzzy logic, the authors present new results of their own in chapters on fuzzy neurocomputation and fuzzy evolutionary computation. In so doing, they reflect a recent trend to view fuzzy logic (FL), neurocomputing (NC), genetic computing (GC), and probabilistic computing (PC) as an association of computing methodologies falling under the rubric of so-called soft computing. The essence of soft computing is that its constituent methodologies are for the most part complementary and synergistic rather than competitive. A concomitant of the concept of soft computing is that in many situations it is advantageous to employ FL, NC, GC, and PC in combination rather than isolation. At this juncture, so-called neurofuzzy systems are the most visible examples of systems of this type. The chapter on fuzzy neurocomputation in this work shows very clearly how a synergistic relation between fuzzy logic and neural network theory can be developed and applied.

Drawing on their mastery of fuzzy logic and incisive understanding of other methodologies for dealing with uncertainty and imprecision, Professors Pedrycz and Gomide provide many insights into the connections between fuzzy set theory, fuzzy logic, multivalued logic, and probability theory. I should like to add to their insights a perception that I have been articulating in some of my recent papers.

Among the basic concepts that underlie human cognition, three stand out in importance: granulation, organization, and causation. Informally, granulation involves a partitioning of a whole into parts; organization involves an integration of parts into a whole; and causation relates to an association of causes with effects.

A granule may be viewed as a clump of points (objects) drawn together by indistinguishability, similarity, or functionality. Modes of information granulation (IG) in which granules are crisp play an important role in many theories, methods and techniques, among them interval analysis, quantization, rough set theory, qualitative process theory, and chunking. What these theories do not reflect, however, is that in much, perhaps most, of human reasoning and concept formation the granules are fuzzy, as are their attributes and attribute values. Nor they reflect the fact that fuzzy IG plays a pivotal role in the remarkable human ability to make rational decisions in an environment of partial knowledge, partial certainty, and partial truth.

In fuzzy logic, fuzzy IG underlies the basic concepts of linguistic variables, fuzzy if-then rules, and fuzzy graphs—concepts that this work discusses very lucidly and in detail. Furthermore, what we see herein is that the machinery of fuzzy IG plays a central role in most applications of fuzzy logic. But what is less obvious is that fuzzy IG is not merely an important part of fuzzy logic, it is its quintessence; and that no methodology other than FL provides machinery for dealing with fuzzy information granulation in ways that parallel human reasoning and decision-making processes.

This perception is reinforced by viewing it in the context of generalization. More specifically, any theory, method, technique, or problem may be fuzzified (or f-generalized) by replacing the concept of a crisp set with that of a fuzzy set. Similarly, any theory, method, technique, or problem can be granulated (g-generalized) by partitioning variables, functions, and relations into granules. Furthermore, we can combine fuzzification with granulation, which gives rise to fuzzy granulation (f-granulation). Fuzzy granulation, then, provides a basis for what might be called f.g-generalization.

The generalization of two-valued logic leads to multivalued logic and parts of fuzzy logic. But fuzzy logic in its wide sense—which is the sense in which it is used today—results from f.g-generalization. This crucial difference between multivalued logic and fuzzy logic explains why fuzzy logic has so many applications, whereas multivalued logic does not.

It is my belief that in coming years, f.g-generalization of various theories, methods, and techniques will have a wide-ranging impact on the evolution of scientific disciplines and their applications. The theories developed and described in this book constitute an important step in this direction.

In sum, Professors Pedrycz and Gomide have produced a text that makes a major contribution to a better understanding of how fuzzy set theory and fuzzy logic address the basic issues of fuzzification and fuzzy granulation. Without exaggeration, *An Introduction to Fuzzy Sets* is must reading for anyone interested in applying fuzzy-set theory, fuzzy logic, and soft computing to the solution of real-world problems and, in particular, to the conception, design, and use of intelligent information and control systems.

Lotfi A. Zadeh
Berkeley, California
January 21, 1997

Preface

Fuzzy sets constitute one of computational intelligence's most fundamental and influential tools. The concept of fuzzy sets is rather new and intellectually stimulating, and their applications diverse and advanced. With the rapidly growing number of industrial endeavors involving fuzzy sets, the need arises for a comprehensive, thoroughly organized and fully updated textbook. This book is intended to fill the wide gap that has developed in recent years between the theory and practice. It provides a highly readable and systematic exposition of the fundamentals of fuzzy sets along with a coherent presentation of sound, comprehensive analysis and design practices. It also explores new insights concerning the areas in which fuzzy set technology has already assumed a sound position, including system modeling, analysis, and design, and several applications of intelligent systems.

Philosophy

The text's main objective is to explain what fuzzy sets are, why they work, when these concepts should (and shouldn't) be used, and how to design systems efficiently using the technology of fuzzy sets. Its underlying didactic philosophy is that the theory should be strongly supported first by applying the reader's intuition to the multitude of linguistic concepts he or she naturally finds and handles in everyday life, and second, by exposing the reader to carefully selected illustrative examples followed by real-world applications. At the same time, the theory should be concise, general and clear enough to allow the reader to appreciate its importance and usefulness.

Because the book is intended mainly to serve as a useful text, numerous illustrative examples at different levels of complexity are prudently distributed across the chapters. Moreover, each chapter contains a series of exercises and problems. The exercises allow and motivate the reader to recall important material and to test his or her basic skills directly concerning the chapter content. The problems consolidate chapter material by requiring the reader to process and apply that material in practice, and propose more advanced topics requiring simulation experiments and careful analysis. In general, some of these experiments could move the reader beyond the core material the chapter presents. In contrast with many past approaches, we are convinced that the reader does not need a flood of definitions and excessive verbosity. Fuzzy sets cry not for a mathematical dignification, but for solid pragmatic fundamentals. They require, as well a sound design approach fully revealing and exploiting their essential properties.

Significant Features

A number of features make the book distinct from others already available:

- a balanced introduction to fuzzy sets emphasizing equally analysis and design aspects of this technology
- a carefully planned organization of the material into three main parts, allowing for easy use of the text
- self-containment of the material, with no prerequisites except for the basic knowledge of introductory calculus and linear algebra
- design-oriented approach toward the utilization of fuzzy sets
- prudently organized exercises distributed throughout the text and more advanced and holistic problems following each chapter
- presentation of formal theory without any unjustified or excessive use of mathematical formalism
- graphical illustrations and intuitive interpretations of the key ideas and concepts.

Chapter Descriptions

The text begins with an introduction that briefly describes the history, main ideas, and concepts of fuzzy set theory and its applications. The subsequent material is divided into three conceptually diverse areas within a cohesive overall framework. Part I deals with the fundamentals of fuzzy sets starting with the basic idea of a fuzzy set, providing the reader with a significant rationale for studying them, and illustrating situations under which the notions of fuzziness start playing a prominent role. It is organized so as to be understandable even to the reader not familiar with fuzzy set technology. Chapter 1 introduces the most basic ideas of fuzzy sets. This introduction is followed by operations on fuzzy sets (chapter 2) and a brief introduction to information theory–based aspects of fuzzy sets (chapter 3). Chapters 4 and 5 are geared toward computing facets, including calculus of fuzzy relations and computing with fuzzy numbers. Chapter 6 reveals some interesting links between fuzzy sets and probability, allowing the reader to appreciate fuzziness and probability as two orthogonal and highly complementary concepts. Subsequently, more advanced materials include linguistic variables (chapter 7), fuzzy logic (chapter 8) and fuzzy measures and fuzzy integrals (chapter 9).

Part II can be used as a continuation of a course on the fundamentals of fuzzy sets. It can be also studied independently as an introduction to the main computational models exploiting the technology of fuzzy sets. The fundamentals of rule-based computation (chapter 10) and fuzzy modeling (chapter 13) are the two key paradigms supporting most of the current applications of fuzzy sets. Similarly, the topics of fuzzy neuro-computation (chapter 11) and evolutionary computation (chapter 12) constitute a core component of computational intelligence.

Problem solving with fuzzy sets, as embraced in part III, deals with many representative studies of fuzzy sets in well-established and practically important fields such as control and optimization (chapter 14), all placed into the framework set up by the paradigms discussed in part II. Finally, chapter 15 exposes the reader, through several carefully selected case studies, to complete and cohesive design practices, a must for both student and practitioner. Part III also unveils the diverse aspects of fuzzy set applications that should be definitely taken into account when dealing with real-would problems.

Intended Readership and Use of the Book

This book is intended for two clearly identified groups: senior undergraduate and graduate students as well as practitioners working in various areas of engineering and science. A wide spectrum of undergraduate students can directly benefit from the book; familiarity with fuzzy set technology becomes a genuine asset in a number of engineering programs (electrical, mechanical, industrial, civil, chemical, agriculture, and computer) as well as other programs, such as computer science, management, applied mathematics, ergonomics, and architecture. As a textbook, the material focuses on the actual needs in this area and takes into account the structure of currently existing undergraduate curricula as well. Currently there are specialized courses on fuzzy sets at a graduate level. The emerging tendency is to apply selected concepts of fuzzy sets in more specialized areas like control, computer vision, software engineering, pattern recognition, and system modeling. For the time being, there are no separate courses at the undergraduate level, but optional courses are often provided to expose students to emerging and advanced topics.

The organization of the material reflects the reader's needs. First, splitting the material into three main parts divides the topics handily and makes the text readily usable as a part of some courses. To live up to the expectations one always has with respect to any sound textbook, we have included a vast number of carefully thought-out exercises throughout the

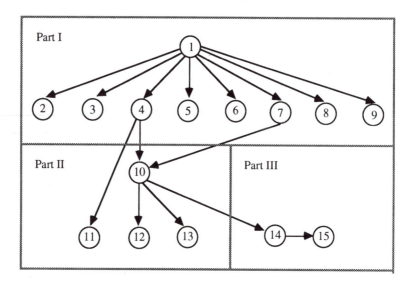

Figure P.1
Introduction to fuzzy sets—analysis and design: a road map

text to motivate continuous learning and thinking, as well as a collection of problems at the end of each chapter. The problems are substantially diverse, starting with some interesting but not highly demanding material, and moving to some problems that are thought-provoking and challenging; they go far beyond the material covered in the text, to the point that their solutions could be easily converted into publishable material. (If you do publish as a response to one of the problems, we would appreciate your acknowledgment.) We have decided to include the most pertinent references; this choice is justifiable, considering the abundant literature on the subject that tends to be somewhat incoherent as far as basic terminology is concerned.

The material in this book may be used in a quarter or semester course or may be extended to two quarters. A ten-week quarter course might adhere to the following outline, which forms a smooth introduction to fuzzy sets: fundamentals of fuzzy sets (chapters 1, 2, 4, 5, and 7) and computational models (chapters 10 and 13). Depending on the character of the course, one might wrap it up by including some elements of chapter 14 and consider eventually one of the case studies in chapter 15. A sixteen-week semester course might simply follow the book outline, with slight modifications depending on the intended focus or audience. In particular, the instructor might select the most pertinent material in chapters 14 and 15; the methodologies and case studies covered there make this choice easier.

Readers who, although familiar with fuzzy sets, would like to pursue further studies in the design of intelligent systems may also find the book's organization beneficial. For such readers, part I could be an useful review, and parts II and III could form an essential focal point of the study. Again, the methodologies and case studies easily address the needs of a broad readership. Readers who identify themselves as practitioners and find that a hands-on approach is a must will benefit from looking at chapters 1, 2, and 7 for a sound introduction and quickly proceeding with chapters 10, 14 and 15 to get into the system development issues.

The road map (figure p.1) visualizes the dependencies among the individual chapters and summarizes the book organization. By examining this diagram, the reader may gain additional flexibility in selecting the material that fits his/her particular interest.

While working on the book, we enjoyed support from various research agencies (Natural Sciences and Engineering Research Council (NSERC) and Research Foundation of the State of Sao Paulo (FAPESP)—W. Pedrycz; Brazilian National Research Council (CNPq) and FAPESP—Fernando Gomide). Our gratitude goes to several dedicated people at The MIT Press: Harry Stanton, Katherine Arnoldi, Jerry Weinstein, and Wendy Drexler—we greatly benefited from their highly professional assistance.

Introduction

Fuzzy sets and fuzzy logic have become one of the emerging areas in contemporary technologies of information processing. Recent studies spread across various areas, from control, pattern recognition, and knowledge-based systems to computer vision and artificial life. A significant number of direct real-world implementations range from home appliances to industrial installations and involve fuzzy sets, both by themselves and hand in hand with other modern approaches, including neural networks.

Fuzzy sets have a rather brief history. The concept itself, coined by L. A. Zadeh in his seminal paper (1965), has emerged as a fundamental and fresh idea. Fuzzy sets involve capturing, representing, and working with linguistic notions—objects with unclear boundaries. What has been conveyed by fuzzy sets and formalized in the resulting framework is definitely essential to many human endeavors and permeates many studies on the role of uncertainty. Two brief excerpts, from Borel and Lukasiewicz, shed light on the essence of the problems and the difficulties arising when looking for eventual solutions.

One seed does not constitute a pile nor two nor three ... from the other side everybody will agree that 100 million seeds constitute a pile. What therefore is the appropriate limit? Can we say that 325 647 seeds don't constitute a pile but 325 648 do? (Borel 1950)

We might assume that a sentence, in the logical sense of the term, might have values other than falsehood or truth. A sentence, of which we do not know whether it is false or true, might have no value determined as truth or falsehood, but might have some third, undetermined, value. We might, for instance, consider that the sentence "In a year from now I shall be in Warsaw" is neither true nor false and has a third, undetermined value, which can be symbolized as "1/2." We might go still further and ascribe to sentences infinitely many values contained between falsehood and truth.... In the logic of infinitely many values, it is assumed that sentences can take on values represented by rational numbers x that satisfy the condition $0 \leq x \leq 1$. (Lukasiewicz 1929)

The first excerpt alludes to the problem of setting up meaningful boundaries when describing even simple concepts. The second one, a vivid manifesto of multiple-valued logic, points out the difficulties one faces in standard two-valued logic, whose semantics may be fundamentally too restrictive in many ways.

Fuzzy sets attempt to examine these issues formally. The original paper by Zadeh set forth the idea in a lucid and highly convincing way within the concise, yet effortlessly comprehended language of mathematics.[1] One of his recent papers (Zadeh 1996) reemphasizes the very nature of fuzzy sets by considering them as a fundamental tool supporting computing with words:

There are two major imperatives for computing with words. First, computing with words is a necessity when the available information is too imprecise to justify the use of numbers, and to achieve tractability, robustness, low solution cost, and better rapport with reality. (L. A. Zadeh 1996)

Fuzzy sets emerged as a new way of representing uncertainty. As such, they naturally got involved in a fervor of philosophical and methodological disputes with proponents of probability and statistics. It was determined quite early, however, that notions of randomness and fuzziness are mostly ortogonal and could eventually coexist. Moreover, there has never been a strong and visible isolation between fuzzy sets regarded as a concept and as a toolbox of useful algorithms. It would be fair to say that by their very nature fuzzy sets are inclined toward multifaceted interaction with certain other methodologies, including neural networks and evolutionary computing. The methods of fuzzy sets have begun to be used as a part of many classic and well-established algorithms. One can refer to operations research, where a number of standard methods of linear programming have already been augmented by fuzzy sets used to handle imprecise objectives and constraints. An interesting synergy has occurred between fuzzy sets and neural networks. The former benefited from neural networks in the form of more profound learning abilities, being inherently associated with neural networks. On the other hand, neural networks were enriched by novel schemes of metalearning expressed in the language of fuzzy sets.

In our opinion, fuzzy sets play a dominant role when it comes to building bridges between symbolic and numeric computation. These two areas constitute essential niches of computing. The latter is dominant in number-crunching scientific computations, the former across all camps of artificial intelligence (AI). Fuzzy sets use symbols (linguistic terms) as does AI, yet fuzzy sets go far beyond this point, for these symbols come equipped with well-defined semantics that, once converted into numeric membership functions, provide a handle for further intensive numeric processing of these concepts. This sort of domain knowledge is omnipresent in various areas. As an illustrative example, let us concentrate on the problem of traffic light control (figure I. 1). Obviously, this is an important problem one is faced with every day, and one whose inefficient solution may cause most of us to suffer a lot. Though very common, the task itself is immensely complex. The growing intensity of traffic along with new traffic patterns contributes to the task's increasing complexity. One must take into consideration in the design a number of additional factors such as the predominant type of vehicles (buses, cars, vans, motocycles, etc.), the proximity of other intersections in the neighborhood, certain types of buildings (hospitals, schools, fire stations), the

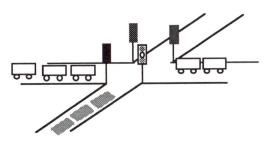

Figure I.1

form of lanes (plain or inclined), and so on. Any realistic model (no matter whether deterministic or stochastic) used to describe the traffic calls for a substantial number of parameters to be estimated; the highly nonstationary control environment may not guarantee that the model designed for a specific situation can perform equally well under variable conditions.

The fuzzy set approach is radically different: We start off by encapsulating all available domain knowledge and organizing it into a manageable format. Because our experience tempts us to generalize, we end up with a collection of "if-then" rules forming a suitable control protocol. For instance, these rules can assume the form

If *many* new arrivals and actual queue is *long*, then the extension of duration time of green light is *very short*.

Most importantly, these rules include linguistic terms (namely, long, very short, etc.) that are inherently associated with the generalization aspect; bear in mind that the rules do not tackle exclusively the single traffic control scenario (one particular crossing) but go far beyond this point by summarizing eventual control policies in similar traffic situations. An important issue is that of calibration of the linguistic terms contained in the rules. Because of the assumed generality, the same control protocol (the underlying control philosophy) can be used successfully in a broad spectrum of traffic situations, once the linguistic terms have been sensibly calibrated. For example, the term *heavy* traffic means something different during rush hours in the streets of Sao Paulo than at the crossing in a small town somewhere in the middle of the prairies. The context is of paramount importance, and fuzzy sets are ready to cope with this conceptual challenge. Fuzzy sets build the necessary conceptual fabric of the control algorithms. The computational details are handled differently, and this is a continuously expanding area of interaction of fuzzy sets with neural networks and genetic computation, which are important suppliers of the optimization tools.

As a discipline, fuzzy sets have roots in set theory and multivalued logic and generalize along these lines. The abolishment of the two-valued (yes-no) dogma has led to a series of interesting mathematical insights and investigations that can easily stand on their own. We are convinced, however, that the term fuzzy mathematics goes a bit too far—as the language of mathematics is universal, so is its rigor, from which the development of the theoretical foundations benefits greatly. The genuine development and progress of fuzzy sets as a technology for the coming century is promoted by its real-world applications and the emerging sound and powerful methodologies resulting there from. We hope, all of these have already begun and the way toward future progress has been paved.

Note

1. The reader should be aware of a number of alternative approaches that are, however, far more mathematically advanced (if not convoluted) and therefore have never gained enough attention. One may refer to the original idea of Klaua (1964) as one among these approaches.

References

Borel, E. 1950. *Probabilite et certitude*, Paris; Press Univevsité de France.

Klaua, D. 1964. *Allgemeine Mengenlehre*. Berlin: Akademie-Verlag.

Lukasiewicz, J. 1929. *Elements of Mathematical Logic*. London: Pergamon Press.

Zadeh, L. A. 1965. Fuzzy sets. *Information and Control* 8:338–53.

Zadeh, L. A. 1996. Fuzzy logic = computing with words. *IEEE Trans. on Fuzzy Systems* 4:103–11.

I FUNDAMENTALS OF FUZZY SETS

1 Basic Notions and Concepts of Fuzzy Sets

This chapter presents the fundamental ideas of fuzzy sets. It introduces and interprets the very notions of the theory and, subsequently, offers them as a solid basis for system analysis and design. Essentially, the fundamental notion of a set as a collection of some objects with a well-defined boundary is generalized to characterize collections whose boundaries are not sharply defined. Since the concept of a set is the backbone of the underlying mathematics, we first proceed with the characteristics and related properties of set theory.

1.1 Set Membership and Fuzzy Sets

The notion of set occurs frequently as we tend to organize, summarize, and generalize knowledge about objects. We can even speculate that the fundamental nature of any human being is to organize, arrange, and systematically classify information about the diversity of any environment. The encapsulation of the objects into a collection whose members all share some general features or properties naturally implies the notion of a set. Sets are used often and almost unconsciously; we talk about a set of even numbers, positive temperatures, fruits, personal computers, and the like. Being intuitively appealing, sets introduce a fundamental notion of *dichotomy*. In its essence, any process of dichotomization imposes a binary, all or none classification decision: either accept or reject an object as belonging to a given collection (category). For instance, consider the set A in universe $\mathbf{X}$, as depicted in figure 1.1. Clearly the object (point) x_1 belongs to the set (collection) A whereas x_2 does not, that is, $x_1 \in A$ and $x_2 \notin A$. In general, if we denote the *accept* decision by 1 and the *reject* decision by 0, for short, then we may express the classification decision through a characteristic (membership) function $A(x), x \in \mathbf{X}$, of the form

$$A(x) = \begin{cases} 1, & \text{if } x \in A \\ 0, & \text{if } x \notin A \end{cases}$$

as illustrated in figure 1.2.

Clearly the empty set $\varnothing$ has a null characteristic function, $\varnothing(x) = 0$, for all x in $\mathbf{X}$, and the universe $\mathbf{X}$ a unity characteristic function $\mathbf{X}(x) = 1, \forall x \in \mathbf{X}$. Also, for a set $A = \{a\}$ with only one element a (that is, a singleton), $A(x) = 1$ if $x = a$ and 0 otherwise.

Actually, the function $A : \mathbf{X} \to \{0, 1\}$ induces a constraint, with a well-defined boundary, on the objects of a universe $\mathbf{X}$ that may be assigned to a set A. The underlying concept of fuzzy sets is to relax this requirement and admit intermediate values of class membership. This allows for an enriched and more realistic interpretation framework accommodating

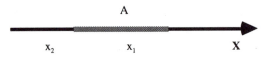

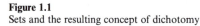

A

x_2 x_1 X

Figure 1.1
Sets and the resulting concept of dichotomy

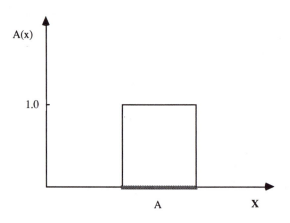

A(x)

1.0

A   X

Figure 1.2
Example of a characteristic function

statements with a partial quantification of belongingness. Most categories used in describing real-world objects do not possess well-defined boundaries. Consider, for instance, notions like *high* temperature, *medium* pressure, *small* neighborhood, *low* speed, *big* car, and so forth, in which the italicized words identify the sources of fuzziness. Whether an object belongs to such a category is a matter of degree, expressed, for example, by a real number in the unit interval [0, 1]. The closer that number is to 1, the higher the grade of the object's membership in particular category.

EXERCISE 1.1   Consider the sets $A = \{x \in \mathbf{R} \,|\, -1 \leq x \leq 2\}$ and $B = \{x \in \mathbf{R} \,|\, 1 \leq x \leq 4\}$ which model intervals on the real line $\mathbf{R}$.

(a) Find the characteristic functions of A and B.

(b) Determine the union, intersection, and the complement of sets A and B in terms of their characteristic functions.

Recall that the union consists of all such elements that belong to at least one set. The intersection is composed of such elements of the universe of discourse that belong to both the sets under discussion, whereas the complement embraces all elements that are not included in the set being considered.

1.2   Basic Definition of a Fuzzy Set

From the inception of the theory (Zadeh 1965, 1981; Kandel 1986; Klir and Folger 1988) a fuzzy set has been defined as a collection of objects with membership values between 0 (complete exclusion) and 1 (complete membership). The membership values express the degrees to which each object is *compatible* with the properties or features distinctive to the collection. Fuzzy sets are formally defined as follows:

DEFINITION 1.1   A fuzzy set is characterized by a membership function mapping the elements of a domain, space, or universe of discourse **X** to the unit interval [0, 1]. (Zadeh 1965) That is,

$$A : \mathbf{X} \to [0, 1].$$

Thus, a fuzzy set A in **X** may be represented as a set of ordered pairs of a generic element $x \in \mathbf{X}$ and its grade of membership: $A = \{(A(x)/x) \mid x \in \mathbf{X}\}$. Clearly, a fuzzy set is a generalization of the concept of a set whose membership function takes on only two values $\{0, 1\}$, as discussed previously.

The value of $A(x)$ describes a degree of membership of x in A. For instance, consider the concept of *high* temperature in, say, an environmental context with temperatures distributed in the interval [0, 50] defined in °C. Clearly 0°C is not understood as a high temperature value, and we may assign a null value to express its degree of compatibility with the *high* temperature concept. In other words, the membership degree of 0°C in the class of high temperatures is zero. Likewise, 30°C and over are certainly *high* temperatures, and we may assign a value of 1 to express a full degree of compatibility with the concept. Therefore, temperature values in the range [30, 50] have a membership value of 1 in the class of high temperatures. The partial quantification of belongingness for the remaining temperature values through their membership values can be pursued as exemplified in figure 1.3, which actually is a membership function $H : \mathbf{T} \to [0, 1]$ characterizing the fuzzy set H of high temperatures in the universe $\mathbf{T} = [0, 50]$.

Fuzzy sets can be regarded as elastic constraints imposed on the elements of a universe or a domain. An interesting analogy that may help the reader grasp this idea can be developed by experimenting with a rubber band as shown in figure 1.4(a). To embrace a new point outside its initial circumference, the band is stretched. The extent of this stretching depends upon the position of the new point, whose membership value can then be conceived as inversely proportional to the force needed to stretch the rubber band and embrace it: the higher the force required, the lower

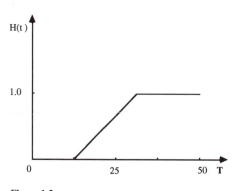

Figure 1.3
Example of a membership function

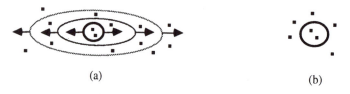

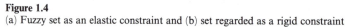

(a)   (b)

Figure 1.4
(a) Fuzzy set as an elastic constraint and (b) set regarded as a rigid constraint

the membership degree. The points the band embraces within its circumference prior to any stretching naturally fit into the category and have a membership value of 1. This example illustrates the essence of fuzzy sets as dealing primarily with a concept of elasticity, or absence of sharply defined boundary (Pedrycz 1995). In contrast, the corresponding model encountered in traditional set theory is formed by rigid bands not amenable to any stretching, as shown in figure 1.4(b). A point is either a member of the set enclosed by the band or it is not: No stretching is permitted.

Two views associated with fuzzy sets should be emphasized. The first concerns the *fuzziness* of a piece of information: the membership value $A(x)$ quantifies how compatible x is with the concept conveyed by A. The second is related to the *uncertainty* about a piece of information: $A(x)$ represents how likely x is to occur given that A (a sort of constraint) is present. This point of view is often referred to as an epistemic view of a fuzzy set representing the state of knowledge about a variable.

Fuzzy sets can be defined in either finite or infinite universes using different notations. If an universe **X** is discrete and finite, with cardinality n, then a fuzzy set is given in a form of an n-dimensional vector whose entries denote grades of membership of the corresponding elements of $\mathbf{X}$. Sometimes a sum notation is used. This allows us to enumerate only elements of $\mathbf{X}$ with nonzero grades of membership in the fuzzy set. For instance, if $\mathbf{X} = \{x_1, x_2, \ldots, x_n\}$, then the fuzzy set $A = \{(a_i/x_i) \mid x_i \in \mathbf{X}\}$, where $a_i = A(x_i), i = 1, \ldots, n$, may be denoted by (Zadeh 1965; Kandel 1986)

$$A = a_1/x_1 + a_2/x_2 + \cdots + a_n/x_n = \sum_{i=1}^{n} a_i/x_i.$$

In this notation the sum should not be confused with the standard algebraic summation; the only purpose of the summation symbol in the above expression is to denote the set of the ordered pairs. Also, note that when $A = \{a/x\}$, that is, when there exists only one point x in a universe for which the membership degree is nonnull, we have a fuzzy singleton. In this sense, we may also interpret the summation symbol as union of singletons. Equivalently, one can summarize A as a vector, meaning that $A = [a_1 \, a_2 \, \ldots \, a_n]$. When the universe $\mathbf{X}$ is continuous, we use, to represent a fuzzy set, the following expression,

$$A(x) = \int_x a/x,$$

where $a = A(x)$ and the integral symbol should be interpreted in the same way as the sum given above.

Points x and p of the real line $\mathbf{R}$ (numerical information) form a very specific form of a fuzzy set described as

$$A(x) = \delta(x - p) = \begin{cases} 1 & \text{if } x = p. \\ 0 & \text{if } x \neq p. \end{cases}$$

Sets and fuzzy sets defined in discrete and finite universes, for example, $\mathbf{X} = \{x_1, x_2, \ldots, x_n\}$, have an interesting and transparent geometrical interpretation (Kosko 1992). Clearly, any set in $\mathbf{X}$ is a member of the power set $P(\mathbf{X})$ of $\mathbf{X}$ or equivalently $2^{\mathbf{X}}$. Therefore, we may associate with each of the 2^n elements of $P(\mathbf{X})$ an n-dimensional vector, containing either one entry or none, that is a corner of the n-dimensional unit hypercube $\{0, 1\}^n$. For instance, if $n = 2$, then $P(\mathbf{X})$ has four elements:

$$P(\mathbf{X}) = \{\varnothing, \{x_1\}, \{x_2\}, \{x_1, x_2\}\}.$$

Using vector notation, $P(\mathbf{X}) = \{[0 \; 0], [1 \; 0], [0 \; 1], [1 \; 1]\}$, whose members are distributed in the four corners of the unit square, as figure 1.5(a) shows.

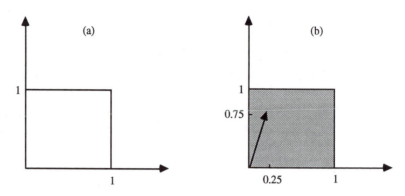

Figure 1.5
Geometric interpretation of (a) sets and (b) fuzzy sets

Interestingly enough, Zadeh was the first to introduce this geometrical interpretation in his 1971 publication (Zadeh, 1971); this, however, is completely overlooked.

Fuzzy sets also form n-dimensional vectors with entries in the $[0, 1]$ range, but, in contrast to the previous construct, the number of different vectors is not finite. Fuzzy sets fill in the entire n-dimensional unit cube, including, in particular, its corners. For $n = 2$ we get a similar geometric illustration, shown in figure 1.5(b), which highlights the fuzzy set $A = [0.25 \ 0.75]$.

EXERCISE 1.2   Consider sets A and B defined in exercise 1.1. How would you relax these rigid interval definitions? Can you describe these constructs?

EXERCISE 1.3   Assume a fuzzy set $A = \{0.1/1 + 0.4/3 + 0.6/4 + 1.0/5 + 1.0/6 + 0.6/7 + 0.5/8 + 0.2/10\}$ defined in $\mathbf{X} = \{1, 2, 3, 4, 5, 6, 7, 8, 9, 10\}$.

(a) Sketch the membership function of this fuzzy set.

(b) What is its representation as a point in an unit hypercube?

(c) Which elements of $\mathbf{X}$ have the lowest and the highest membership degrees?

1.3 Types of Membership Functions

In principle any function of the form $A : \mathbf{X} \to [0, 1]$ describes a membership function associated with a fuzzy set A that depends not only on the concept to be represented, but also on the context in which it is used. The graphs of the functions may have very different shapes, and may have some specific properties (e.g., continuity). Whether a particular shape is suitable can be detemined only in the application context. In certain cases, however, the meaning semantics captured by fuzzy sets is not too sensitive to variations in the shape, and simple functions are convenient. In many practical instances fuzzy sets can be represented explicitly by families of parametrized functions, the most common being the following:

1. Triangular Functions:

$$A(x) = \begin{cases} 0, & \text{if } x \leq a \\ \dfrac{x-a}{m-a}, & \text{if } x \in [a,m] \\ \dfrac{b-x}{b-m}, & \text{if } x \in [m,b] \\ 0, & \text{if } x \geq b, \end{cases}$$

where m is a modal value, and a and b denote the lower and upper bounds, respectively, for nonzero values of $A(x)$. Sometimes it is more convenient to use the notation explicitly highlighting the membership function's parameters; in this case, we obtain

$$A(x; a, m, b) = \max\{\min[(x-a)/(m-a), (b-x)/(b-m)], 0\}.$$

2. Γ-function:

$$A(x) = \begin{cases} 0, & \text{if } x \leq a \\ 1 - e^{-k(x-a)^2}, & \text{if } x > a \end{cases}$$

or

$$A(x) = \begin{cases} 0, & \text{if } x \leq a \\ \dfrac{k(x-a)^2}{1 + k(x-a)^2}, & \text{if } x > a, \end{cases}$$

where $k > 0$.

3. S-function:

$$A(x) = \begin{cases} 0, & \text{if } x \leq a \\ 2\left(\dfrac{x-a}{b-a}\right)^2, & \text{if } x \in [a,m] \\ 1 - 2\left(\dfrac{x-b}{b-a}\right)^2, & \text{if } x \in [m,b] \\ 1, & x > b. \end{cases}$$

The point $m = a + b/2$ is known as a crossover of the S-function.

4. Trapezoidal function:

$$A(x) = \begin{cases} 0, & \text{if } x < a \\ \dfrac{x-a}{m-a}, & \text{if } x \in [a,m] \\ 1, & \text{if } x \in [m,n] \\ \dfrac{b-x}{b-n}, & \text{if } x \in [n,b] \\ 0, & \text{if } x > b. \end{cases}$$

Using equivalent notation, we obtain

$$A(x; a, m, n, b) = \max\{\min[(x-a)/(m-a), 1, (b-x)/(b-n)], 0\}.$$

5. Gaussian function:

$$A(x) = e^{-k(x-m)^2},$$

where $k > 0$

6. Exponential-like function:

$$A(x) = \frac{1}{1 + k(x-m)^2},$$

$k > 1$, or

$$A(x) = \frac{k(x-m)^2}{1 + k(x-m)^2},$$

$k > 0$.

Figure 1.6 shows the example membership functions for types 2 through 6.

1.4   Characteristics of a Fuzzy Set

We can characterize fuzzy sets in more detail by referring to the features used in characterizing the membership functions that describe them (Kandel 1986; Dubois, Prade, and Yager 1993; Yager et al. 1987). This brings us to the concepts of normality, height, support, convexity, concavity, and cardinality, as well as to simple operations used to modify membership functions.

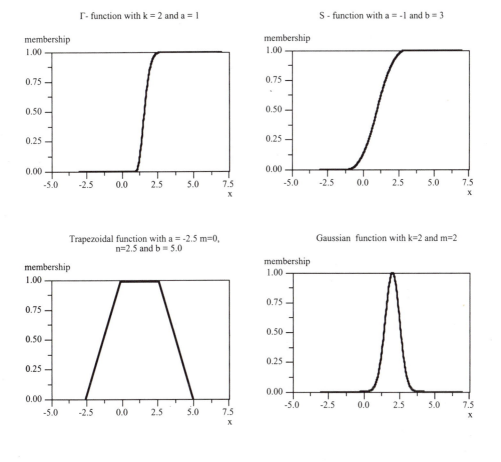

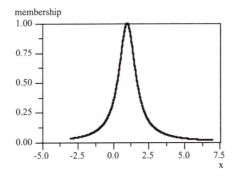

Figure 1.6
Examples of selected membership functions; exponential membership function is described by the first of the two formulas provided in the text

DEFINITION 1.2 A fuzzy set A is *normal* if its membership function attains 1, that is,

$$\sup_x A(x) = 1.$$

If the supremum is less than 1, then A is called *subnormal*. The supremum above is usually referred to as the *height* of A, hgt(A). Hence saying that a particular fuzzy set is normal is equivalent to saying its height is equal to 1.

DEFINITION 1.3   By the *support* of a fuzzy set A, denoted by Supp(A), we mean all elements of **X** that belong to A to a nonzero degree,

$$\text{Supp}(A) = \{x \in \mathbf{X} \,|\, A(x) > 0\}.$$

Alternatively, the *core* of a fuzzy set A is the set of all elements of **X** that exhibit a unit level of membership in A. More formally,

$$\text{Core}(A) = \{x \in \mathbf{X} \,|\, A(x) = 1\}.$$

The support and core of fuzzy sets may be viewed as closely related concepts in the sense that they identify elements belonging to the fuzzy set. Obviously, they are both sets. All elements of core are subsumed by the support.

Set A is unimodal if its membership is a unimodal function. A related concept is that of convexity.

DEFINITION 1.4   A fuzzy set A is convex if its membership function is such that

$$A[\lambda x_1 + (1 - \lambda)x_2] \geq \min[A(x_1), A(x_2)]$$

for any $x_1, x_2 \in \mathbf{X}$, and $\lambda \in [0, 1]$. Similarly, a fuzzy set A is concave if the corresponding membership function satisfies the relationship

$$A[\lambda x_1 + (1 - \lambda)x_2] \leq \max[A(x_1), A(x_2)]$$

for any $x_1, x_2 \in \mathbf{X}$, and $\lambda \in [0, 1]$.

Note that convex and concave fuzzy sets are unimodal, but the converse is not true.

EXERCISE 1.4   Clearly, the trapezoidal membership function $A(x; 1, 2, 4, 5)$ in $\mathbf{X} = [0, 10]$ is normal.

(a) Find its support and core.

(b) Is it convex?

DEFINITION 1.5   Given a fuzzy set A in a finite universe $\mathbf{X}$, its cardinality, denoted by $\mathrm{Card}(A)$, is defined as

$$\mathrm{Card}(A) = \sum_{x \in \mathbf{X}} A(x).$$

Often, $\mathrm{Card}(A)$ is referred to as the scalar cardinality or the sigma count of A. For example, the fuzzy set $A = 0.1/1 + 0.3/2 + 0.6/3 + 1.0/4 + 0.4/5$ in $\mathbf{X} = \{1, 2, 3, 4, 5, 6\}$ is such that $\mathrm{Card}(A) = 2.4$.

A number of simple, yet useful operations may be performed on fuzzy sets. These are one-argument mappings, because they apply to a single membership function. (We will study two- and many-argument operations in chapter 2).

1. Normalization: This operation converts a subnormal, nonempty fuzzy set into its normal version by dividing the original membership function by the height of A:

$$\mathrm{Norm\_}A(x) = \frac{A(x)}{\mathrm{hgt}(A)}.$$

2. Concentration: When fuzzy sets are *concentrated*, their membership functions take on relatively smaller values. That is, the membership function becomes more concentrated around points with higher membership grades as, for instance, being raised to power two,

$$\mathrm{Con\_}A(x) = A^2(x).$$

An amplified effect is obtained by using any power $p > 1$,

$$\mathrm{Con\_}A(x) = A^p(x).$$

3. Dilation: Dilation has the opposite effect from concentration and is produced by modifying the membership function through the transformation

$$\mathrm{Dil\_}A(x) = A^{0.5}(x)$$

or

$$\mathrm{Dil\_}A(x) = 2A(x) - A^2(x).$$

As before one can generalize this operation by admitting any power $r \in (0, 1.0)$

$$\mathrm{Dil\_}A(x) = A^r(x).$$

4. Contrast intensification: As the name suggests, in contrast intensification, the membership values less that $1/2$ are diminished while the grades of membership above this threshold are elevated. This intensifies contrast once it reduces a set's fuzziness. The operation is defined by

$$\text{Int\_}A(x) = \begin{cases} 2A^2(x), & \text{if } 0 \leq A(x) \leq 0.5 \\ 1 - 2(1 - A(x))^2, & \text{otherwise,} \end{cases}$$

or, to make the intensification more pronounced, for $p > 1$,

$$\text{Int\_}A(x) = \begin{cases} 2^{p-1}A^p(x), & \text{if } 0 \leq A(x) \leq 0.5 \\ 1 - 2^{p-1}(1 - A(x))^p, & \text{otherwise.} \end{cases}$$

5. Fuzzification: The effect of fuzzification is complementary to that of intensification and is produced by altering the membership function as follows:

$$\text{Fuzz\_}A(x) = \begin{cases} \sqrt{A(x)/2}, & \text{if } A(x) \leq 0.5 \\ 1 - \sqrt{(1 - A(x))/2}, & \text{otherwise.} \end{cases}$$

More complex operations may be built from the elementary operations on fuzzy sets by combination or composition. They will be essential in chapter 7 in dealing with linguistic variables. In particular, linguistic hedges such as *very, slightly, and more or less* may be regarded as operators that act on the fuzzy set representing the meaning of its operand. For instance, in the composite statement *very high* temperature, *very* can be viewed as an operator acting on the fuzzy meaning of *high* temperature. Intuitively, it seems reasonable to interpret this operator as a concentrator, because concentration reduces the magnitude of the grade of membership. The reduction of membership grades in related to the idea of inclusion and fits our intuition that *very high* temperature is a concept which is embraced within the concept of *high* temperature. In other words, if H is a fuzzy set characterizing *high* temperature, then the fuzzy set Con_H is a reasonable approximation of *very high* temperature. Note that, by definition, Con_$H(t) \leq H(t)$ and thus Con_$H \subset H$; that is, Con_H is included in H, as the next section shows.

Figure 1.7 exemplifies the effect of concentration, dilation, contrast intensification, and fuzzification on a Gaussian membership function.

EXERCISE 1.5   Let A be a fuzzy set with the triangular membership function $A(x; -1, 0, 1)$ in $\mathbf{X} = [-2, 2]$. Perform the five operations discussed above, (normalization, concentration, dilation, contrast identification, and fuzzification) show them graphically, and compare the results.

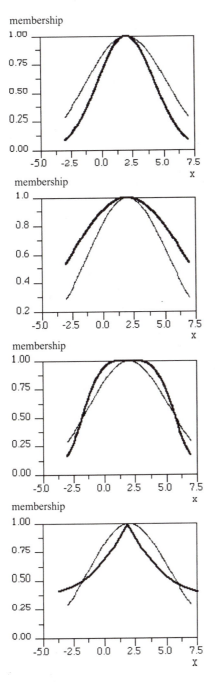

Figure 1.7
Operations of concentration $(p = 2.0)$, dilation $(p = 0.5)$, contrast intensification and fuzzification applied to Gaussian membership function with $k = 0.05$ and $m = 2$

1.5   Basic Relationships between Fuzzy Sets: Equality and Inclusion

As in set theory, we can define generic relations between two fuzzy sets, such as equality and inclusion. We say that two fuzzy sets, A and B, defined in the same space $\mathbf{X}$ are equal if and only if (iff) their membership functions are identical. That is, for each element of X,

$A = B$ iff $A(x) = B(x)$.

Analogously, A is said to be included in B if any only if the membership function of A is less than that of B for each x in $\mathbf{X}$:

$A \subseteq B$ iff $A(x) \leq B(x)$.

When the fuzzy sets A and B are defined in a finite universe $\mathbf{X}$, and the requirement that for each x in $\mathbf{X}$, $A(x) \leq B(x)$, is relaxed, we may define the degree of subsethood (Kosko 1992) as

$$S(A, B) = \frac{1}{\text{Card}(A)} \left\{ \text{Card}(A) - \sum_{x \in \mathbf{X}} \max[0, A(x) - B(x)] \right\}.$$

$S(A, B)$ provides a normalized measure of the degree to which the inequality $A(x) \leq B(x)$ is violated.

EXERCISE 1.6   Assume fuzzy sets A, with the trapezoidal membership function $A(x; 0, 1, 3, 5)$, and B, with the triangular membership function $B(x; 1, 2, 4)$, in $\mathbf{X} = [0, 10]$.

(a) Is $A \subset B$?

(b) Is $B \subset A$?

1.6   Fuzzy Sets and Sets: The Representation Theorem

Any fuzzy set can be regarded as a family of fuzzy sets. This is the essence of an identity principle known also as the representation theorem. To explain this construction, we need to define the notion of an α-cut of a fuzzy set. The α-cut of A, denoted by A_α, is a set consisting of those elements of the universe $\mathbf{X}$ whose membership values exceed the threshold level α,

$A_\alpha = \{x | A(x) \geq \alpha\}$.

In other words, A_α consists of elements of $\mathbf{X}$ identified with A to a degree of at least α. In particular, the highest level, $\alpha = 1$, determines a set of $\mathbf{X}$

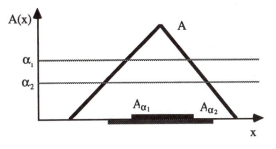

Figure 1.8
Fuzzy set as a family of its α-cuts

totally belonging to A. Clearly, the lower the level of α, the more elements are admitted to the corresponding α-cut, that is,

if $\alpha_1 > \alpha_2$ then $A_{\alpha_1} \subset A_{\alpha_2}$.

The representation theorem states that any fuzzy set A can be decomposed into a series of its α-cuts,

$$A = \bigcup_{\alpha \in [0,1]} (\alpha A_\alpha),$$

or, equivalently, as illustrated in figure 1.8,

$$A(x) = \sup_{\alpha \in [0,1]} [\alpha A_\alpha(x)].$$

Conversely, any fuzzy set can be "reconstructed" from a family of nested sets (assuming that they satisfy the constraint of consistency: if $\alpha_1 > \alpha_2$ then $A_{\alpha_1} \subset A_{\alpha_2}$). This theorem's importance lies in its underscoring of the very nature of the generalization provided by fuzzy sets. Furthermore, the theorem implies that problems formulated in the framework of fuzzy sets (such as fuzzy optimization, programming, decision making, etc.) can be solved by transforming these fuzzy sets into their corresponding families of nested α-cuts and determining solutions to each using standard, nonfuzzy techniques. Subsequently, all the partial results derived in this way can be merged, reconstructing a solution to the problem in its original formulation based on fuzzy sets. By increasing the number of quantization levels of the membership values (that is, the α-cuts), the reconstruction can be made more detailed.

EXERCISE 1.7   Find all α-cuts of $A = 0.1/2 + 0.3/3 + 1.0/4 + 0.3/5$ in $\mathbf{X} = \{1, 2, 3, 4, 5\}$. Show how to recover A from these α-cuts.

1.7   The Extension Principle

The extension principle plays a fundamental role in translating set-based concepts into their fuzzy-set counterparts. Typical examples include arithmetic operations with fuzzy numbers, discussed in chapter 5, and computing induced constraints. Essentially, the extension principle is used to transform fuzzy sets via functions (Kandel 1986; Klir and Folger 1988; Yager et al. 1987). Let $\mathbf{X}$ and $\mathbf{Y}$ be two sets and f a mapping from $\mathbf{X}$ to $\mathbf{Y}$:

$$f : \mathbf{X} \to \mathbf{Y}.$$

Let A be a fuzzy set in $\mathbf{X}$. The extension principle states that the image of A under this mapping is a fuzzy set $B = f(A)$ in $\mathbf{Y}$ such that, for each $y \in \mathbf{Y}$,

$$B(y) = \sup_{x} A(x), \text{ subject to } x \in \mathbf{X} \text{ and } y = f(x),$$

as illustrated in figure 1.9.

The extension principle easily generalizes to functions of many variables as follows. Let $\mathbf{X}_i, i = 1, \ldots, n$, and $\mathbf{Y}$ be universes of discourse, and $\mathbf{X} = \mathbf{X}_1 \times \mathbf{X}_2 \times \cdots \times \mathbf{X}_n$ constitute the Cartesian product of the $\mathbf{X}_i$s. Consider fuzzy sets A_i in $\mathbf{X}_i$, $i = 1, \ldots, n$, and a mapping $y = f(x)$, $x = (x_1, x_2, \ldots, x_n), x \in \mathbf{X}$. Fuzzy sets $A_1, A_2, \ldots, A_n$ are then transformed via f producing the fuzzy set $B = f(A_1, A_2, \ldots, A_n)$ in $\mathbf{Y}$ such that, for

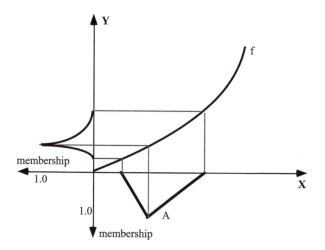

Figure 1.9
The idea of the extension principle

each $y \in \mathbf{Y}$,

$$B(y) = \sup_{x}\{\min[A_1(x_1), A_2(x_2), \ldots, A_n(x_n)]\},$$

subject to $x \in \mathbf{X}$ and $y = f(x)$.

Actually, in the expression above, the min operator is just a choice within a family of operators called triangular norms introduced in chapter 2.

EXERCISE 1.8   Show that if $\mathbf{X} = \{1, 2, 3, 4\}$, $\mathbf{Y} = \{1, 2, 3, 4, 5, 6\}$, $y = x + 2$, and $A = 0.1/1 + 0.2/2 + 0.7/3 + 1/4$, then $B = f(A) = 0.1/3 + 0.2/4 + 0.7/5 + 1/6$.

1.8   Membership Function Determination

Six classes of experimental methods help determine membership functions: horizontal approach, vertical approach, pairwise comparison, inference based on problem specification, parametric estimation, and fuzzy clustering. Which method to select depends heavily on the specifics of an application, in particular, the way the uncertainty is manifested and captured during the experiment.

1.8.1   Horizontal Method of Membership Estimation

The underlying idea of the *horizontal method* is to gather information about membership values of the concept at some selected elements of the universe of discourse $x_1, x_2, \ldots, x_n$. (These need not be evenly distributed; on the contrary, their distribution might be very uneven depending on the form of the concepts to be captured). The method relies on some experimental findings collected under the following scenario. A group of testees (experts) are asked to answer the question: Can x_i be accepted as compatible with the concept A? only the answers "yes" or "no" are permitted (although one could eventually enlarge this repertoire to include a third option, "unknown"). The estimated value of the membership function at x_i is taken as a ratio of the number of positive replies $P(x_i)$ to the total number N of responses,

$$A(x_i) = \frac{P(x_i)}{N},$$

$i = 1, 2, \ldots, n$. This very simple experiment, when prudently arranged and completed, can deliver reliable and significant estimates. Moreover, the results (estimated membership values) can be evaluated with respect to their statistical relevance by associating with each membership value its standard deviation taken from the results of the experiments. Assuming that the results comply with a certain binomial distribution, the standard deviation of the estimate $A(x_i)$ can be computed as

$$\sigma(x_i) = \sqrt{\frac{A(x_i)(1 - A(x_i))}{N}}.$$

In fact, the horizontal method provides an interval-valued fuzzy set with the bounds defined by

$$\left[A(x_i) - \sqrt{\frac{A(x_i)(1 - A(x_i))}{N}}, A(x_i) + \sqrt{\frac{A(x_i)(1 - A(x_i))}{N}}\right].$$

EXERCISE 1.9   The replies of ten experts regarding the linguistic term *acceptable* (*low*) power consumption with reference to a certain electronic device is shown in a tabular format:

Power consumption [mW]	5	10	20	30	40	50	60	70	80	90
Number of positive answers	10	9	6	5	4	3	2	2	1	0

Determine this membership function and compute the bounds of membership values. Discuss their distribution over the universe of discourse. Sketch the membership function. Which analytic expressions would you suggest to represent this notion?

1.8.2   Vertical Method of Membership Estimation

The *vertical method* takes advantage of the identity principle and "reconstructs" a fuzzy set via identifying its α-cuts. After several levels of α are selected, testees are requested to identify the corresponding subset of **X** whose elements belong to A to a degree not less than α. The fuzzy set is built by stacking up the successive α-cuts.

In comparison to the horizontal approach, the factor of uncertainty emerging in this class of experiment is distributed along the membership axis. Observe in figure 1.10 that in the vertical method the uncertainty component resides along the universe of discourse (α-levels). Thus the methods could be regarded as highly orthogonal. To some degree, they are also complementary, since the estimation results can be merged by selecting several x_is and α-levels, thus achieving some cross validation of the results derived.

EXERCISE 1.10   We are concerned with an experimental determination of the membership function of the linguistic term *safe* speed. Obviously, any driver who is either too slow or too fast could pose a substantial hazard

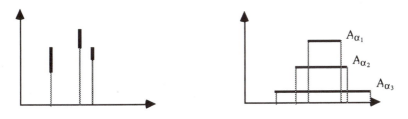

Figure 1.10
Horizontal (left) and vertical (right) methods of membership function estimation

on a road. An expert is interviewed to identify plausible intervals of safe speed assuming a certain level of confidence associated with them. The results are given below (all values in km/h):

$\alpha = 0.05$: $[50, 120]$
$\alpha = 0.10$: $[60, 105]$
$\alpha = 0.20$: $[65, 100]$
$\alpha = 0.30$: $[70, 100]$
$\alpha = 0.40$: $[70, 95]$
$\alpha = 0.90$: $[80, 90]$
$\alpha = 1.00$: $[90, 90]$

Reconstruct a fuzzy set from the results and model its membership function.

The main advantage of these two methods lies in their conceptual clarity. On the other hand, the most prominent shortcoming stems from the "local" (as opposed to "global") character of the experiments set up in each of these environments—the expert becomes exposed to a single element either in the universe of discourse or the membership scale. This makes the individual experiments quite isolated, which goes against the fundamental concept of membership continuity inherent in the very concept of fuzzy sets. Because the elements are treated discretely, the outcomes of the experiment may be scattered and inconsistent.

1.8.3   Pairwise-Comparison Method of Membership Function Estimation

The pairwise-comparison method proposed by Saaty (1980) mitigates one of the deficiencies of the horizontal and vertical approaches by determining the membership function through a sequence of pairwise comparisons of the individual objects in a finite universe of discourse. To explain the rationale behind the method, let us assume for the time being that the membership function is given and its values at $x_1, x_2, \ldots, x_n$ are known and are equal to $A(x_1), A(x_2), \ldots, A(x_n)$, respectively. Consider the ratios

$A(x_i)/A(x_j), i, j = 1, 2, \ldots, n$ and arrange them in the form of a square matrix $\mathscr{A} = [a_{ij}] = [A(x_i)/A(x_j)]$, namely,

$$\mathscr{A} = [a_{ij}] = \begin{bmatrix} \dfrac{A(x_1)}{A(x_1)} & \dfrac{A(x_1)}{A(x_2)} & & \dfrac{A(x_1)}{A(x_n)} \\[2ex] \dfrac{A(x_2)}{A(x_1)} & \dfrac{A(x_2)}{A(x_2)} & & \\[2ex] & & \dfrac{A(x_i)}{A(x_j)} & \\[2ex] \dfrac{A(x_n)}{A(x_1)} & & & \dfrac{A(x_n)}{A(x_n)} \end{bmatrix}$$

$\mathscr{A}$ is usually referred to as a *reciprocal matrix*. Note that

1. All diagonal elements of $\mathscr{A}$ are equal to unity, $A(x_i)/A(x_i) = 1$.

2. $\mathscr{A}$ satisfies a property of reciprocality since $a_{ij} = 1/a_{ji}$ (or $a_{ij}a_{ji} = 1$), $i, j = 1, 2, \ldots, n$.

3. $\mathscr{A}$ is transitive in the sense that $a_{ik}(a_{kj}) = a_{ij}$. This property can be easily verified by noting that

$$a_{ik}a_{kj} = \frac{A(x_i)}{A(x_k)} \frac{A(x_k)}{A(x_j)} = \frac{A(x_i)}{A(x_j)} = a_{ij}.$$

Now multiplying $\mathscr{A}$ by the vector $\mathbf{a} = [A(x_1)A(x_2)\ldots A(x_n)]^T$ one obtains,

$$\mathscr{A}\mathbf{a} = n\mathbf{a},$$

namely,

$$(\mathscr{A} - n\mathbf{I})\mathbf{a} = \mathbf{0}.$$

From basic matrix algebra ($\mathscr{A}$ is a positive definite matrix; $\mathbf{I}$ is the identity matrix) one concludes that n denotes the largest eigenvalue of $\mathscr{A}$. Moreover the corresponding eigenvector associated with n is simply equal to $\mathbf{a}$.

Now let us reverse the problem (as this usually happens in reality) and consider that the entries of $\mathscr{A}$ are not known and need to be estimated through a series of pairwise comparisons. In designing the experiment we can easily preserve reciprocality. However, maintaining the transitivity of the entries of $\mathscr{A}$ becomes practically impossible.

As the elements of $\mathscr{A}$ are collected experimentally, one selects a suitable ratio scale (that usually involves 7 ± 2 quantization levels) for comparing the objects pairwise in the context of $\mathscr{A}$. The level of preference of x_i over x_j is quantified numerically: the more x_i is preferred over x_j, the higher the numerical level associated with this pair. The level of preference of x_i over x_i is always equal to 1, as shown by the diagonal elements of $\mathscr{A}$. If x_i is not preferred over x_j, then one considers the level of preference attached to the swapped pair of the elements. The number of necessary comparisons is therefore $n(n-1)/2$. As the transitivity property cannot always be strictly enforced, the maximal eigenvalue is no longer n. As shown in Saaty (1980), it exceeds n. Interestingly, the higher the value of the maximal eigenvalue, the more significant the transitivity inconsistencies within the collected data. This furnishes us with a useful tool for monitoring the quality of estimates; if it is too low, the experiments may need to be repeated.

1.8.4 Problem Specification–Based Membership Determination

The *problem specification–based method* of determining a membership function relies directly on computations that take into account a certain numerical objective function studied in the problem. For illustrative purposes, let us study a problem in which a nonlinear function $y = f(x)$ must be linearly approximated around a certain point, say x^*. It is well known that the linear approximation $y = a(x - x^*)$ holds only in a *small* neighborhood of $x = x^*$. How *small* this neighborhood can be is quantified by introducing a fuzzy set describing an acceptable error of approximation (or briefly, the quality of linear approximation). The approximation error can be quantified as $F(x)$. In particular, one considers

$$F(x) = |f(x) - f(x^*)|,$$

forming a suitable measure quantifying the approximation effect. This function, in turn, can be used to construct a membership function of the term A describing an *acceptable* linearization error. The membership function of A at x characterizes the extent to which the linearization of the function is acceptable. Assuming that the expression

$$M = \max_x F(x)$$

is finite, the membership function of A is defined as

$$A(x) = 1 - \frac{F(x) - F(x^*)}{M - F(x^*)}.$$

In general, we can admit as a model of A any monotonically increasing function

$\phi : [0, 1] \to [0, 1]$ such that $\phi(0) = 0$ and $\phi(1) = 1$, and

$$A(x) = \phi\left(1 - \frac{F(x) - F(x^*)}{M - F(x^*)}\right).$$

EXERCISE 1.11   The function $g(x) = \sin(0.5x)$ needs to be linearized around $x = 0$. Propose a model of a suitable membership function describing the quality (or relevance) of this linearization.

1.8.5   Membership Estimation as a Problem of Parametric Optimization

In a nutshell, the *parametric membership estimation* method is not aimed at constructing a certain membership function from scratch. The intent here is to fit a standard curve to some experimental data consisting of ordered pairs (element, membership value) denoted by $(x_k, M(x_k))$. The form of the parameterized membership function is given in advance and should reflect the nature of the concept to be described. (Several examples of commonly used membership functions were provided in section 1.3.) The parametric membership estimation procedure is as follows.

Let us assume a parameterized membership function $A(x; \mathbf{p})$ where $x \in \mathbf{X}$, and $\mathbf{p}$ is a vector of its parameters in the appropriate parameter space $\mathbf{P}$. For given pairs of data $(x_k, M(x_k)), k = 1, \dots, N$, the vector of parameters $\mathbf{p}$ of the assumed membership function $A(x; \mathbf{p})$ needs to be determined. A commonly used procedure exploits the mean squared errors as the estimation criterion. Therefore the problem reads

$$\min_{\mathbf{p} \in \mathbf{P}} \sum_{k=1}^{N} [M(x_k) - A(x_k; \mathbf{p})]^2,$$

which is actually a nonlinear optimization problem and can be solved by choosing, for instance, an appropriate iterative, gradient-based algorithm (Bazaraa and Shett 1979).

1.8.6   Membership Estimation via Fuzzy Clustering

Fuzzy clustering forms another important class of membership estimation methods and is algorithmic in nature. Its primary objective is to partition a collection of numerical data into a series of overlapping clusters whose degrees of belongingness are interpreted as membership values. The method is concerned with a fuzzy partition (a family of fuzzy sets) of the universe of discourse rather than a single membership function. Various algorithms for fuzzy clustering are availabe, fuzzy isodata (Bezdek 1981) being among the most dominant. A description of the fuzzy isodata algorithm follows.

Consider N vectors $x_1, x_2, \ldots, x_N$ viewed as elements of an n-dimensional Euclidean space $\mathbf{R}^n$, equipped with a distance measure between any two elements x_i and x_j, denoted by $\|x_i - x_j\|$. The focus in the clustering method is the fuzzy partition matrix $F = [f_{ij}]$ consisting of c rows and N columns whose entries satisfy the following properties:

$$0 \le f_{ij} \le 1,$$

$$\sum_{i=1}^{c} f_{ij} = 1 \quad \forall j = 1, 2, \ldots, N,$$

$$0 < \sum_{j=1}^{N} f_{ij} < N \quad \forall i = 1, 2, \ldots, c.$$

The ith row of F provides a discrete membership function of the ith cluster, and therefore each f_{ij} denotes the grade of membership of the jth element x_j to the ith cluster. Let $\mathscr{F}$ be a family of all matrices fulfilling the properties above. The fuzzy isodata algorithm comprises the following steps:

Step 1. Find an initial partiton matrix $F(0) \in \mathscr{F}$. Set $k = 0$.

Step 2. Compute centroids $v_i(k)$ of the clusters using $F(k)$ as follows:

$$v_i(k) = \frac{\sum_{j=1}^{N} f_{ij}^2(k) x_j}{\sum_{j=1}^{N} f_{ij}^2(k)}$$

Step 3. Update the partition matrix to obtain $F(k+1) \in \mathscr{F}$ according to

$$(f_{ij}(k+1))^{-1} = \sum_{l=1}^{c} \left(\frac{\|x_j - v_i\|}{\|x_j - v_l\|} \right)^2$$

Step 4. Compare $F(k)$ with $F(k+1)$. If they are sufficiently alike, then stop. Otherwise, go to step 2.

It can be shown (Bezdek 1981) that the fuzzy isodata algorithm provides local solutions (or saddle points) to the following nonlinear optimization problem:

$$\min_{v_i, f_{ij}} \sum_{j=1}^{N} \sum_{i=1}^{c} f_{ij}^2 \|x_j - v_i\|^2$$

subject to

$$0 \le f_{ij} \le 1,$$

$$\sum_{i=1}^{c} f_{ij} = 1 \quad \forall j = 1, 2, \ldots, N,$$

$$0 < \sum_{j=1}^{N} f_{ij} < N \quad \forall i = 1, 2, \ldots, c.$$

1.9   Generalizations of Fuzzy Sets

A number of distinct generalizations have been made from the generic concept of partial membership conveyed by fuzzy sets. Although some of these extensions are primarily mathematical, others arise as an effect of difficulties in representing the concept of membership through single numerical values. The two outlined below fall under the second category, conveying application-oriented constructs. Their role in application of fuzzy sets will be explained through several specific constructs, particularly in fuzzy modeling, chapter 13.

1.9.1   Interval-Valued Fuzzy Sets and Second-Order Fuzzy Sets

Sambuc (1975) discusses a conceptually convincing idea: As there are difficulties in assigning single membership values, we admit numerical intervals defined in the unit interval. Thus an interval-valued fuzzy set $A^\wedge$ defined in $\mathbf{X}$ is a pair,

$$A^\wedge = (A_-, A_+),$$

where A_-, and A_+ are the lower and upper membership functions such that the inequality

$$A_-(x) \le A_+(x)$$

holds for each $x \in \mathbf{X}$; see figure 1.11. Here x_1 is assigned to the concept with degrees between 0.5 and 0.7, x_2 to any with a degree between 0.4 and 0.6, and so forth. Technically $A^\wedge$ can be treated and represented in any application as a pair of two fuzzy sets forming the bounds of $A^\wedge$.

The concept of second-order fuzzy set $A\sim$ exploits the above idea further by using fuzzy sets defined in the unit interval to capture uncertainty about membership grades. Figure 1.12 illustrates how this concept is implemented using triangular membership functions defined in the unit interval.

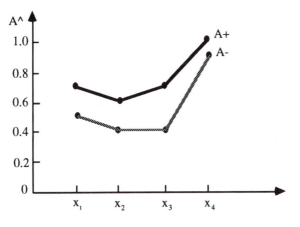

Figure 1.11
Interval fuzzy set $A^\wedge$

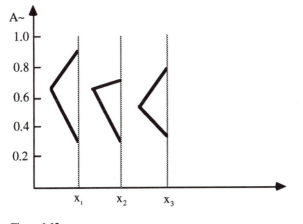

Figure 1.12
Second-order fuzzy set $A\sim$

1.9.2 Type-Two Fuzzy Sets

In type-two fuzzy sets, one assumes that the concept of membership is captured by some fuzzy set defined in the unit interval, instead of single membership values or intervals. In other words, for finite $\mathbf{X}$ we can look at the type-two fuzzy set as a collection of individual fuzzy sets. This type of generalization is convenient in organizing information about the concept under consideration. A brief example may help clarify this point. Consider traffic on a highway, which is usually a mixture of several categories of vehicles: trucks, buses, automobiles, motocycles, and so forth. To characterize the traffic, we specify an intensity for each of the categories of vehicles, the intensities being characterized by fuzzy sets. For instance we may have

traffic $= \{heavy/\text{trucks}, light/\text{motorcycles}, moderate/\text{automobiles},$
$\qquad light/\text{buses}\},$

where *heavy, light, moderate*, are relevant fuzzy sets in the space of traffic intensity attached to the corresponding category of the vehicles.

Although one could easily unleash the imagination and generalize even further, caution is a must. First, any "generalization" could prove less innovative than anticipated, and the resulting concept could be quite well represented in the already existing framework of fuzzy sets. Secondly, an innovation could lead to structures that would be almost impossible to construct effectively, for example, determining their membership functions (or the like). This happens with fuzzy sets of n-order ($n > 2$). For example, a third-order fuzzy set, $A\sim\sim$, is described by a family of membership functions defined in $[0, 1]$ whose values are again fuzzy sets in $[0, 1]$; this is fairly unrealistic. The process of building these sets is infinite through an elegant inductive definition, yet the obtained structures are, at least so far, out of any practical reach.

1.10 Chapter Summary

This chapter introduced the fundamental notions and concepts of fuzzy sets and showed that the main idea behind fuzzy sets is the graded distinction among elements of a universe of discourse. Interpretations at conceptual and geometrical levels were given to provide insights in application settings. Typical examples of membership functions and characteristics of fuzzy sets were presented as well. The idea that a fuzzy set can be viewed as a family of fuzzy sets was demonstrated by the representation theorem. An important mechanism to extend pointwise concepts to concepts involving fuzzy sets was conveyed be the extension principle. Since membership function estimation is a central issue in analyzing, designing, and implementing fuzzy systems, experimental and algorithmic methods for estimation were provided. Extensions of fuzzy sets were also discussed.

1.11   Problems

1. What is the characteristic function of the set $E = \{x \mid x \neq x\}$?

2. Assume you were told that the room temperature is *around* 20°C. How would you represent this piece of information, or concept, by (a) a set and (b) a fuzzy set? Ask someone else to answer this question independently and compare his answer with yours.

3. The middle point of a line segment is, at the same time, *close to* and *far from* its extreme points. How would you geometrically depict this idea through (a) a set and (b) a fuzzy set, both in the unit square?

4. Given the fuzzy set A with the following membership function

$$A(x) = \begin{cases} x - 5, & \text{if } 5 \leq x \leq 6 \\ -x + 7, & \text{if } 6 < x \leq 7 \\ 0, & \text{elsewhere.} \end{cases}$$

   (a) Sketch the graph of the function. What is its type?

   (b) How would you attach a linguistic description to the concept conveyed by A?

5. Consider the fuzzy set A whose membership functions is

$$A(x) = \begin{cases} 0.5(x/3 - 5/3), & \text{if } 5 \leq x \leq 8 \\ -0.5(x/3 - 11/3), & \text{if } 8 < x \leq 11 \\ 0, & \text{elsewhere.} \end{cases}$$

   (a) Sketch the graph of the function. What is its height? Is it normal?

   (b) Perform the operations of concentration, dilation, contrast intensification, and fuzzification for the normalized (if A is not normal) fuzzy set. After performing these operations, what can you say about them and the normalized (again, if A is not normal) fuzzy set regarding the inclusion and equality relationships.

6. Is $S(A, B) = S(B, A)$? Illustrate your answer with an example.

7. Assume fuzzy set $A = 1.0/1 + 0.8/2 + 0.5/3 + 0.1/4$ defined in $\mathbf{X} = \{1, 2, 3, 4, 5\}$. Find all of its α-cuts. Show how A can be expressed in terms of the family composed of all of its α-cuts.

8. Let $\mathbf{X} = \{1, 2, 3, 4\}, \mathbf{Y} = \{1, 2, 3, 4, 5, 6\}$, and $y = f(x) = x^2 + 1$. Given the fuzzy set A, with $A = 0.2/1 + 0.1/2 + 1.0/3 + 0.3/4$, compute the image $B = f(A)$. Sketch this transformation graphically.

9. One can sometimes come up with a more qualitative motivation behind the use of some families of the membership functions. Let us consider that the rate of membership change at point x, say $\partial A / \partial x$, must rise if our confidence that x belongs to A is strengthened. Expressing this fact analytically, we produce the relation

$$\frac{\partial A}{\partial x} = kA(x)[1 - A(x)],$$

where $k > 0$ is a certain constant. Find the membership function that follows this rationale.

10. The geometric representation of a fuzzy set as a point in a unit hypercube has already been discussed. How could you go about making a similar interpretation of the generalized version of fuzzy sets, such as interval-valued fuzzy sets and second-order fuzzy sets?

References

Bazaraa, M., and C. Shetty. 1979. *Nonlinear Programming*. New York: John Wiley and Sons.

Bezdek, J. 1981. *Pattern Recognition with Fuzzy Objective Function Algorithms*. New York: Plenum Press.

Dubois, D., H. Prade, and R. R. Yager, eds. 1993. *Readings in Fuzzy Sets for Intelligent Systems*. San Mateo, CA: Morgan Kaufmann.

Kandel, A. 1986. *Fuzzy Mathematical Techniques with Applications*. Reading, MA: Addison-Wesley.

Klir, G. J., and Folger, T. A. 1988. *Fuzzy Sets, Uncertainty and Information*. Englewood Cliffs, NJ: Prentice Hall.

Kosko, B. 1992. *Neural Networks and Fuzzy Systems: A Dynamical Systems Approach to Machine Intelligence*. Englewood Cliffs, NJ: Prentice Hall.

Pedrycz, W. 1995. *Fuzzy Sets Engineering*. Boca Raton, FL: CRC Press.

Saaty, T. L. 1980. *The Analytic Hierarchy Processes*. New York: McGraw Hill.

Sambuc, R. 1975. *Fonctions d'F-flous: Application a l'aide au diagnostic en pathologie thyroidienne*. Ph. D. thesis, Universite de Marseille.

Yager, R., S, Ovchinnikov, R. Tong, and H. Nguyen, eds. 1987. *Fuzzy Sets and Applications: Selected Papers by L. A. Zadeh*, New York: John Wiley and Sons.

Zadeh, L. A. 1965. Fuzzy sets. *Information and Control* 8:338–53.

Zadeh, L. A. 1971. Toward a theory of fuzzy systems. In *Aspects of Network and System Theory*, ed. R. E. Kalman and N. De Claris. New York: Holt, Rinehart and Winston.

Zadeh, L. A. 1981. Possibility theory and soft data analysis. In *Mathematical Frontiers of Social and Policy Sciences*, ed. L. Cobb and R. Thrall, 69–129. Boulder, CO: Westview Press.

2

Fuzzy Set Operations

To perform operations on sets means to combine, compare, or aggregate sets. Set operations allow constructs that are of an utmost importance in any situation involving information and data processing. This chapter introduces the essential fuzzy set operations, emphasizing the key issues of these operations and their proper and context-dependent interpretations. Because of the richness of the fuzzy sets, we study several general classes of operations, of which set operations are merely special cases.

2.1 Set Theory Operations and Their Properties

Before getting into the operations on fuzzy sets themselves, it is instructive to review the basic operations encountered in set theory and highlight the relationships between them and their fuzzy set relatives. The main properties of these operations form a suitable reference one can apply when analyzing various models of fuzzy set operations: in particular, the properties of commutativity, idempotency, associativity, distributivity, and transitivity. Because characteristic functions are equivalent representations of sets, the basic intersection, union and complement operations are conveniently represented by taking the minimum, maximum, and one-complement of the corresponding characteristic functions for all $x \in \mathbf{X}$:

$$(A \cap B)(x) = \min(A(x), B(x)) = A(x) \wedge B(x),$$

$$(A \cup B)(x) = \max(A(x), B(x)) = A(x) \vee B(x),$$

$$\bar{A}(x) = 1 - A(x),$$

where A and B are sets defined in a universe $\mathbf{X}$ (universe of discourse), and $(A \cap B)(x)$ and $(A \cup B)(x)$ denote the membership functions of the sets resulting from the intersection and union of A and B, respectively.

EXERCISE 2.1 Perform the operations of intersection, union, and complement in terms of characteristic functions, considering $A = \{x \in \mathbf{R} \mid 1 \leq x \leq 3\}$ and $B = \{x \in \mathbf{R} \mid 2 \leq x \leq 4\}$. Show the results graphically .

Though the following features have been expressed in standard set notation, they can also be expressed in the language of characteristic functions:

1. Commutativity:

$A \cup B = B \cup A$

$A \cap B = B \cap A$

2. Associativity:

$$A \cup (B \cup C) = (A \cup)B \cup C = A \cup B \cup C$$

$$A \cap (B \cap C) = (A \cap)B \cap C = A \cap B \cap C$$

3. Idempotency:

$$A \cup A = A$$

$$A \cap A = A$$

4. Distributivity:

$$A \cap (B \cup C) = (A \cap B) \cup (A \cap C)$$

$$A \cup (B \cap C) = (A \cup B) \cap (A \cup C)$$

5. Boundary Conditions:

$$A \cup \varnothing = A, \quad A \cup \mathbf{X} = \mathbf{X}$$

$$A \cap \varnothing = \varnothing, \quad A \cap \mathbf{X} = A$$

6. Involution:

$$\bar{\bar{A}} = A$$

7. Transitivity:

$$A \subset B \text{ and } B \subset C \text{ implies } A \subset C$$

The following relationships express properties of containment and cardinality and are directly implied by the properties outlined above:

$$(A \cap B) \subset A \subset (A \cup B)$$

If $A \subset B$, then $A = A \cap B$ and $B = A \cup B$

$$\text{Card}(A) + \text{Card}(B) = \text{Card}(A \cap B) + \text{Card}(A \cup B)$$

$$\text{Card}(A) + \text{Card}(\bar{A}) = \text{Card}(\mathbf{X}).$$

2.2 Triangular Norms

The concept of triangular norms comes from the ideas of so-called probabilistic metric spaces originally proposed by Menger (1942) and Schweizer and Sklar (1983) and actively pursued by Weber (1983) and Butnariu and Klement (1993), to name a few. In these spaces, a geometric concept of a triangular inequality was placed in the setting of probability theory. In fuzzy set theory, triangular norms play a key role by providing generic models for intersection and union operations on fuzzy sets, which must possess the properties of commutativity, associativity and monotonicity. Boundary conditions must also be satisfied, to assure that they behave as set operations. Therefore, triangular norms form general classes of intersection and union operators. Formally they are defined as follows.

DEFINITION 2.1 A *triangular norm* (*t-norm*) is a binary operation $t : [0, 1]^2 \to [0, 1]$ satisfying the following requirements:

- Commutativity: $x \, t \, y = y \, t \, x$
- Associativity: $x \, t \, (y \, t \, z) = (x \, t \, y) t \, z$
- Monotonicity: If $x \leq y$ and $w \leq z$, then $x \, t \, w \leq y \, t \, z$
- Boundary conditions: $0 \, t \, x = 0, 1 \, t \, x = x$

As a formal construct, s-norms are dual to t-norms.

DEFINITION 2.2   An *s-norm*, also known as a *triangular co-norm* is a binary operation $s : [0, 1]^2 \to [0, 1]$ satisfying the following requirements:

- Commutativity: $x \, s \, y = y \, s \, x$
- Associativity: $x \, s \, (y \, s \, z) = (x \, s \, y) \, s \, z$
- Monotonicity: if $x \leq y$ and $w \leq z$ then $x \, s \, w \leq y \, s \, z$
- Boundary conditions: $x \, s \, 0 = x, x \, s \, 1 = 1.$

Clearly, the min operator ($\wedge$) is a t-norm, whereas the max operator ($\vee$) is an s-norm. They correspond to set intersection and union operators, respectively, when membership degrees are constrained to the two-element set of grades of belongingness $\{0, 1\}$. Thus, they may be regarded as natural extensions to fuzzy sets of set intersection and union operations (figure 2.1). As further examples, selected instances of triangular norms are assembled below. Some are frequently encountered in the literature.

Examples of t-norms:

- $x \, t_1 \, y = \dfrac{1}{1 + \sqrt[p]{((1 - x)/x)^p + ((1 - y)/y)^p}}, \quad p > 0$
- $x \, t_2 \, y = \max(0, (1 + p)(x + y - 1) - pxy), \quad p \geq -1$
- $x \, t_3 \, y = 1 - \min(1, \sqrt[p]{(1 - x)^p + (1 - y)^p}), \quad p > 0$
- $x \, t_4 \, y = xy$
- $x \, t_5 \, y = \dfrac{xy}{p + (1 - p)(x + y - xy)}, \quad p \geq 0$

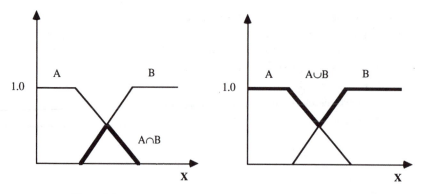

Figure 2.1
Intersection and union of fuzzy sets based on min and max operators

- $x \, t_6 \, y = \dfrac{1}{\sqrt[p]{1/x^p + 1/y^p - 1}}$

- $x \, t_7 \, y = \sqrt[p]{\max(0, x^p + y^p - 1)}, \quad p > 0$

- $x \, t_8 \, y = \dfrac{xy}{\max(x, y, p)}, \quad p \in [0, 1]$

- $x \, t_9 \, y = \log_p \left[1 + \dfrac{(p^x - 1)(p^y - 1)}{p - 1} \right], \quad p > 0, p \neq 1$

- $x \, t_{10} \, y = \dfrac{1}{1 + \sqrt[p]{((1 - x)/x)^p + ((1 - y)/y)^p}}, \quad p > 0$

- $x \, t_{11} \, y = \begin{cases} x, & \text{if } y = 1 \\ y, & \text{if } x = 1 \\ 0, & \text{otherwise} \end{cases}$

Figure 2.2 plots some of these triangular norms as a function of a single variable (y has been fixed equal to 0.6).

Examples of s-norms:

- $x \, s_1 \, y = \dfrac{1}{1 + \sqrt[p]{(x/1 - x)^p + (y/1 - y)^p}}, \quad p > 0$

- $x \, s_2 \, y = \min(1, x + y + pxy), \quad p \geq 0$

- $x \, s_2 \, y = \min(1, \sqrt[p]{x^p + y^p}), \quad p > 0$

- $x \, s_4 \, y = x + y - xy$

- $x \, s_5 \, y = \dfrac{x + y - xy - (1 - p)xy}{1 - (1 - p)xy}, \quad p \geq 0$

- $x \, s_6 \, y = 1 - \dfrac{1}{\sqrt[p]{1/(1 - x)^p + 1/(1 - x)^p - 1}}$

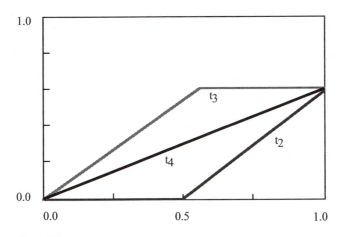

Figure 2.2
Selected t-norms; $p = 0.5$

- $x\,s_7\,y = 1 - \max(0, \sqrt[p]{((1-x)^p + (1-y)^p - 1)}), \quad p > 0$

- $x\,s_8\,y = 1 - \dfrac{(1-x)(1-y)}{\max((1-x),(1-y),p)}, \quad p \in [0,1]$

- $x\,s_9\,y = \log_p\left[1 + \dfrac{(p^{1-x}-1)(p^{1-y}-1)}{p-1}\right], \quad p > 0, p \neq 1$

- $x\,s_{10}\,y = \dfrac{1}{1 - \sqrt[p]{(x/1-x)^p + (y/1-y)^p}}, \quad p > 0$

- $x\,s_{11}\,y = \begin{cases} x, & \text{if } y = 0 \\ y, & \text{if } x = 0 \\ 1, & \text{otherwise} \end{cases}$

As figure 2.2 did for t-norms, figure 2.3 plots selected s-norms (again, y is equal to 0.6).

Triangular norms cannot be linearly ordered. However, their bounds are easily identified, namely:

$x\,t_{11}\,y \le x\,t\,y \le \min(x,y),$

that is, the min operator is the biggest t-norm, while t_{11}, known also as a *drastic product*, forms the lower bound on the family of t-norms. Similarly, the bounds for the family of s-norms are identified as

$\max(x,y) \le x\,s\,y \le x\,s_{11}\,y,$

where now the maximum operation constitutes the lowest bound. The *drastic sum*, s_{11}, is the upper bound of the family of all s-norms.

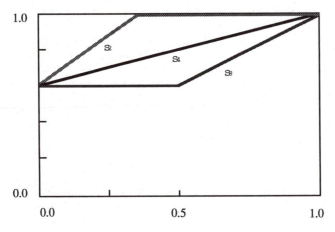

Figure 2.3
Selected s-norms; $p = 0.5$

For each t-norm there exists a dual s-norm; this means that

$$x \, s \, y = 1 - (1 - x) \, t \, (1 - y),$$

and alternatively,

$$x \, t \, y = 1 - (1 - x) \, s \, (1 - y).$$

Once these two relationships are rewritten in the form

$$1 - x \, s \, y = (1 - x) \, t \, (1 - y), \quad 1 - x \, t \, y = (1 - x) \, s \, (1 - y),$$

it becomes obvious that they are just De Morgan's laws, commonly encountered in set theory:

$$\overline{A \cup B} = \bar{A} \cap \bar{B},$$

$$\overline{A \cap B} = \bar{A} \cup \bar{B}.$$

The complement, $\bar{A}$, of a fuzzy set A is defined analogously to its definition in set theory, namely:

$$\bar{A}(x) = 1 - A(x).$$

According to this definition, the complement is *involutive*, meaning that

$$\bar{\bar{A}}(x) = A(x),$$

so it behaves analogously to its well-known counterpart in set theory.

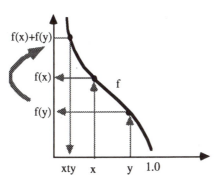

Figure 2.4
Construction of t-norms based on additive generators

2.2.1   Several Classes of Triangular Norms

Following the general definitions outlined above, it is worth specifying some specific classes of triangular norms.

DEFINITION 2.3   A t-norm is Archimedean if it is continuous and $x \, t \, x < x$ for all $x \in (0, 1)$. Moreover an Archimedean t-norm is strict if t is strictly increasing on $(0, 1) \times (0, 1)$.

For strict Archimedean t-norms an interesting representation is provided in terms of so-called additive generators, namely, continuous and decreasing function from [0,1] onto $\mathbf{R}_+$:

$$f : [0, 1] \rightarrow [0, \infty).$$

Any strict Archimedean t-norm can be generated in the form

$$x \, t \, y = f^{-1}[f(x) + f(y)],$$

and, vice versa, any t-norm from the above class implies the corresponding additive generator. Figure 2.4 illustrates the construction of t-norms using additive generators.

The product $x \, t \, y = xy$ is an example of a strict Archimedean t-norm. Its additive generator is obtained by solving the following functional equation (that is, a relationship whose unknown is a function):

$$xy = f^{-1}[f(x) + f(y)].$$

One can verify that the function $f(x) = -\log x$ satisfies the above relationship. Indeed, we have

$$f(x) + f(y) = -\log x - \log y = -[\log x + \log y] = -\log xy.$$

Therefore

$$f^{-1}(u) = e^{-u},$$

and finally

$$f^{-1}[f(x) + f(y)] = e^{\log xy} = xy.$$

Observe that $f(x) = -c \log x, c > 0$, is also the additive generator, as verified through simple computation:

$$f(x) + f(y) = -c \log xy$$

$$f^{-1}(u) = e^{-u/c}$$

$$f^{-1}[f(x) + f(y)] = e^{c \log xy/c} = xy.$$

Interestingly, any additive generator is unique up to a positive multiplicative constant.

Similarly, a t-conorm (s-norm) is said to be Archimedean if it satisfies

$$x \, s \, x > x$$

for all $x \in (0, 1)$. The additive generator of any continuous Archimedean s-norm is a continuous and strictly increasing function $g : [0, 1] \to [0, \infty)$ with $g(0) = 0$ such that

$$x \, s \, y = g^{-1}[g(1) \wedge (g(x) + g(y))].$$

If $g(1)$ is finite, $g(1) < \infty$, then the resulting s-norm is nilpotent. Again, the additive generators of s-norms are unique up to a positive constant.

2.2.2 Triangular Norms as Models of Operations on Fuzzy Sets

In general, triangular norms do not satisfy the laws of contradiction and excluded middle, $A \cap \bar{A} \neq \varnothing$ and $A \cup \bar{A} \neq \mathbf{X}$, respectively. This can be easily checked via the min and max operations; more specifically, we obtain a so-called overlap and underlap effect where the intersection and union of the fuzzy set and its complement do not produce the empty set and the universe, respectively. These two contradictions of the laws arise through operating on membership values between 0 and 1: Observe that the most profound deviation occurs for the membership value $1/2$. Consider A with the constant membership value of $1/2$ over $\mathbf{X}$. In this situation, the minimum and maximum operations return the same value for intersection and union of A and its complement,

$$\min[A(x), 1 - A(x)] = \min(1/2, 1/2) = 1/2,$$

$$\max[A(x), 1 - A(x)] = \max(1/2, 1/2) = 1/2,$$

for each x in **X**. One can say that A is the most fuzzy set among all fuzzy sets defined in **X**. In other words, the degree to which a particular set fails to meet the requirements of excluded middle and contradiction can serve as a useful measure of how fuzzy that set is. We will return to this way of evaluating fuzziness in chapter 3. An exception is a bounded sum (t_2 with $p = 0$) and bounded product (s_2 with $p = 0$); they always produce

$$(A \cap \bar{A})(x) = \max[0, A(x) + (1 - A(x)) - 1] = 0,$$

$$(A \cup \bar{A})(x) = \min[1, A(x) + (1 - A(x))] = 1.$$

These two operations were introduced by J. Lukasiewicz (1920) in his model of three-valued and many-valued logics; although evident extensions of the model of two-valued logic, they were still somewhat conservative because they retained the excluded middle and contradiction—the two fundamental principles of two-valued logic.

As far as the remaining properties of set theory are concerned, the operators defined through triangular norms have both commutativity and associativity owing to the definitions themselves. The set requirement of idempotency, as expressed in terms of t-norms and s-norms, states that

$x \, t \, x = x,$

$x \, s \, x = x.$

Because the property of associativity holds true, we get

$(x \, t \, x) \, t \, x = x,$

$(x \, s \, x) \, s \, x = x.$

In general,

$x \, t \, x \, t \ldots t \, x = x,$

$x \, s \, x \, s \ldots s \, x = x.$

In other words, no matter how many identical fuzzy sets are combined, the result is not affected. The idempotency property, as outlined above, does not hold for all triangular norms. On the contrary: The only meaningful idempotent triangular norms are the maximum and minimum operations. It should also be noted that when t-norms are iterated, the sequence of obtained values decreases as the number of arguments involved in the pertinent computation increases:

$x \, t \, x \ldots x \, t \, x \leq x \, t \, x \ldots x \, t \, x.$
n times $\qquad (n - 1)$ times

On the other hand, the s-norms generate an increasing sequence of membership values when iterated, meaning that

$$\underbrace{x \, s \, x \ldots x \, s \, x}_{n \text{ times}} \geq \underbrace{x \, s \, x \ldots x \, s \, x}_{(n-1) \text{times}}.$$

In general, the distributivity property is not satisfied for triangular norms. An exception is noted when we consider the pair min and max norms to model the intersection and union, respectively.

A point worth noting here concerns the interpretation of set operations in terms of logical connectives. Supported by the isomorphism between set theory and propositional two-valued logic, the intesection and union operations can be identified with the conjunctive (AND) and disjunctive (OR) connectives, respectively. This is also the case when triangular norms are viewed as general conjunctive and disjunctive connectives within the framework of a multivalued logic, as is explored in chapter 8.

2.3   Aggregation Operations on Fuzzy Sets

Elements of a collection of fuzzy sets can be combined to produce a single fuzzy set through aggregation operations. This is the case, for instance, in the intersection and union of any number of fuzzy sets. In general, many other types of aggregation may be performed on fuzzy sets, the operations being characterized formally as follows.

DEFINITION 2.4   An aggregation is an n-ary operation $A : [0,1]^n \to [0,1]$ satisfying the following requirements:

- Boundary conditions: $A(0,\ldots,0) = 0$ and $A(1,\ldots,1) = 1$
- Monotonicity: $A(x_1,\ldots,x_n) \geq A(y_1,\ldots,y_n)$ if $x_i \geq y_i$, $i = 1,\ldots,n$

Thus, triangular norms clearly are aggregation operators.

2.3.1   Compensatory Operators

The plain set-theoretic operations or their corresponding logic operations, as determined by experiments (see Greco and Rocha 1987), may not model well in some experimental findings. For instance, Zimmermann and Zysno (1980) revealed through experiments that the minimum operator does not work well as a model of the AND connective, producing too conservative (low) results: The experiments suggested a sort of averaging effect instead. In the same experiment, the maximum operator performed much better. To alleviate this deficiency, Zimmermann and

Zysno proposed a family of so-called compensatory operators in which the operation of aggregation arises as a sort of combination of "pure" logical AND and OR connectives providing the required mechanism of compensation. The operation defined in Zimmermann and Zysno (1980) is expressed as

$$(A \ominus B)(x) = [(A \cap B)(x)]^{1-\gamma}[(A \cup B)(x)]^{\gamma},$$

where $\cap$ and $\cup$ are intersection and union operations, whereas γ stands for the compensation factor, $\gamma \in [0, 1]$, indicating where the actual operator is located between AND and OR. There are also other ways to come up with a similar effect of compensation, say

$$(A \otimes B)(x) = (1 - \gamma)[(A \cap B)(x)] + \gamma[(A \cup B)(x)].$$

The intersection and union operations can be modeled using either standard min and max or, in general, triangular norms.

As an example, we study the data set coming from Zimmermann and Zysno (1980), as presented in table 2.1. It consists of 24 pairs of membership values $A(x)$ and $B(x)$ along with the experimentally collected results of their aggregation. As seen in figures 2.5 and 2.6, neither min nor algebraic product operations perform well on this data. Also, the max

Table 2.1
Zimmermann-Zysno data

	$A(x)$	$B(x)$	Target
1	0.426	0.241	0.215
2	0.352	0.662	0.427
3	0.109	0.352	0.221
4	0.630	0.052	0.212
5	0.484	0.496	0.486
6	0.000	0.000	0.000
7	0.270	0.403	0.274
8	0.156	0.130	0.119
9	0.790	0.284	0.407
10	0.725	0.193	0.261
11	1.000	1.000	1.000
12	0.330	0.912	0.632
13	0.949	0.020	0.247
14	0.202	0.826	0.500
15	0.744	0.551	0.555
16	0.572	0.691	0.585
17	0.041	0.975	0.355
18	0.534	0.873	0.661
19	0.674	0.587	0.570
20	0.440	0.450	0.418
21	0.909	0.750	0.789
22	0.856	0.091	0.303
23	0.974	0.164	0.515
24	0.073	0.788	0.324

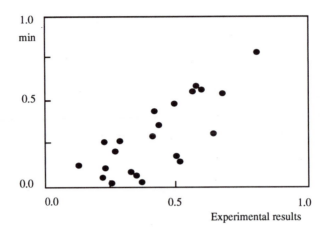
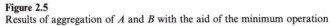

Figure 2.5
Results of aggregation of *A* and *B* with the aid of the minimum operation

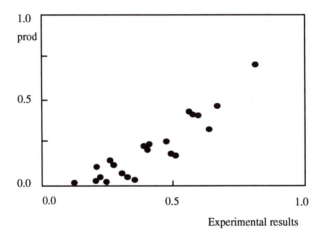
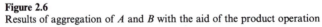

Figure 2.6
Results of aggregation of *A* and *B* with the aid of the product operation

operator is not suitable, as visualized in table 2.1. Observe that an ideal operator should distribute all these points along a straight line.

2.3.2   Symmetric Sums

Symmetric sums provide another option for aggregating membership values. More formally, these are n-argument functions, denoted by S_sum, such that in addition to fulfilling boundary conditions and monotonicity, they are continuous, commutative, and auto-dual:

$$S\_sum(x_1, \ldots, x_n) = 1 - S\_sum(1 - x_1, \ldots, 1 - x_n).$$

Dubois and Prade (1980) showed that any symmetric sum can be represented in the form

$$S\_sum(x_1, \ldots, x_n) = \left[1 + \frac{\rho(1 - x_1, \ldots, 1 - x_n)}{\rho(x_1, \ldots, x_n)} \right]^{-1},$$

with ρ being any increasing continuous function with $\rho(0, \ldots, 0) = 0$.

2.3.3   Averaging Operation

An averaging (generalized mean) operator defined for n arguments is idempotent and commutative, in addition to fulfilling monotonicity and boundary conditions. As discussed in Dyckhoff and Pedrycz (1984), the generalized mean takes on the form

$$A(x_1, \ldots, x_n) = \sqrt[p]{\frac{1}{n} \sum_{i=1}^{n} (x_i)^p}, \quad p \in \mathbf{R}, p \neq 0.$$

It is important to note that the family of generalized means subsumes some well-known cases of fuzzy-set operators. In particular, we obtain

1. Arithmetic Mean ($p = 1$) $A(x_1, \ldots, x_n) = \frac{1}{n} \sum_{i=1}^{n} (x_i).$

2. Geometric Mean ($p \to 0$) $A(x_1, \ldots, x_n) = (x_1, x_2 \ldots, x_n)^{1/n}.$

3. Harmonic Mean ($p = -1$) $A(x_1, \ldots, x_n) = \dfrac{n}{\sum_{i=1}^{n} 1/x_i}.$

4. Minimum ($p \to -\infty$) $A(x_1, \ldots, x_n) = \min(x_1, x_2, \ldots, x_n).$

5. Maximum ($p \to \infty$) $A(x_1, \ldots, x_n) = \max(x_1, x_2, \ldots, x_n).$

The parameter p can be referred to as a compensation factor.

Similarly, as discussed in Dyckhoff and Pedrycz (1984), we introduce a grade of compensation that becomes a core of the compensative operators. It is viewed as a strictly monotone increasing mapping from $\mathbf{R}$ onto

$[0, 1], \gamma : \mathbf{R} \to [0, 1]$. Two examples of this mapping are

$$\gamma = 0.5\left(1 + \frac{p}{1 + |p|}\right),$$

$$\gamma = 0.5\left(1 + \frac{2}{\pi}\arctan(p)\right).$$

These forms of the compensation grade yield

- $p = -\infty,\quad \gamma = 0$ (minimum)
- $p = -1,\quad \gamma = -0.25$ (harmonic mean)
- $p = 0,\quad \gamma = 0.50$ (geometric mean)
- $p = 1,\quad \gamma = 0.75$ (arithmetic mean)
- $p = \infty,\quad \gamma = 1$ (maximum)

A certain generalization of the above operations is a family of so-called quasi-arithmetic means of the form

$$A(x_1, \ldots, x_n) = f\left(\sqrt[p]{[f^{-1}(x_1, \ldots, x_n)]^p}\right),$$

where f is any continuous, strictly monotonic function. The operators of this form constitute a class of all decomposable continuous averaging operators (cf. Butnariu and Klement 1993). In general, we can introduce a class of averaging operators as a family of functions

$$v : [0, 1] \times [0, 1] \to [0, 1],$$

that satisfy the following requirements (Berg and Schwarz 1995):

1. Boundary Conditions: $\min(x, y) \le Av(x, y) \le \max(x, y)$ and $Av(0, 0) = 0$ and $Av(1, 1) = 1$.

2. Idempotency: $Av(x, x) = x$.

3. Commutativity: $Av(x, y) = Av(y, x)$.

4. v is increasing and continuous.

2.3.4   Ordered Weighted Averaging Operations

The family of ordered weighted averaging (OWA) operations proposed in Yager (1988) can be regarded as a class of weighted operators. Let $\mathbf{w} = (w_1, w_2, \ldots, w_n)$ be a vector of weights such that

$$\sum_{i=1}^{n} w_i = 1.$$

Let the sequence of membership values $\{A(x_i)\}$ be ordered: $A(x_1) \leq A(x_2) \leq \cdots \leq A(x_n)$. Thus,

$$\text{OWA}(A, \mathbf{w}) = \sum_{i=1}^{n} w_i A(x_i).$$

Essentially, the OWA is a weighted sum whose arguments are ordered. Let us analyze the behavior of the OWA operator for some selected weight vectors.

1. If $\mathbf{w} = (1\ 0 \ldots 0)$ owing to the organization of the arguments, we immediately obtain

$$\text{OWA}(A, (1, 0, \ldots, 0)) = A(x_1) = \min[A(x_1), \ldots, A(x_n)].$$

2. If $\mathbf{w} = (0 \ldots 0\ 1)$ then OWA is the maximum operation:

$$\text{OWA}(A, (0, 0, \ldots, 1)) = A(x_n) = \max[A(x_1), \ldots, A(x_n)].$$

3. If $\mathbf{w} = (1/n, \ldots, 1/n)$ then the OWA becomes a standard mean:

$$\text{OWA}(A, \mathbf{w}) = \frac{1}{n} \sum_{i=1}^{n} A(x_i).$$

2.4 Sensitivity of Fuzzy Set Operators

Briefly stated, the *sensitivity* of a fuzzy connective characterizes the extent to which the changes in the values of membership affect the result of aggregation. As usual, this impact can be quantified with the aid of partial derivatives. For the time being, let any differentiable t-norm be used as a model of the logical operator. Then the derivatives $\partial(x\ t\ y)/\partial x$ and $\partial(x\ t\ y)/\partial y$ are viewed as two local measures of sensitivity—they pertain to certain specific membership values, x and y. To express a global measure of sensitivity, $S_t(x, y)$, one can integrate the above derivatives over the unit square by considering the expressions of the form

$$S_t(x, y) = \int_0^1 \int_0^1 \left[\left| \frac{\partial(x\ t\ y)}{\partial x} \right| + \left| \frac{\partial(x\ t\ y)}{\partial y} \right| \right] dx\, dy$$

or

$$S_t(x, y) = \int_0^1 \int_0^1 \left[\left(\frac{\partial(x\ t\ y)}{\partial x} \right)^2 + \left(\frac{\partial(x\ t\ y)}{\partial y} \right)^2 \right] dx\, dy.$$

For example, consider the min operator and product, commonly used examples of t-norms. We have

$$\frac{\partial[\min(x,y)]}{\partial x} = \frac{\partial}{\partial x}\begin{cases} x, & \text{if } x \le y \\ y, & \text{if } x > y \end{cases} = \begin{cases} 1, & x \le y \\ 0, & \text{otherwise,} \end{cases}$$

and the above derivative becomes an indicator function describing the inclusion condition $x \le y$, $\mathbf{I}(x - y)$, for y fixed. (Here $\mathbf{I}$ stands for the unit function.) The product operator yields

$$\frac{\partial(xy)}{\partial x} = y,$$

so its local sensitivity measure depends linearly with respect to the values of y. These local measures depend very much on the triangular norms used. Observe, however, that the integrals for these two t-norms,

$$\int_0^1 \frac{\partial(xy)}{\partial x}\, dx = \int_0^1 y\, dx = y,$$

$$\int_0^1 \frac{\partial[\min(x,y)]}{\partial x}\, dx = \int_0^1 \begin{cases} 1, & x \le y \\ 0, & \text{otherwise} \end{cases} dx = \int_0^1 y\, dx = y,$$

are equal. In other words, the local sensitivities can differ significantly despite the fact that globally the sensitivities are the same. Completing the integration as indicated by $S_t(x,y)$, we get

1. for the minimum operation:

$$\int_0^1\int_0^1 \frac{\partial[\min(x,y)]}{\partial x}\, dx\, dy + \int_0^1\int_0^1 \frac{\partial[\min(x,y)]}{\partial y}\, dx\, dy$$

$$= \int_0^1\int_0^1 \begin{cases} 1, & \text{if } x \le y \\ 0, & \text{otherwise} \end{cases} dx\, dy + \int_0^1\int_0^1 \begin{cases} 1, & \text{if } y \le x \\ 0, & \text{otherwise} \end{cases} dx\, dy$$

$$= \frac{1}{2} + \frac{1}{2} = 1.$$

2. for the product operator:

$$\int_0^1\int_0^1 \frac{\partial(xy)}{\partial x}\, dx\, dy + \int_0^1\int_0^1 \frac{\partial(xy)}{\partial y}\, dx\, dy$$

$$= \int_0^1\int_0^1 y\, dx\, dy + \int_0^1\int_0^1 x\, dy\, dx$$

$$= \frac{1}{2} + \frac{1}{2} = 1.$$

Obviously, the sensitivity of s-norms as well as negations (see section 2.5) can be defined in the same way as outlined above.

2.5   Negations

The notion of the complement of A can be generalized by studying a so-called negation operation,

$$N : [0, 1] \rightarrow [0, 1],$$

fulfilling the following conditions:

- Monotonicity: N is nonincreasing
- Boundary conditions: $N(0) = 1, N(1) = 0$

Further properties may be also imposed as requirements, if necessary. Additional features that may be deemed essential involve

- Continuity: N is a continuous function
- Involution: $N(N(x)) = x$ for $x \in [0, 1]$

For instance, if N is a threshold function,

$$N(x) = \begin{cases} 1, & \text{if } x < a \\ 0, & \text{if } x \geq a, \end{cases}$$

$a \in [0, 1]$, then it satisfies monotonicity and boundary conditions; however, as can be easily verified, this function is not involutive.

Some other examples of negations are

$$N(x) = \begin{cases} 1, & \text{if } x = 0 \\ 0, & \text{if } x > 0 \end{cases}$$

$$N(x) = \frac{1 - x}{1 + \lambda x}, \lambda \in (-1, \infty)$$

$$N(x) = \sqrt[w]{1 - x^w}, \quad w \in (0, \infty)$$

The last two examples are both involutive; moreover, for $\lambda = 0$ and $w = 1$, they become the original complement function. Figure 2.7 illustrates the negation operator for various values of λ.

We can also develop a formal system of logic operations (formed by triangular norms and negations). In the system (t, s, N), we call a t-norm and an s-norm dual with respect to N iff

$$x \text{ s } y = N(N(x) \text{ t } N(y)),$$

or, equivalently,

$$x \text{ t } y = N(N(x) \text{ s } N(y)).$$

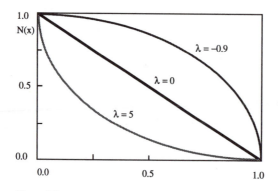

Figure 2.7
Negation operator $N(x) = \dfrac{1-x}{1+\lambda x}$, $\lambda \in (-1, \infty)$, for selected values of λ

Observe that the property of duality with respect to N differs from the property of duality between t-norms and s-norms previously discussed. Two examples of (t, s, N) systems are

1. $x \text{ t } y = \min(x, y)$,

   $x \text{ s } y = \max(x, y)$,

   $N(x) = 1 - x$.

2. $x \text{ t } y = \max\left(0, \dfrac{x + y - 1 + \lambda xy}{1 + \lambda}\right)$,

   $x \text{ s } y = \min(1, x + y - 1 + \lambda xy)$

   $N(x) = \dfrac{1-x}{1+\lambda x}$, $\lambda > 1$.

EXERCISE 2.2   Does the function $N(x) = \frac{1}{2}\{1 + \sin((2x + 1)\pi/2)\}$ qualify as a negation operation?

2.6   Comparison Operations on Fuzzy Sets

Fuzzy sets, as defined by membership functions, can be compared in different ways. Although the primary intent in comparison is to express the extent to which two fuzzy sets match, it is next to impossible to come up with a single method. Instead, we can enumerate several classes of methods available today for satisfying this objective. We review main classes of methods and highlight the aspects of matching supported by them.

2.6.1 Distance Measures

Distance measures consider a distance function between membership functions of fuzzy sets A and B and treat it as an indicator of their closeness. Because the computations involve two functions, they emphasize the functional facet of fuzzy sets. Comparing fuzzy sets via distance measures does not place the matching procedure in the set-theoretic perspective. In general, the distance between A and B, defined in the same universe of discourse $\mathbf{X}, \mathbf{X} \subseteq \mathbf{R}$, can be defined using the Minkowski distance

$$d(A, B) = \left[\int_{\mathbf{X}} |A(x) - B(x)|^p \, dx \right]^{1/p},$$

where $p \geq 1$; we assume that this integral exists. Several specific cases are typically encountered in applications:

1. Hamming distance ($p = 1$): $d(A, B) = \int_{\mathbf{X}} |A(x) - B(x)| \, dx$.
2. Euclidean distance ($p = 2$): $d(A, B) = [\int_{\mathbf{X}} [A(x) - B(x)]^2 \, dx]^{1/2}$.
3. Tchebyschev distance ($p = \infty$): $d(A, B) = \sup_{x \in \mathbf{X}} |A(x) - B(x)|$.

For discrete universes of discourse, the integration is replaced by the summation. The more similar the two fuzzy sets, the lower the distance function between them. For this reason, sometimes it is more convenient to normalize the distance function and denote it $d_n(A, B)$, and use this version to express similarity as a straight complementation, $1 - d_n(A, B)$.

2.6.2 Equality Indexes

Equality indexes concentrate on the logic-oriented expression of similarity (or difference) between fuzzy sets. As a prerequisite, let us recall that two sets are equal if A is included in B and, at the same time, B is included in A. This is captured by the two-valued predicate

truth$(A \subset B$ and $B \subset A) = true$.

As we are now concerned with fuzzy sets, the property of being equal can hold to a certain, if not necessarily the highest, degree. Admitting the above observation as a starting point of our construct, we introduce the following definition,

$$(A \equiv B)(x)$$

$$= 0.5\{[A(x) \varphi B(x)] \wedge [B(x) \varphi A(x)] + [\bar{A}(x) \varphi \bar{B}(x)] \wedge [\bar{B}(x) \varphi \bar{A}(x)]\},$$

where the conjunction of the conditions in the original predicate is modeled by the minimum operation while the inclusion is represented by

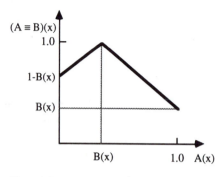

Figure 2.8
Equality index realized with aid of t_2 t-norm, $p = 0$

the φ-operator (referred to as a residuation) and induced by some continuous t-norm:

$$A(x) \, \varphi \, B(x) = \sup_{c \in [0,1]} [A(x) \, t \, c \le B(x)].$$

The intent of averaging the two terms in $A \equiv B$ is to make the equality index a more symmetrical function of the membership values under comparison. Note that the second term involves complements of the respective fuzzy sets. Figure 2.8 shows the plot of the equality index defined above. We have assumed the t-norm to be specified as t_2 with $p = 0$. This yields

$$A(x) \, \varphi \, B(x) = \begin{cases} 1, & \text{if } A(x) < B(x) \\ B(x) - A(x) + 1, & \text{if } A(x) \ge B(x), \end{cases}$$

and, consequently

$$(A \equiv B)(x) = \begin{cases} A(x) - B(x) + 1, & \text{if } A(x) < B(x) \\ B(x) - A(x) + 1, & \text{if } A(x) \ge B(x). \end{cases}$$

The above definition embraces a single element of the universe of discourse. To come up with the global measure of equality, the previous values need to be aggregated over **X**. Three basic mechanisms are of interest:

1. Optimistic equality index: $(A \equiv B)_{\text{opt}} = \sup_{x \in \mathbf{X}} (A \equiv B)(x)$.
2. Pessimistic equality index: $(A \equiv B)_{\text{pes}} = \inf_{x \in \mathbf{X}} (A \equiv B)(x)$.
3. Averaged equality index: $(A \equiv B)_{\text{av}} = \int_{\mathbf{X}} (A \equiv B)(x) \, dx. \; 1/\text{Card}(\mathbf{X})$

The following relationship obviously holds:

$$(A \equiv B)_{\text{pes}} \le (A \equiv B)_{\text{ave}} \le (A \equiv B)_{\text{opt}}.$$

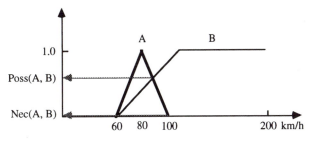

Figure 2.9
Fuzzy sets representing linguistic terms of *high* speed and *around* 80 km/h

2.6.3 Possibility and Necessity Measures

Consider a freeway with a speed limit of 100 km/h. In this specific context, the concept of *high* speed can be represented by a fuzzy set B defined in the space of velocities, as figure 2.9 shows. The membership function of a vehicle moving at a speed of 80 km/h is triangular. Under this scenario, several questions arise: What is the likelihood that the vehicle is traveling at a *high* speed? What grade of *high* speed means "*around* 80 km/h"? These questions can be naturally discussed in the framework of possibility calculus, since it concentrates on processing flexible constraints of the type envisioned in the problem.

The possibility measure (Zadeh 1978; Dubois and Prade 1980) of fuzzy set A with respect to fuzzy set B, denoted by $\mathrm{Poss}(A, B)$, is defined as

$$\mathrm{Poss}(A, B) = \sup_{x \in \mathbf{X}} [\min(A(x), B(x))].$$

The necessity measure of A with respect to B, $\mathrm{Nec}(A, B)$, is defined as

$$\mathrm{Nec}(A, B) = \inf_{x \in \mathbf{X}} [\max(A(x), 1 - B(x))].$$

An interesting interpretation arises from these measures. The possibility measure quantifies the extent to which A and B overlap. By virtue of the definition introduced, the measure is symmetric: $\mathrm{Poss}(A, B) = \mathrm{Poss}(B, A)$. In the above example,

$$\mathrm{Poss}(around\ 80, high\ \text{speed}) = 0.6.$$

The necessity measure describes the degree to which B is included in A. In the case illustrated in figure 2.9, this equals 0. As seen from definition, the necessity measure is asymmetrical, $\mathrm{Nec}(A, B) \neq \mathrm{Nec}(B, A)$. Moreover, the following relationship is valid:

$$\mathrm{Nec}(A, B) + \mathrm{Poss}(\bar{A}, B) = 1.$$

The straightforward generalization of the definitions uses triangular norms in place of min and max, respectively.

EXERCISE 2.3   Verify the following identities:

(a) $\text{Poss}(A \cup B, C) = \max[\text{Poss}(A, C), \text{Poss}(B, C)]$.

(b) $\text{Nec}(A \cap B, C) = \min[\text{Nec}(A, C), \text{Nec}(B, C)]$.

(Hint: recall that t-norms and s-norms are distributive with respect to maximum and minimum.)

Suppose now that B is taken as the universe of discourse $\mathbf{X}$. We get

$$\text{Poss}(A, \mathbf{X}) = \sup_{x \in \mathbf{X}} [\min(A(x), 1)] = \sup_{x \in \mathbf{X}} A(x).$$

We introduce the notation

$$\Pi(A) = \text{Poss}(A, \mathbf{X}).$$

$\Pi(A)$ is referred to as the possibility of A. Clearly, $\Pi(\varnothing) = 0$ and $\Pi(\mathbf{X}) = 1$.

In general, the possibility Π is a function of a collection $\mathscr{F}$ of fuzzy sets in $\mathbf{X}$, $\mathscr{F}(\mathbf{X}) = \{A_1, \ldots, A_n\}$,

$$\Pi : \mathscr{F}(\mathbf{X}) \to [0, 1],$$

associating with a fuzzy set a number in the unit interval such that

$$\Pi(\varnothing) = 0,$$

$$\Pi(\mathbf{X}) = 1,$$

$$\Pi\left(\bigcup_i A_i\right) = \sup_i \Pi(A_i), \quad i = 1, \ldots, n.$$

Intuitively, if A_i represents a concept (or event) in $\mathbf{X}$, then the possibility of $A_i, \Pi(A_i)$, characterizes the extent to which this event is possible. Clearly, the possibility of the null event is zero; similarly, the possibility of the universal event (universe of discourse) is one. The possibility of the occurrence of A or B is the possibility of the more possible event between the two.

Suppose that the linguistic concepts of speed of *around* 80 and *around* 60 are given by two triangular fuzzy sets A and B, as in figure 2.10. Clearly $\Pi(A \cup B) = 1$, since either A or B may happen. However, what is the possibility of having A and B occur at the same time?

It should be noted that for any sets A and B defined in the same space, we have

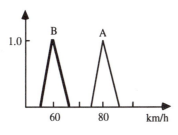

Figure 2.10
Two linguistic concepts

$$\max[\Pi(A), \Pi(\bar{A})] = 1,$$

$$\Pi(A) + \Pi(\bar{A}) \geq 1, \text{and}$$

$$\Pi(A \cap B) \leq \min[\Pi(A), \Pi(B)].$$

EXERCISE 2.4   Verify that $\Pi(A \cup B) = \max[\Pi(A), \Pi(B)]$. (Hint: See exercise 2.3.)

The possibility measure provides information about the occurrence of the event A with respect to $\mathbf{X}$. However, the possibility measure alone does not suffice to describe the uncertainty concerning this event: If $\Pi(A) = 1$, the event A may happen, but if at the same time $\Pi(\bar{A}) = 1$, its certainty is indeterminate. If on the other hand $\Pi(\bar{A}) = 0$, then the occurrence of A is certain. To obtain complete information about A, it becomes indispensable to indicate how certain it is that A will occur. This auxiliary information is provided by the necessity measure.

Let us now assume $B = \mathbf{X}$ and use a shorthand notation,

$$N(A) = \text{Nec}(A, \mathbf{X}).$$

We have

$$\text{Nec}(A, \mathbf{X}) = \inf_{x \in \mathbf{X}} [\max(A(x), 0)] = \inf_{x \in \mathbf{X}} A(x).$$

$N(A)$ is called the necessity of A. As before, $N(\varnothing) = 0$, and $N(\mathbf{X}) = 1$.

EXERCISE 2.5   Show that $N(A \cap B) = \min[N(A), N(B)]$. Hint: See exercise 2.3.

More formally, the necessity N is a function of fuzzy sets

$$N : \mathscr{F}(\mathbf{X}) \to [0, 1],$$

associating with a fuzzy set of $\mathscr{F}(\mathbf{X})$ a number in the unit interval such that

$N(\varnothing) = 0,$

$N(\mathbf{X}) = 1, \text{and}$

$N\left(\bigcap_i A_i\right) = \inf_i N(A_i), \quad i = 1, \dots, n.$

The following relationships are also valid for sets A and B in $\mathbf{X}$:

$\min[N(A), N(\bar{A})] = 0,$

$N(A) + N(\bar{A}) \leq 1, \text{ and}$

$N(A \cup B) \geq \max[N(A), N(B)].$

Suppose that A is a certain fuzzy set in $\mathbf{X}$. Then

$N(A) = \inf_{x \in \mathbf{X}} A(x) = 1 - \sup_{x \in \mathbf{X}} [1 - A(x)] = 1 - \Pi(\bar{A}),$

or equivalently,

$N(A) = 1 - \Pi(\bar{A}).$

This identity reveals why the necessity measure completes the information concerning the uncertainty underlying an event A: the higher the necessity measure, the less likely is the complementary event A. In other words, the occurrence of A is certain. $N(A) = 1$, iff the occurrence of its complement $\bar{A}$ is impossible, meaning that $\Pi(\bar{A}) = 0$, and therefore $\Pi(A) = 1$.

EXERCISE 2.6   Assume that $A \subseteq B$. How are $N(A)$ and $N(B)$ related?

2.6.4 Compatibility Measures

Compatibility measures quantify the extent to which X is compatible with another fuzzy set defined in the same space. As before, let X and A be expressed in $\mathbf{X}$. Formally, the compatibility of X with A, say $\mathrm{Comp}(X, A)$, is defined as (Zadeh 1975)

$\mathrm{Comp}(X, A)(u) = \sup_{u=A(x)} X(x), \quad u \in [0, 1].$

Conceptually, there is an evident difference between this measure and the approaches presented previously. First, compatibility is not a single numerical quantity but a mapping between two unit intervals. In other words, we can think of $\mathrm{Comp}(X, A)$ as a fuzzy set defined in $[0,1]$—we can think of a fuzzy set of compatibility. Second, compatibility is asymmetrical, for it quantifies the truth of the statement

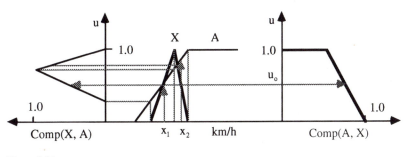

Figure 2.11
Computations of Comp(X, A) and Comp(A, X)

Comp(X, A) = Compatibility $(X$ is $A)$,

and this is inherently distinct form saying that A is compatible with X,

Comp$(X, A) \neq$ Comp(A, X).

(See figure 2.11.) Evidently, the fuzzy set A plays the role of a certain reference (conceptual reference) with respect to which the compatibility is to be computed. For instance, one can think of compatibility of speed *about* 80 km/h with the concept of *high* speed,

Comp(*about* 80 km/h, *high* speed) = Comp(X, A).

Figure 2.11 illustrates the computations of fuzzy sets of compatibility. Consider, for example, u_0 in [0,1] when computing Comp(A, X). Two arguments of X are associated with this membership value, x_1 and x_2. The supremum (maximum) taken over the corresponding membership values $A(x_1)$ and $A(x_2)$ eliminates ambiguity, since the higher of the values is accepted as the membership of the compatibility at u_0.

Inspection reveals several basic properties, assuming that both X and A are normal:

1. If $X \subset X'$, then Comp$(X, A)(u) \leq$ Comp$(X', A)(u)$.

2. If $X = A$, then the Comp(X, A) assumes a linear membership function,

Comp$(X, X)(u) = u$.

3. If X is a single numerical entity, namely, a fuzzy singleton, then the compatibility function assumes a value of 0 over the entire unit interval except at $u = A(x_0)$, where the compatibility assumes a value of 1:

$$\text{Comp}(X, A) = \begin{cases} 1, & \text{if } u = A(x_0) \\ 0, & \text{otherwise.} \end{cases}$$

The possibility and necessity measures are subsumed in the compatibility

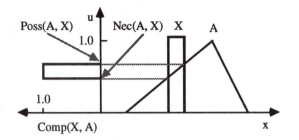

Figure 2.12
Relationships between $\text{Comp}(X, A)$, $\text{Poss}(A, X)$, and $\text{Nec}(A, X)$

measure as included in the support of $\text{Comp}(X, A)$. In particular, as shown in figure 2.12, where X is a set, the upper bound of the support of the compatibility is the possibility value of A with respect to X, whereas the lower bound corresponds to the necessity value.

EXERCISE 2.7   Consider A and X, characterized by triangular membership functions $A(x; a, m, b)$ and $X(x; c, n, d)$. Find the general expression of the membership functions of the compatibility measures $\text{Comp}(X, A)$ and $\text{Comp}(A, X)$.

2.7   Chapter Summary

This chapter introduced the essential operations involving fuzzy sets, including intersection, union, and complement viewed as general classes of operations modeled by triangular norms, as well as one-complement and negations. These operations were analyzed to provide a sense of their potential and flexibility for developing applications. It was also shown that triangular norms as well as averaging operators are aggregation operators. Operations aimed at comparing fuzzy sets were also discussed, including distance measures, possibility- and necessity-based matching computations, equality indexes, and compatibility measures.

2.8   Problems

1. Consider two fuzzy sets with triangular membership functions $A(x; 1, 2, 3)$ and $B(x; 2, 2, 4)$.

    (a) Find their intersection and union and express them analytically, using the min and max operators.

    (b) Exercise the intersection and union by choosing different t-norms and s-norms.

    (c) Find the complement of A and B and intersect them with the original sets using several t-norms. Repeat the same with the union operation using several different s-norms.

2. Show that the drastic sum and drastic product satisfy the law of excluded middle and the law of contradiction.

3. Verify that De Morgan's laws are satisfied if we take the standard union (max) and intersection (min) with the negation defined by

    (a) $N(x) = 1 - x/1 + \lambda x, \lambda \in (-1, \infty)$.
    (b) $N(x) = \sqrt[w]{1 - x^w}, w \in (0, \infty)$.

4. Show that any symmetric sum can be represented by

    $$S\_\text{sum}(x_1, \ldots, x_n) = \left[1 + \frac{\rho(1 - x_1, \ldots, 1 - x_n)}{\rho(x_1, \ldots, x_n)} \right]^{-1},$$

    ρ being any increasing continuous function with $\rho(0, \ldots, 0) = 0$.

5. Assume any idempotent aggregation operator $A(x_1, \ldots, x_n)$. Show that it satisfies the inequalities $\min(x_1, x_2, \ldots, x_n) \leq A(x_1, \ldots, x_n) \leq \max(x_1, x_2, \ldots, x_n)$. Is the converse true?

6. Consider an OWA operator defined by the weights $w_k = 1$ and $w_j = 0$ for all $j \neq k$. What is the aggregated value provided by this form of aggregation?

7. Let $\mathbf{X} = \{1, 2, 3, 4, 5, 6, 7, 8, 9, 10\}$ and the fuzzy sets $A = \{0.1/1 + 0.2/2 + 0.5/3 + 1/4 + 0.4/5 + 0.2/6\}$ and $B = \{0.1/3 + 0.2/4 + 0.5/5 + 1/6 + 0.4/7 + 0.2/8\}$ in $\mathbf{X}$. Compute their Hamming, Euclidean and Tchebyschev distances. Compare the results with their possibility and necessity measures.

8. Prove that if A is a normal fuzzy set, then $\text{Poss}(B, A) \geq \text{Nec}(B, A)$. Illustrate your proof.

9. A scheme for providing an index that summarizes the possibility and necessity measures in an averaged form is to define $C(A, B) = 1/2(\text{Poss}(A, B) + \text{Nec}(A, B))$. Show that the following equality then holds: $C(A, B) + C(\bar{A}, B) = 1$.

10. Assume that $A \subseteq B$. What is the relationship between $\Pi(A)$ and $\Pi(B)$?

11. Let X be a fuzzy set defined in $\mathbf{R}$ whose membership function is equal to 1 over the entire universe of discourse. Determine the compatibility measure of X with respect to any fuzzy set defined in the same universe of discourse.

References

Berg, M., and H. Schwarz. 1995. Logical valuation of connectives for fuzzy control by partial differential equations. *Applied Mathematics and Computer Science* 5:597–614.

Butnariu, D., and E. P. Klement. 1993. Triangular Norm–Based Measures and Games with Fuzzy Coalitions. Dordrecht: Kluwer Academic Publishers.

Dubois, D., and H. Prade. 1980. Fuzzy Sets and Systems: Theory and Applications. New York: Academic Press.

Dyckhoff, H., and W. Pedrycz. 1984. Generalized means as a model of compensative connectives. *Fuzzy Sets and Systems* 14:143–54.

Greco, D., and A. F. Rocha. 1987. The fuzzy logic of text understanding. *Fuzzy Sets and Systems* 23:347–60.

Lukasiewicz, J. 1920. O logice trojwartosciowej. *Ruch Filozoficzny* 5:169–170.

Menger, K. 1942. Statistical metric spaces. *Proceedings of the National Academy of Sciences (USA)* 28:535–37.

Pedrycz, W. 1993. Fuzzy Control and Fuzzy Systems (2nd extended edition). Taunton, UK: Research Studies Press/John Wiley.

Schweizer, B., and A. Sklar. 1983. Probabilistic Metric Spaces. Amsterdam: North Holland.

Weber, S. 1983. A general concept of fuzzy connectives, negations and implications based on t-norms. *Fuzzy Sets and Systems* 11:115–34.

Yager, R. 1988. On ordered weighted averaging aggregation operations in multicriteria decision making. *IEEE Transactions on Systems, Man, and Cybernetics* 18:183–90.

Zadeh, L. A. 1975. The concept of a linguistic variable and its application to approximate reasoning, *Information Sciences* 8:199–249 (part I); 8:301–57 (part II); 9:43–80 (part III).

Zadeh, L. A. 1978. Fuzzy sets as a basis for a theory of possibility. *Fuzzy Sets and Systems* 1:3–28.

Zimmermann, H., and P. Zysno. 1980. Latent connectives in human decision making. *Fuzzy Sets and Systems* 4:37–51.

3

Information-Based Characterization of Fuzzy Sets

Fuzzy sets are collections of elements with various grades of belongingness. In addition to describing them individually through the corresponding membership values or membership functions, we may be interested in a global characterization of fuzzy sets in terms of some scalar indices such as entropy and energy measures of fuzziness as well as a specificity measure. After discussing these measures, we concentrate on families of fuzzy sets that generate the concept of a frame of cognition. The notion of frame of cognition, which embeds the idea of information granulation, is fundamental in many engineering problems such as fuzzy modeling, fuzzy control, fuzzy classifiers, and fuzzy information processing systems, to name a few.

3.1 Entropy Measures of Fuzziness

To review briefly the notion of entropy and explain its meaning, let us consider an experiment involving a variable X with a finite number of outcomes $x_1, x_2, \ldots, x_n$ that occur with probabilities $p_1, p_2, \ldots, p_n$; obviously one has

$$0 \le p_i \le 1, \quad \sum_{i=1}^{n} p_i = 1,$$

where $p_i = p(x_i)$, with $p(x)$ a probability distribution on a finite set $\mathbf{X}$. The notion of entropy, as originally introduced by Shannon and Weaver (1949), reads as

$$H(p_1, \ldots, p_n) = -\sum_{i=1}^{n} p_i \log_2 p_i. \tag{3.1}$$

To get to the essence of this idea, let us analyze three highly illustrative situations:

1. When $n = 2$, the outcomes of the experiment are $p_1 = p$ and $p_2 = 1 - p$. Hence

$$H(p_1, p_2) = -p \log_2 p - (1 - p) \log_2 (1 - p).$$

Now let $p = 1/2$, meaning that x_1 and x_2 are equiprobable. This implies that

$$H(p_1, p_2) = -1/2 \log_2 (1/2) - 1/2 \log_2 (1/2) = 1 \text{ bit}.$$

Moreover, the entropy attains its maximum at this particular value of probability.

2. As a straightforward generalization of the two-outcome experiment, one can verify that the expression $H(p_1, \ldots, p_n)$ reaches its maximum for equal probabilities assigned to the results of the experiment

$$H(p_1, \ldots, p_n) \leq H(1/n, 1/n, \ldots, 1/n).$$

3. If one of the probabilities equals 1, then the entropy becomes equal to 0.

$$H(0, \ldots, 1, \ldots, 0) = 0.$$

Here we define $p_i \log_2 p_i = 0$, if $p_i = 0$, by extending $-\log_2 p$ to the origin by continuity. As clearly indicated by the above observations, entropy quantifies uncertainty that stems from a lack of predictability of the results of the experiment because of its probabilistic nature. This uncertainty vanishes if a single outcome occurs (with probability 1) and assumes its maximal value if all the outcomes are equiprobable (occur with the same probability).

Referring to the original definition (3.1), we can express entropy as an expected value of the function $-\log_2 p(x)$, $x \in \mathbf{X}$, say,

$$H(p_1, \ldots, p_n) = H(p(x)) = \text{Expected\_Value}[-\log_2 p(x)] = H(X).$$

The definition easily generalizes to the continuous case; here the sum is replaced by the integral with p being the probability density function:

$$H(X) = \int_x p(x) \log_2 p(x) \, dx,$$

where X is a random variable defined in $\mathbf{X}$ characterized by the probability density function $p(x)$.

Another useful generalization of the entropy function pertains to a so-called weighted entropy:

$$H(p_1, \ldots, p_n) = -\sum_{i=1}^{n} w_i p_i \log_2 p_i \tag{3.2}$$

where all the weight factors are greater than zero, $w_i > 0$. Assuming in particular that

$$w_i = -p_i / \log_2 p_i,$$

the weighted entropy is expressed as a sum of squared probabilities:

$$H(p_1, \ldots, p_n) = \sum_{i=1}^{n} p_i^2.$$

In the following discussion, we confine ourselves to a finite universe of discourse, $\mathbf{X} = \{x_1, x_2, \ldots, x_n\}$. De Luca and Termini (1972, 1974) introduced the notion of *entropy measures of fuzziness*. Further generalizations and refinements can be found in Knopfmacher 1975; Trillas and Riera 1978; and Czogala, Gottwald, and Pedrycz 1982. We start by defining a functional, $h: [0, 1] \rightarrow [0, 1]$, with the following properties (Ebanks 1983):

1. Sharpness: $h(A(x_i))$ assumes a value of 0 iff the membership value $A(x_i)$ takes on the value 0 or 1 (complete exclusion or complete membership).

2. Maximality: $h(A(x_i))$ assumes its maximal value iff $A(x_i) = 1/2$.

3. Resolution: $h(A(x_i))$ is greater than or equal to $h(A^*(x_i))$, where A^* denotes any sharpened version of A, meaning that

$$A^*(x_i) \geq A(x_i), \quad \text{if } A(x_i) \geq 1/2, \text{ and}$$

$$A^*(x_i) < A(x_i), \quad \text{if } A(x_i) < 1/2.$$

4. Symmetry: $h(A(x_i)) = h(A(1 - x_i))$.

5. Monotonicity: $h(x)$ is monotonically increasing over $[0, 1/2]$ and decreasing over $[1/2, 1]$; moreover $h(1/2) = 1$.

6. Valuation: $h[\max(A(x_i), A(x_j))] + h[\min(A(x_i), A(x_j))] = h(A(x_i)) + h(A(x_j))$.

Several commonly encountered examples of this functional include (see also figure 3.1):

1. Shannon functional: $h(u) = -u \log u - (1 - u) \log(1 - u)$.

2. Quadratic functional: $h(u) = 4u(1 - u)$.

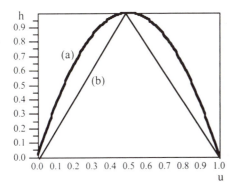

Figure 3.1
Examples of entropy functionals $h(u)$: (a) quadratic functional (b) piecewise linear functional

3. Piecewise linear functional: $h(u) = \begin{cases} 2u, & \text{if } u \in [0, 1/2) \\ 2(1 - u), & \text{if } u \in [1/2, 1]. \end{cases}$

The entropy of A is then defined as the sum of the functionals of the membership function of A:

$$H(A) = \sum_{i=1}^{n} h(A(x_i)). \tag{3.3}$$

Ebanks (1983) has shown that the above additive form (3.3) satisfies the requirements of sharpness, maximality, resolution, symmetry, and valuation. Moreover, the functional h is symmetrical,

$$H(A) = H(\bar{A}).$$

The following example reveals some interesting links between fuzzy entropy and the concept of distance between fuzzy sets.

EXAMPLE 3.1   Let us study a piecewise linear form of the functional h. Then the entropy of A computes accordingly:

$$H(A) = \sum_{i:A(x_i)<1/2}^{n} h(A(x_i)) + \sum_{i:A(x_i)\geq1/2}^{n} h(A(x_i))$$

$$= \sum_{i:A(x_i)<1/2}^{n} 2A(x_i) + \sum_{i:A(x_i)\geq1/2}^{n} 2(1 - A(x_i))$$

$$= 2\left[\sum_{i:A(x_i)<1/2}^{n} A(x_i) + \sum_{i:A(x_i)\geq1/2}^{n} (1 - A(x_i)) \right].$$

Denote by $A_{1/2}$ the 1/2-cut set of A,

$$A_{1/2} = \{x_i \,|\, A(x_i) \geq 1/2\}.$$

Then

$$H(A) = 2\sum_{i=1}^{n} |A(x_i) - A_{1/2}(x_i)|$$

and is nothing but a Hamming distance between A and its 1/2-cut multiplied by the constant 2. This, in fact, emphasizes what has been said about entropy: The more A resembles $A_{1/2}$, the higher its entropy. Geometrically, this means A is approaching the middle of the unit hypercube.

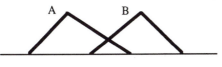

Figure 3.2
Two triangular fuzzy sets

EXERCISE 3.1   Consider the quadratic form of h, $h(u) = 4u(1 - u)$. Show that the entropy of A is equal (up to a certain constant factor) to the Euclidean distance between this fuzzy set and its 1/2-cut.

In the case where $\mathbf{X}$ is the line of reals, the entropy measure of fuzziness is modified; namely, the summation is replaced by integration. (Obviously we assume that the resulting integral does make sense.)

EXERCISE 3.2   Compute entropies of the two triangular fuzzy sets A and B shown in figure 3.2. Use in your calculations several forms of the functional h. Now consider the union and intersection of A and B and calculate entropies of the fuzzy sets obtained.

It should be stressed that the uncertainty effect captured by the fuzzy entropy pertains to the effect of partial rather than binary membership values conveyed by A. Undoubtedly, this is uncertainty of an inherently different type than the probabilistic (statistical) uncertainty represented previously in (3.1).

Because fuzziness and randomness may emerge within the same problem, the uncertainty originating from these two sources can be aggregated. One possible way of merging is to use the additive form:

$$H(A) = -\sum_{i=1}^{n} p_i \log_2 p_i + H(A).$$

Note that if events (occurrences) represented by x_is are equiprobable, the second component of the equation remains, whereas if they are sets, the formula reduces to the probability-oriented first part. One should be aware that because fuzziness and probability are orthogonal concepts, the additive model described above can be regarded as one among feasible alternatives aimed at aggregation. For a more detailed discussion, refer to Pal and Bezdek 1994.

3.2   Energy Measures of Fuzziness

An energy measure of fuzziness (De Luca and Termini 1972, 1974), $E(A)$, is defined as the quantity

$$E(A) = \sum_{i=1}^{n} e(A(x_i)), \tag{3.4}$$

where $e \colon [0,1] \to [0,1]$ is a functional increasing over the entire domain with the boundary conditions $e(0) = 0$ and $e(1) = 1$. In particular, the above functional can be regarded as an identity mapping, $e(u) = u$, for all u in the unit interval. In this case the energy measure of fuzziness is referred to as a cardinality of the fuzzy set,

$$\mathrm{Card}(A) = \sum_{i=1}^{n} e(A(x_i)). \tag{3.5}$$

Some other types of the mapping e include

$$e(u) = u^p, \quad p > 0, \tag{3.6}$$

$$e(u) = \sin\left(\frac{\pi}{2} u\right), \tag{3.7}$$

and so forth.

In general, the energy measure of fuzziness is aimed at expressing a total mass of the fuzzy set. If $e(u) = u^p$, we can regard the energy measure of fuzziness as the distance of A from the origin (empty set):

$$E(A) = d(A, \varnothing).$$

EXERCISE 3.3 Calculate the energy measure of fuzziness of the fuzzy set defined as a Gaussian membership function $A(x) = \exp(-x^2/2)$. Discuss several forms of the mapping e and compare the obtained results.

EXERCISE 3.4 Using the same triangular fuzzy numbers as in exercise 3.2, calculate the energy measure of fuzziness for $e(u) = u^p$. Plot $E(A)$ as a function of p by carrying out the respective computations.

EXERCISE 3.5 Consider triangular membership functions as in exercise 3.2. Determine their union and intersection and compute their energy measure of fuzziness for $e(u)$, given by (3.7).

If each x_i appears with probability p_i, then the energy of A can include this probabilistic information by assuming the form

$$\mathsf{E}(A) = \sum_{i=1}^{n} p_i e(A(x_i)). \tag{3.8}$$

Essentially, $\mathsf{E}(A)$ becomes an expected valued of the functional $e(A)$.

As in the case of the fuzzy entropies, the definition of energy can be extended to the continuous case,

$$\mathsf{E}(A) = \int_x p(x)e(A(x))\,dx. \tag{3.9}$$

3.3   Specificity of a Fuzzy Set

It is often helpful to be able to quantify how difficult it is to pick up a single point in the universe of discourse as a reasonable representative of the fuzzy set. Let us consider some specific cases.

1. If the fuzzy set is of a degenerated form, namely, it is already a single element, $A = \{x_0\}$, then there is no hesitation in selecting x_0 as an excellent (in fact, the only) representative of A.

2. If A covers almost the entire universe of discourse and embraces many elements with a membership grade equal to 1, then the choice of only one should cause a great deal of hesitation.

As seen from these considerations, in the first instance the fuzzy set is very specific, whereas its specificity in the second situation is zero.

Prompted by this problem, we can define the *specificity* of a fuzzy set as follows (Yager 1982, 1983): Let $\mathbf{X}$ be a finite universe of discourse. The specificity of A defined in $\mathbf{X}$, denoted by $\mathrm{Sp}(A)$, assigns to a fuzzy set A a nonnegative number such that

- $\mathrm{Sp}(A) = 1$, if and only if there exists only one element of $\mathbf{X}$ for which A assumes a value of 1 while the remaining membership values equal 0;
- $\mathrm{Sp}(A) = 0$, if $A(x) = 0$ for all elements of $\mathbf{X}$; and
- $\mathrm{Sp}(A_1) \leq \mathrm{Sp}(A_2)$, if $A_1 \supset A_2$.

Yager 1982 defines the specificity measure as the integral

$$\mathrm{Sp}(A) = \int_0^{\alpha_{\max}} \frac{1}{\mathrm{Card}(A_\alpha)}\,d\alpha \tag{3.10}$$

where $\alpha_{\max} = \mathrm{hgt}(A)$. For a finite universe of discourse (implying a finite number of membership values), the integration is replaced by a summation,

$$\mathrm{Sp}(A) = \sum_{i=1}^n \frac{1}{\mathrm{Card}(A_{\alpha_i})}\,\Delta\alpha_i, \tag{3.11}$$

where $\Delta\alpha_i = \alpha_i - \alpha_{i-1}$ and $\alpha_0 = 0$ whereas n denotes a number of the membership values of A.

EXERCISE 3.6 Show that the specificity measure (3.10) satisfies the requirements of the basic definition.

As a straightforward computational exercise, let us determine the specificity measure of A for the membership function $A = [0.2\ 0.4\ 1.0\ 0.8\ 0.3\ 0.0]$. Following (3.11) we enumerate the α-cuts and their cardinalities:

α	α-cut	Cardinality
0.0	[1 1 1 1 1 1]	6
0.2	[1 1 1 1 1 0]	5
0.3	[0 1 1 1 1 0]	4
0.4	[0 1 1 1 0 0]	3
0.8	[0 0 1 1 0 0]	2
1.0	[0 0 1 0 0 0]	1

Thus $Sp(A) = (1/5)(0.2) + (1/4)(0.1) + (1/3)(0.1) + (1/2)(0.4) + (1/1)(0.2) = 0.04 + 0.025 + 0.033 + 0.2 + 0.2 = 0.498$.

3.4 Frames of Cognition

So far we have discussed a single fuzzy set and proposed several conceptually different scalar characterizations of it. What really matters in most applications of fuzzy sets technology are families of fuzzy sets. We usually refer to them as frames of cognition.

3.4.1 Basic Definition

The notion of frames of cognition emerges in fuzzy modeling, fuzzy controllers, classifiers, and the like: primarily, in any use of fuzzy sets that calls for some form of interfacing with any real world process. Generally speaking, the frame consists of several normal fuzzy sets, also called linguistic labels or linguistic landmarks (to emphasize their focal role in this processing), used as reference points for fuzzy information processing. When the aspects of fuzzy information processing need to be emphasized, we may refer to these fuzzy sets as a fuzzy codebook, a concept widely exploited in information coding and its transmission. By adjusting the specificity of the labels and their granularity (Zadeh 1979), one can easily implement the principle of incompatibility. In particular, this allows one to cover a broad spectrum of information granularity spreading from that of a qualitative form (symbols) to that of the numerical character with the highest granularity possible.

Let us now examine a more formal definition. A frame of cognition (Pedrycz 1990, 1992),

$$A = \{A_1, A_2, \ldots, A_n\},$$

is a collection of all fuzzy sets defined in the same universe of discourse X and satisfying the following conditions:

1. Coverage: A covers X, that is, any element of $x \in X$ belongs to at least one label of A. More precisely, this requirement can be written down in the form

$$\forall_{x \in X} \exists_{i=1,\ldots,n} A_i(x) > 0.$$

Being more stringent, we may demand an ε-level of coverage of X that formalizes in the form

$$\forall_{x \in X} \exists_{i=1,\ldots,n} A_i(x) > \varepsilon,$$

where $\varepsilon \in [0, 1]$ stands for the coverage level. This simply means that any element of X belongs to at least one label to a degree not less than ε. Expressed another way, we can regard this label as a representative of this element to a nonzero extent. The condition of coverage assures us that each element of X is sufficiently represented by A.

2. Semantic Soundness of A: This condition translates into a general requirement of a linguistic interpretability of its elements. We may pose a few more detailed conditions in particular characterizing this notion (see de Oliveira 1993 and Pedrycz and de Oliveira 1993):

• A_is are unimodal and normal fuzzy sets; in this way they identify the regions of X that are semantically equivalent to the linguistic terms.

• A_is are sufficiently disjointed; this requirement assures that the terms are sufficiently distinct and therefore become linguistically meaningful.

• The number of elements of A is usually quite small; some psychological findings suggest 7 ± 2 linguistic terms as an upper limit for the cardinality of the frame of cognition.

The above features are described rather than formally defined and should therefore be treated as a collection of useful guidelines rather than strict definitions. In particular, some threshold levels (like ε) may need to be specified numerically.

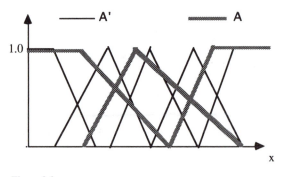

Figure 3.3
Two frames of cognition with different levels of granularity

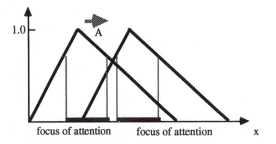

focus of attention   focus of attention   x

Figure 3.4
Focus of attention (scope of perception) realized through A

3.4.2   Main Properties

Considering the family of linguistic labels encapsulated in a frame of cognition, several properties are worth emphasizing.

3.4.2.1   Specificity

The frame of cognition A' is more specific than A if all the elements of A' are more specific than the elements of A. The specificity of the fuzzy sets encompassed in A can be conveniently evaluated using the specificity measure defined by Yager (1982, 1983). Figure 3.3 shows an example of A and A'. Note the granularity of A and A'. Clearly A' is finer than A.

3.4.2.2   Focus of Attention

A focus of attention (scope of perception) set up by a fuzzy set $A = A_i$ in A is defined as an α-cut of this fuzzy set. By moving A along $\mathbf{X}$ while not changing its membership function, we can focus attention on a certain region of $\mathbf{X}$. Figure 3.4 illustrates this phenomenon.

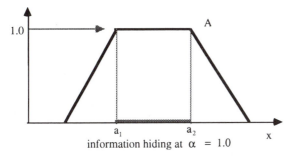

Figure 3.5
Effect of information hiding realized with fuzzy sets

3.4.2.3 Information Hiding

The idea of information hiding is directly linked to the notion of focus of attention. By modifying the membership function of $A_i = A$, an element of A, we can achieve an equivalence of the elements lying within some regions of **X**. Consider a trapezoidal fuzzy set A in **R** with its 1-cut distributed between a_1 and a_2, as depicted in figure 3.5. Observe that all the elements falling within this interval are now made nondistinguishable (equivalent) once expressed via A; beyond doubt $A(x) = 1$ for $x \in [a_1, a_2]$. Thus at the assumed level of granularity, a processing module does not distinguish between any two elements in the 1-cut of A. Hence the more detailed information (namely, that concerning the position of x within this interval) becomes hidden. By modulating (namely, increasing or decreasing) the level of the α-cut, we can accomplish an α-information hiding.

3.5   Information Encoding and Decoding Using Linguistic Landmarks

In current way uses of fuzzy sets, the main question is that of representing any datum, whether numeric or not, in terms of the codebook **A** and then transforming it back into its original format. These two fundamental tasks are, in communication theory, commonly referred to as encoding and decoding. The encoding activities occur at the transmitter, while the decoding takes place at the receiver. Fuzzy set literature has traditionally used the terms fuzzification and defuzzification to denote encoding and decoding, respectively. These are, unfortunately, quite misleading and meaningless terms, because they mask the very nature of the processing that takes place and neither address any design criteria nor introduce any measures aimed at characterizing the quality of encoding and decoding information completed by the fuzzy channel.

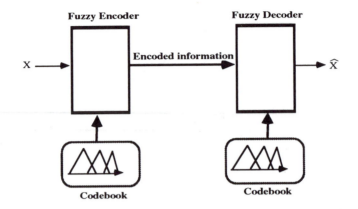

Figure 3.6
Fuzzy communication channel with fuzzy encoding and decoding

Figure 3.6 illustrates encoding and decoding with the use of the code-book **A**. The channel functions as follows. Any input information, despite its nature, is encoded (represented) in terms of the elements of the code-book. In this internal format (encoded information) it is sent across the channel. Using the same codebook, the message is decoded at the receiver.

3.5.1   Encoding Schemes in the Fuzzy Communication Channel

We can consider several encoding mechanisms. Here we focus on the mechanism exploiting the use of possibility and necessity measures. As discussed in section 2.6.3, the possibility measure expresses the degree to which an input datum X overlaps (intersects) with any component of the codebook. Similarly, the necessity of X taken with respect to the elements of **A** describes a degree of inclusion of the message in the successive elements of the codebook. After this type of encoding, the internal form of X arises as a vector of possibilities and necessities (Dubois and Prade 1988):

$$[\text{Poss}(A_1, X)\, \text{Poss}(A_2, X) \ldots \text{Poss}(A_n, X)\, \text{Nec}(A_1, X)\, \text{Nec}(A_2, X) \ldots$$

$$\text{Nec}(A_n, X)].$$

It is instructive now to look at encoding and relationships between the possibility and necessity values and granularity of **A** and X. To carry out this analysis, let A be one of the elements of the frame of cognition and X an input datum. Denote by λ the possibility of A computed with respect to X,

$$\lambda = \text{Poss}(A, X),$$

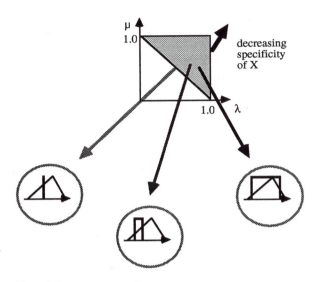

Figure 3.7
Ignorance-conflict plane

and by μ complement of the necessity measure,

$$\mu = 1 - \text{Nec}(A, X).$$

Note that $\mu = \text{Poss}(\bar{A}, X)$, as seen in section 2.6.

Here we are in a position to introduce the notions of conflict and ignorance as they emerge in the representation put in context of A. As we look at the values of λ and μ, three characteristic cases emerge:

1. $\lambda + \mu = 1$:   no uncertainty

2. $\lambda + \mu > 1$:   conflict (X becomes a conflicting piece of evidence, because it invokes both A and its complement.)

3. $\lambda + \mu < 1$:   ignorance (X indicates a lack of a sufficient support, making difficult to make any decision either in favor of A or against it.)

Let us quantify these statements as

$$\lambda + \mu = 1 + \alpha$$

or

$$\lambda + \mu = 1 - \beta$$

where α and β are used in the evaluation of the level of conflict or ignorance, respectively. A convenient graphical illustration can be formed in terms of the so-called ignorance-conflict plane, as shown in figure 3.7.

The higher the values of these indices (α or β), the higher the uncertainty (conflict or ignorance) associated with X. If X is a singleton (or is perceived as such when processed in context of A), the corresponding element in the plane moves along the diagonal. Progressively decreasing specificity of X corresponds to points progressively more distant from the main diagonal of the plane in the ignorance-conflict plane.

3.5.2 Decoding Mechanisms

The decoding mechanism should reconstruct (decode) the transmitted information. The fundamental, ideal requirement is that the decoded result should equal X. Denoting by F and F^{-1} the encoding and decoding, respectively, used by the channel, we are interested in satisfying the relationship

$$F^{-1}(F(X)) = X.$$

Despite this concise formulation, we should not be misled: The solution to this problem is not straightforward. Aside from any other consideration, the above requirement is difficult to meet. Reading it more carefully, this condition states: "Develop such encoding (F) and (F^{-1}) mechanisms that the reconstruction condition holds for any type (character) of input datum X." We distinguish two characteristic scenarios:

1. Pointwise data: $X = \{x\}$.
2. Nonpointwise data: Intervals (sets) and fuzzy sets.

In the first instance, we are primarily concerned with the procedure that has usually been referred to as defuzzification. In this scheme, the above requirement is specified as

$$F^{-1}(F(\{x\})) = x.$$

3.6   Decoding Mechanisms for Pointwise Data

By confining ourselves to this restricted model, involving pointwise data only, we can restrict the transmitted information to the possibility values (the necessities are the same). A number of algorithms produce the decoded result. We start with the center of gravity and then review some other options that can be placed in two general categories, depending on whether the decoding (1) involves modal values of the fuzzy sets of the codebook or (2) is based on computations of the membership area of the codebook's elements.

3.6.1   Decoding Based on Modal Values of the Codebook

3.6.1.1   Center-of-Gravity Decoding

The decoded result using center-of-gravity decoding reads as

$$F^{-1}(x) = \frac{\sum_{i=1}^{n} A_i(x)a_i}{\sum_{i=1}^{n} A_i(x)},$$

where a_i denotes a modal value of A_i. In general, using center-of-gravity decoding provides only an approximation of the actual value x. A remarkable exception comes with the family of triangular membership functions with an overlap of 1/2 between each of the successive linguistic terms (fuzzy sets). These circumstances entail ideal reconstruction. Unfortunately, any change in the overlap of the labels or modifications in their functional form leads to a nonzero reconstruction error; refer to figure 3.8 for the case when the overlap falls below 1/2, producing a stairlike effect in the reconstructed values.

3.6.1.2   Polynomial Expansion

In polynomial expansion, the modal values of the codebook's fuzzy sets are weighted not only by the simple possibility values $A_i(x)$ but also their powers:

$$\hat{x} = \frac{\sum_{i=1}^{n}[p_{i0} + p_{i1}A_i(x) + p_{i2}A_i^2(x) + \cdots]a_i}{\sum_{i=1}^{n}[p_{i0} + p_{i1}A_i(x) + p_{i2}A_i^2(x) + \cdots]}.$$

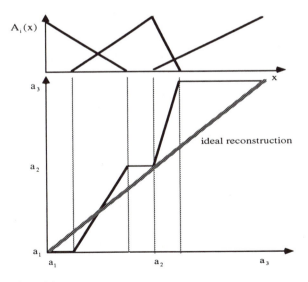

Figure 3.8
A stairlike reconstruction effect

3.6.1.3 Linguistic Expansion

In linguistic expansion, the computations use A_i modified by the linguistic modifiers very (A_i^2) and more or less $(A_i^{1/2})$:

$$\hat{x} = \frac{\sum_{i=1}^{n}[p_{i1}A_i(x) + p_{i2}A_i^2(x) + p_{i3}A_i^{1/2}(x)]a_i}{\sum_{i=1}^{n}[p_{i1}A_i(x) + p_{i2}A_i^2(x) + p_{i3}A_i^{1/2}(x)]}.$$

Considering even simple triangular membership functions in A, one can easily verify that polynomial and linguistic expansion methods cannot assure error-free decoding.

Despite the obvious diversity of the existing decoding algorithms, two essential points must be made:

1. Decoding should be carried out in the framework of a particular codebook.

2. A clearly defined criterion should guide the choice of the decoding algorithm and its eventual optimization.

3.7 Decoding Using Membership Functions of the Linguistic Terms of the Codebook

Yet another category of the decoding algorithms deals with membership functions of the codebook. More precisely, a new fuzzy set A is formed based on the information transmitted through the communication channel,

$$A = \bigcup_{i=1}^{n}[A_i \cap \Lambda_i],$$

where

$$\Lambda_i = \text{Poss}(A_i, X).$$

The main versions of the decoding include:

1. Mean-of-maxima method (MoM): The values corresponding to the highest membership functions of A are selected and averaged, giving rise to the result of decoding.

2. Center-of-gravity method (CoG): Here the result of decoding is equal to the center of gravity of A:

$$\hat{x} = \frac{\int_x A(x)x\,dx}{\int_x A(x)\,dx}.$$

3. Center-of-area method (CoA):   Here $\hat{x}$ arises as a result of the balance between the two areas of A delimited by $\hat{x}$, that is, $\hat{x}$ is such that

$$\int_{-\infty}^{\hat{x}} A(x)\,dx = \int_{\hat{x}}^{\infty} A(x)\,dx.$$

There are numerous modifications of these basic methods. For instance, instead of MoM, we can choose either the least or the greatest value for which the membership function A attains its maximum. We can, in addition, reduce any impact coming from long tails in the membership function, particularly if the corresponding membership values in these regions are low. The CoG modification incorporating this modification reads as

$$\hat{x} = \frac{\int_{x:A(x)\geq\beta} A(x)x\,dx}{\int_{x:A(x)\geq\beta} A(x)\,dx}.$$

We can consider a similar modification to the CoA method,

$$\int_{x:A(x)\geq\beta}^{\hat{x}} A(x)\,dx = \int_{\hat{x}}^{x:A(x)\geq\beta} A(x)\,dx,$$

where β is the threshold level eliminating influence from low membership functions. Another modification of the CoG method involves an extra weight factor δ that occurs in the expression

$$\hat{x} = \frac{\int_x A^{\delta}(x)x\,dx}{\int_x A^{\delta}(x)\,dx}$$

and makes this decoding scheme more flexible. In particular, if $\delta = 1$, then this modification is the standard CoG method, whereas for $\delta \to 0$, it performs like the MoM scheme. Because they require computations of A, these methods are more demanding than the decoding algorithms based on modal values discussed in section 3.6.1.

3.8   General Possibility-Necessity Decoding

In general, if the encoded (and decoded) information is not numerical (pointwise), the methods presented in sections 3.6 and 3.7 are not suitable as a vehicle to process it, for two reasons in particular. First, the result of decoding nonpointwise information should be a fuzzy set, an interval or the like. None of the methods previously discussed returns more than a single numerical quantity. Second, it is not possible to quantify the uncertainty of X (expressed with respect to the elements of the codebook) in nonpointwise information using only possibility measures. To quantify the uncertainty of X, both possibility and necessity values need to be transmitted through the channel; the decoding is therefore based on both measures, and none of the methods previously discussed includes necessity measures. Although the methods introduced in the previous sections can be modified to accommodate these extensions, the result of decoding using these modified methods could be very complicated and intuitively troublesome.

We can consider an alternative approach by treating the decoder as solving an inverse problem to that arising at the level of the encoder. In the following discussion, we confine ourselves to the possibility and necessity computations used at the point of encoding. To grasp the essence of the inverse task, let us discuss only a single fuzzy set of the codebook, say A. Then the transmitted information consists of two numbers,

$$\lambda = \sup_{x \in X} [A(x) \wedge X(x)], \quad \text{and} \quad \mu = \inf_{x \in X} [A(x) \vee \overline{X}(x)].$$

By treating these two relationships as fuzzy relational equations to be solved with respect to X, we can immediately take advantage of the analytical results. (See chapter 4.) In fact, we obtain bounds of the membership values of the transmitted datum X. The maximal solution to the possibility condition (namely, the one with the highest membership value) is

$$X'(x) = \lambda \varphi A(x) = \begin{cases} 1, & \text{if } A(x) \leq \lambda \\ \lambda, & \text{if } A(x) > \lambda. \end{cases}$$

X' constitutes the upper bound (maximal fuzzy set) of this decoding. The necessity constraint of decoding determines the lowest bound and is equal to

$$\overline{X''}(x) = \mu \beta A(x) = \begin{cases} 0, & \text{if } A(x) \geq \mu \\ \mu, & \text{if } A(x) < \mu. \end{cases}$$

Let us rewrite the last expression in the equivalent form,

$$X'' = \mu \beta A(x) = \begin{cases} 1, & \text{if } A(x) \geq \mu \\ 1 - \mu, & \text{if } A(x) < \mu. \end{cases}$$

Figure 3.9 illustrates the decoding of X for two combinations of the values of λ and μ. Note that the decoded object is a fuzzy set whose membership values come from a limited repertoire of four admissible values: $0, 1, \lambda$, and μ.

Quite obviously, the result of decoding becomes an interval-valued fuzzy set. The width of the bounds identifies the location of the most

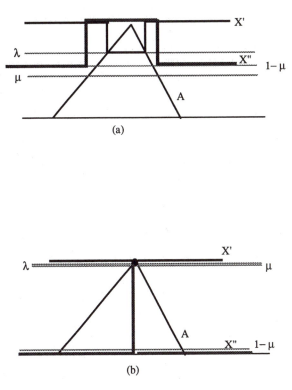

Figure 3.9
Decoding with possibility and necessity measures

essential part of the result of decoding. For instance, for λ and μ approaching 1, represented in figure 3.9(b), only in a narrow region around the modal value of A does the decoded fuzzy set show very similar bounds; in the rest of the universe, these bounds are very broad. As λ and μ decrease, the bounds of the fuzzy set tend to expand, as exemplified in figure 3.9(a).

EXERCISE 3.7   Consider some other characteristic combinations of the possibility and necessity values: high possibility, low necessity; low possibility, low necessity; possibility $= 1/2$, necessity $= 1/2$. Comment on the quality of decoding in terms of the interval-valued fuzzy set that is produced.

Because the codebook consists of several fuzzy sets, the decoding problem calls for a solution of a system of fuzzy relational equations of the form

$$\lambda_k = \sup_{x \in X} [X(x) \wedge A_k(x)],$$

$$\mu_k = \inf_{x \in X} [\overline{X}(x) \vee A_k(x)],$$

$k = 1, 2, \ldots, n$. The solution to the possibility constraints comes as an intersection of the individual solutions (recall that these were the maximal solutions to the individual constraints),

$$X'(x) = \bigwedge_{k=1}^{n} [X'_k(x)\varphi\lambda_k].$$

For the necessity constraints, the union of the results is produced:

$$\overline{X''}(x) = \bigvee_{k=1}^{n} [\overline{X''}(x)\beta\mu_k].$$

EXERCISE 3.8   Derive detailed decoding formulas for the product and Lukasiewicz t-norm.

3.9   Distance between Fuzzy Sets Based on Their Internal, Linguistic Representation

The selection of the codebook, though guided to some degree by intuition, should also reflect the qualitative domain knowledge about the variable to be quantified linguistically and its relationship to the input-output mapping to be modeled. The fundamental question arises: How can these linguistic labels be distributed, and how can granularity and specificity be decided upon?

To illustrate the essence of the problem, let us consider a single input–single output mapping manifested through a series of experimental data, as shown in figure 3.10. Our intent is to approximate these data by some rules of the form

If x is A_1, then z is z_1,

If x is A_2, then z is z_2,

$\ldots$

If x is A_n, then z is z_n,

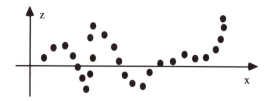

Figure 3.10
Experimental data obtained for unknown nonlinear mapping $z = f(x)$

where $A_1, A_2, \ldots, A_n$ are linguistic labels (frame of cognition) to be distributed along the universe of discourse, and $z_1, z_2, \ldots, z_n$ denote the
corresponding constants in the output space. Roughly speaking, the
idea of this approximation is to represent the mapping through a series
of representatives whose relevance is restricted to the selected regions
of the universe of discourse identified by the corresponding linguistic
labels. As intuitively appealing as this is, more linguistic terms are necessary
to model these regions of the function where its variability (changes) are
profound. Moreover, these terms should be of finer granularity. On the
other hand, in the region in which the mapping is almost constant, it is
enough to assign only a single linguistic term (like the one with the trapezoidal membership function). This pattern of allocating the membership functions either keeps some of the data almost similar at the internal
(linguistic) level (despite their distant allocation in the original space) or
makes their linguistic images as distinct as possible (moved apart as far as
possible). For instance, as figure 3.11(a) illustrates, two distinct points in
the real line are mapped onto the same element in the linguistic space. In

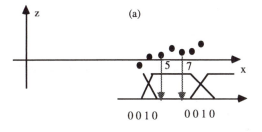

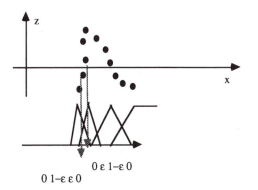

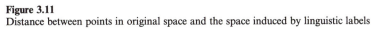

Figure 3.11
Distance between points in original space and the space induced by linguistic labels

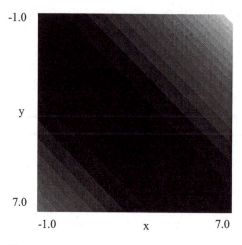

Figure 3.12
Distance in induced linguistic space

other words, the distance $d(5,7) = |5 - 7| = 2$. At the same time, the Hamming distance, determined now in the space of the linguistic labels, equals zero:

$$d([0\,0\,1\,0\,0], [0\,0\,1\,0\,0]) = 0.$$

On the other hand, figure 3.11(b) highlights the case when some *small* changes in the position of the elements in the original space are linearly amplified in the induced linguistic space. It then becomes evident that a suitable allocation of the linguistic terms allows us to modify the distance function and produce the desired transformation effect. Figures 3.12 and 3.13 contrast the Euclidean distance computed within the original space with the one computed in the space of the linguistic labels—the darker the shading, the greater the distance. Figure 3.14 summarizes the linguistic labels for which these computations have been carried out.

3.10   Chapter Summary

In contrast to the ideas presented in the previous chapters, this chapter focused on global characterization of fuzzy sets in terms of entropy and energy measures of fuzziness and on specificity measures as well. These concepts provide a key to capturing the meaning of fuzzy sets and linguistic labels within an application context and an important background in achieving solutions for engineering problems. The chapter also introduced the important concept of frame of cognition, which embeds the idea of granulation, the fuzzy-set counterpart of quantization. Besides

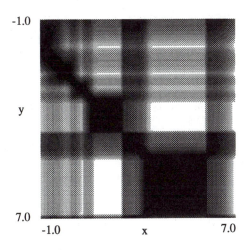

Figure 3.13
Euclidean distance computed in the original space

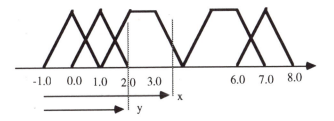

Figure 3.14
Euclidean distance computed in the linguistic space (space of the linguistic labels)

playing an essential role in developing system interfaces, here called fuzzy encoders and decoders in a general sense, the concept of frame of cognition, coupled with the informational characterization of fuzzy sets, is also an important issue in fuzzy modeling, fuzzy control, and fuzzy algorithms and systems.

3.11   Problems

1.   Compare the two classes of decoding methods (using modal values and membership functions) with respect to computational complexity and optimization transparency of the decoding and encoding algorithms. (By *optimization transparency*, we mean how easily one can update or optimize the codebook for the channel.) Consider the following optimization problem. For given encoding and decoding schemes, modify the codebook's fuzzy sets so that the losses of information described as

$$Q = \sum_{k=1}^{N} (\hat{x}_k - x_k)^2$$

become minimized; the series of input values $\{x_k\}$ is given. The codebook modifications may concern modal values of the fuzzy sets, their membership functions, or both. Assuming that we are interested in modifying the modes of the fuzzy sets, these can be governed by the well-known gradient-based scheme,

$$a_i(\text{new}) = a_i(\text{old}) - \alpha \frac{\partial Q}{\partial a_i},$$

$i = 1, 2, \ldots, n, \alpha \in (0, 1]$. Derive detailed formulas depending on the specific form of decoding and comment on their suitability in supporting these computations. Comment on any difficulties the decoding experiences in exploiting the codebook's complete membership functions—this can be referred to as a lack of optimization transparency of the decoding mechanism.

2. Let X be encoded as a numeric interval. Let the codebook consist of five fuzzy sets with triangular membership functions and 1/2 overlap between successive linguistic terms, and let the fuzzy sets be distributed uniformly across the universe $\mathbf{X} = [-5, 5]$. Discuss the results of decoding if the width of X is

    (a) significantly smaller than the support of the codebook's elements;

    (b) approximately equal to this support;

    (c) significantly higher than the support.

    Try to generalize your findings by commenting on the required relationship between the codebook's granularity of the codebook and the granularity of transmitted information.

3. For the image in figure 3.15, compute its entropy and energy measure of fuzziness. (Refer also to its tabular format below.) Modify the brightness level of each pixel according to the formula (contrast intensification)

$$\text{Int\_Image}(x) = \begin{cases} 2x^2, & \text{if } 0 \leq x \leq 0.5 \\ 1 - 2(1 - x)^2, & \text{otherwise.} \end{cases}$$

Calculate entropy and energy of the resulting image. Obtain a series of images by iteratively intensifying contrast. What happens to their entropy and energy measure?

	1	2	3	4
1	0.7	0.9	1	1
2	0.0	0.1	1	0.2
3	0.0	0.3	1	0.1
4	0.0	0.6	1	0.1

Figure 3.15
An example image

4. Prove that the weighted entropy defined by (3.2) attains its maximum for equal probabilities. Hint: Treat this as a constrained optimization problem and use the method of Lagrange multipliers.

5. Let us assume that for some reason, the center-of-gravity method based on the complete membership function is your favorite decoding scheme. Your goal is to emulate it, namely, to propose an equivalent decoding method based on some numerical prototypes, $b_1^*, b_2^*, \ldots, b_n^*$, defined in **X**. (These need not necessarily be the existing codebook's modal value.) The optimization can be completed in the form outlined in figure 3.16 Propose a suitable performance index and organize the entire optimization experiment. It should be stressed that problem has been formulated in this way with the intent of having you model a certain decoding scheme using another; the aspect of lossless transmission of information is not taken into account.

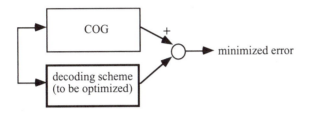

Figure 3.16
CoG decoding scheme and its emulation

6. Assume a frame of cognition **A** whose elements are triangular membership functions with the same form of the membership function and distributed uniformly in **X**.

    (a) Calculate

    1. the entropy measure of fuzziness,

    $$H(\mathbf{A}) = \sum_{i=1}^{n} H(A_i).$$

    2. the energy measure of fuzziness,

    $$E(\mathbf{A}) = \sum_{i=1}^{n} E(A_i).$$

    (b) Calculate the entropy and energy measures of fuzziness of the union of the fuzzy sets. Compare the results with those obtained in (a).

    (c) Now modify **A** by changing the overlap between successive fuzzy sets leaving their modal values intact (figure 3.17). Repeat (a) and (b), and try to represent the entropy and energy measures of fuzziness as functions of the overlap level.

Figure 3.17
Fuzzy sets of **A** with different spreads

7.   Yager (1982) proposed a method of determining a numerical representative of a fuzzy set defined in the unit interval. The concept is based on the following expression:

$$\text{Rep}(A) = \int_0^{\text{hgt}(A)} M(A_\alpha)\,d\alpha$$

where A_α is an α-cut of A, and M is a mean value of the arguments appearing in this operation, say $M(0.1, 0.2, 0.9) = (0.1 + 0.2 + 0.9)/3 = 0.4$.

(a) Find the numerical representative of $A = [0.4\ 0.7\ 0.5\ 0.6\ 1.0\ 0.7\ 0.3\ 0.2\ 0.1\ 0.0\ 0.0]$ defined in the equidistantly spaced elements of the unit interval $\{0.0, 0.1, \ldots, 0.9, 1.0\}$.

(b) What is the numerical representative of linguistically modified A, say *very A* and *more or less A*? Hint: see chapter 7.

(c) How can you go about calculating Rep (A) for A defined in **R** rather than the unit interval?

References

Czogala, E., S. Gottwald, and W. Pedrycz. 1982. Contribution to application of energy measure of fuzzy sets. *Fuzzy Sets and Systems* 8:205–14.

De Luca, A., and S. Termini. 1972. A definition of nonprobabilistic entropy in the setting of fuzzy sets. *Information and Control* 20:301–12.

De Luca, A., and S. Termini. 1974. Entropy of L-fuzzy sets. *Information and Control* 24:55–73.

de Oliveira, J. V. 1993. On optimal fuzzy systems with I/O interfaces. *Proc. 2nd Int. Conf. on Fuzzy Systems*, San Francisco.

Dubois, D., and H. Prade. 1988. Possibility Theory: An Approach to Computerized Processing of Uncertainty. New York: Plenum Press.

Dubois, D., and H. Prade. 1985. A note on measures of specificity for fuzzy sets. *Int J. General Systems* 10:279–83.

Ebanks, B. R. 1983. On measures of fuzziness and their representations. *J. Math. Analysis and Applications* 94:24–37.

Knopfmacher, J. 1975. On measure of fuzziness. *Journ. Math. Analysis and Applications* 49:529–34.

Pal, N. R., and J. C. Bezdek. 1994. Measuring fuzzy uncertainty. *IEEE Trans. on Fuzzy Systems* 2:107–18. 1994.

Pedrycz, W. 1990. Fuzzy sets framework for development of perception perspective. *Fuzzy Sets and Systems* 37:123–37.

Pedrycz, W. 1992. Selected issues of frame of knowledge representation realized by means of linguistic labels. *Int. J. Intelligent Systems* 7:155–70.

Pedrycz, W., and J. V. de Oliveira. 1993. Optimization of fuzzy relational models. *Proc. 5th IFSA World Congress*, Seoul, South Korea, July 4–9, vol. 2:1187–90.

Shannon, C. E., and W. W. Weaver. 1949. The Mathematical Theory of Communication. Urbana: University of Illinois Press.

Trillas, E., and R. Riera. 1978. Entropies for finite fuzzy sets. *Information Sciences* 15:159–68.

Yager, R. 1983. Entropy and specificity in a mathematical theory of evidence. *Int. J. General Systems* 9:249–60.

Yager, R. 1982. Measuring tranquility and anxiety in decision making: An application of fuzzy sets. *Int. J. General Systems* 8:139–46.

Zadeh, L. A. 1979. Fuzzy sets and information granularity. In *Advances in Fuzzy Set Theory and Applications*, ed. M. M. Gupta, R. K. Ragade, R. R. Yager, 3–18. Amsterdam: North Holland.

4

Fuzzy Relations and Their Calculus

This chapter has two main objectives. First, it provides the reader with the fundamentals of fuzzy relations as a generic concept used in describing relationships between linguistic terms. Second, it examines how the calculus of fuzzy relations permeates and supports the applied facet of fuzzy sets and becomes a key algorithmic platform. With these two objectives in mind, it discusses fuzzy-relational equations as one of the advanced vehicles of computing with fuzzy relations.

4.1   Relations and Fuzzy Relations

Let $\mathbf{X}$ and $\mathbf{Y}$ be two universes of discourse. A relation R defined in $\mathbf{X} \times \mathbf{Y}$ is any subset of the Cartesian product of these two universes:

$$R : \mathbf{X} \times \mathbf{Y} \rightarrow \{0, 1\}.$$

If the value of the relation for some x and y in $\mathbf{X}$ and $\mathbf{Y}$ equals 1,

$$R(x, y) = 1,$$

these two elements are *related*. Otherwise, if $R(x, y) = 0$, we say that x and y are *unrelated*.

EXAMPLE 4.1   The relation "equal to" defined in $\mathbf{X} = \{0, 1, 2, 3\}$ identifies the following set in the Cartesian product of $\mathbf{X} \times \mathbf{X}$:

$$\text{equal}(x, y) = \{(x, y) \mid x = y\} \subset \mathbf{X} \times \mathbf{X}.$$

The equivalent explicit listlike notation summarizes all the pairs (x, y) of the Cartesian product related in the sense indicated above, namely,

$$\text{equal}(0, 0) = 1, \quad \text{equal}(1, 1) = 1, \quad \text{equal}(2, 2) = 1, \quad \text{equal}(3, 3) = 1.$$

For the remaining pairs of the Cartesian product $\mathbf{X} \times \mathbf{X}$, this relation assumes the value 0.

Another way of defining relations is through relevant analytical formulas.

EXAMPLE 4.2   The relation

$$R_d = \{\mathbf{x} \in \mathbf{R}^2 \mid (\mathbf{x} - \mathbf{x}_0)^T (\mathbf{x} - \mathbf{x}_0) \leq 9\}$$

describes all points $\mathbf{x} = (x, y)$ located in the plane $\mathbf{R}^2$ such that their Euclidean distance from the point $\mathbf{x}_0 = (2, 3)$ is less than 9. (Refer to figure 4.1.) In fact, the relation represents a disk with a center at $(2, 3)$.

Elementary geometry can easily provide some other examples of relations in $\mathbf{R}^2$, as

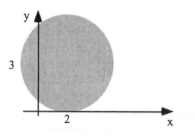

Figure 4.1
Relation representing a disk centered at $(2, 3)$

- Square: $R_c(x, y) = \begin{cases} 1, & \text{if } |x| < 1 \text{ and } |y| < 1 \\ 0, & \text{otherwise,} \end{cases}$

- Circle: $R_c(x, y) = \begin{cases} 1, & \text{if } x^2 + y^2 = r \\ 0, & \text{otherwise.} \end{cases}$

Similarly, the expression

$$R_1 = \{(x, y) \mid y \leq x\}$$

defines a relation *less than or equal to* (y is less than or equal to x).

Overall, depending on the form of the universe of of discourse (finite or infinite), relations are defined in a tabular form or given analytically. The listlike format used in example 4.1 represents the relation as a (4×4) matrix whose entries are either 1 or 0:

$$R = \begin{bmatrix} 1 & 0 & 0 & 0 \\ 0 & 1 & 0 & 0 \\ 0 & 0 & 1 & 0 \\ 0 & 0 & 0 & 1 \end{bmatrix}.$$

It is important to stress that relations subsume functions but not the other way around; that is, all functions are relations, but not all relations are functions. A relation is a function from $\mathbf{X}$ to $\mathbf{Y}$ if and only if

$$\forall_x \in \mathbf{X} \quad \exists! \, y \in \mathbf{Y} \quad R(x, y) = 1.$$

Observe also that functions are directional constructs implying that a certain direction (here, from $\mathbf{X}$ to $\mathbf{Y}$) has been specified:

$$f : \mathbf{X} \to \mathbf{Y}.$$

Obviously, if the mapping $f : \mathbf{X} \to \mathbf{Y}$ is a function, there is no guarantee that the mapping $f^{-1} : \mathbf{Y} \to \mathbf{X}$ is also a function (except in some cases when f^{-1} is said to exist). In contrast, relations are direction free (have no specific direction identified). They can be "accessed" from any direction. This makes a profound conceptual and computational difference. Relations are regarded as key notions in relational databases and non-procedural languages such as PROLOG. Although one can easily compute f for any x, given $f(x)$, this does not automatically imply that $f^{-1}(y)$ can be determined in the same way. The computations with relations are very distinct. Given the table (relation) R,

$$R = \begin{array}{c} \\ a \\ b \\ c \end{array} \begin{array}{cccc} 1 & 2 & 3 & 4 \\ \left[\begin{array}{cccc} 1 & 1 & 0 & 0 \\ 0 & 1 & 0 & 1 \\ 0 & 1 & 1 & 0 \end{array}\right] \end{array},$$

where $\mathbf{X} = \{a, b, c\}$, $\mathbf{Y} = \{1, 2, 3, 4\}$, one can determine all values of x for which $R(x, 2)$ holds just as readily as one can find the values of $\mathbf{Y}$ for $x = c$, $R(c, y)$. In the first instance, we derive

$$R(a, 2) = 1, \quad R(b, 2) = 1, \quad R(c, 2) = 1.$$

In the latter case, one obtains

$$R(c, 1) = 0, \quad R(c, 2) = 1, \quad R(c, 3) = 1, \quad R(c, 4) = 0.$$

The n-ary relation defined in $\mathbf{X}_1 \times \mathbf{X}_2 \ldots \mathbf{X}_n$ is regarded as any subset of the Cartesian product of these spaces:

$$R : \mathbf{X}_1 \times \mathbf{X}_2 \times \cdots \times \mathbf{X}_n \to \{0, 1\}.$$

Fuzzy relations generalize the generic concept of relations by admitting the notion of partial membership (association) between the points in the universe of discourse. The examples are fuzzy relations defined in $\mathbf{R}^2$ such as *much smaller* than, *approximately* equal, and *similar*, with the membership functions defined as

- x *much smaller* than y: $R(x, y) = \exp(-|y - x|)$

- x and y *approximately* equal: $R(x, y) = \exp\left\{\dfrac{|x - y|}{\alpha}\right\} \alpha > 0$

- x and y *similar*: $R(x, y) = \begin{cases} \exp(-(x - y)^2/\beta) & \text{if } |x - y| \le 5 \\ 0 & \text{if } |x - y| > 5 \end{cases}, \quad \beta > 0.$

EXERCISE 4.1   Propose a membership function of a two-dimensional torus.

When they involve discrete universes of discourse, fuzzy relations are represented as matrices or, equivalently, directed graphs. The nodes of the graph are the elements of $\mathbf{X}$ and $\mathbf{Y}$, whereas the edges (links) correspond

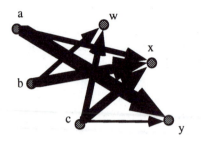

Figure 4.2
Graph of the fuzzy relation in example 4.3

to the entries of the matrix. If a certain entry is equal to 0, this connection does not show up on the graph.

EXAMPLE 4.3   Consider the fuzzy relation

$$R = \begin{array}{c} \\ a \\ b \\ c \end{array} \begin{array}{ccc} w & x & y \\ \begin{bmatrix} 0.0 & 0.4 & 1.0 \\ 0.3 & 0.9 & 0.0 \\ 0.5 & 0.7 & 0.2 \end{bmatrix} \end{array}.$$

Figure 4.2 provides the corresponding graph. Figure 4.3 provides a similar graphic notation is given in figure 4.3, where the corresponding entries of the table are shadowed according to the membership values of the fuzzy relation.

The domain of R, dom(R), is defined as

$$\text{dom}(R)(x) = \sup_{y \in \mathbf{Y}} R(x, y).$$

The codomain of R, co(R), is expressed by finding the maximal value of R along $\mathbf{X}$,

$$\text{co}(R)(y) = \sup_{x \in \mathbf{X}} R(x, y).$$

For discrete universes of discourse with card($\mathbf{X}$) and card($\mathbf{Y}$) $< \infty$, the domain and range are regarded as the height of the rows and columns of the fuzzy relation.

Similarly, as fuzzy sets, fuzzy relations can be represented by their α-cuts, meaning that

$$R = \bigcup_{\alpha} \alpha R_{\alpha},$$

$$R(x, y) = \bigvee_{\alpha} [\min(\alpha, R_{\alpha}(x, y))].$$

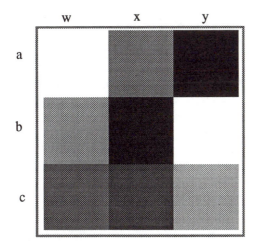

Figure 4.3
Fuzzy relation in example 4.3 as an array of associations between elements (the stronger the link, the darker the color of the corresponding entry of R)

Multidimensional fuzzy relations are generalized in a straightforward manner:

$$R : \mathbf{X}_1 \times \mathbf{X}_2 \times \cdots \times \mathbf{X}_n \to [0,1].$$

EXERCISE 4.2   For the given fuzzy relation

$$R = \begin{bmatrix} 1.0 & 0.8 & 0.6 & 0.2 \\ 0.1 & 0.5 & 1.0 & 0.4 \\ 0.2 & 0.3 & 0.5 & 0.9 \end{bmatrix}$$

determine the domain and range and the α-cuts for $\alpha = 0.1, 0.5, 0.7$.

4.2   Operations on Fuzzy Relations

The definitions of basic operations on fuzzy relations follow very closely the corresponding operations in fuzzy sets. For illustrative purposes, the definitions are specified for two-argument fuzzy relations, R, W, $P, \ldots$, defined in $\mathbf{X} \times \mathbf{Y}$. As in the case of fuzzy sets, all are defined pointwise.

- Union: $(R \cup W)(x, y) = R(x, y) \, \mathrm{s} \, W(x, y).$
- Intersection: $(R \cap W)(x, y) = R(x, y) \, \mathrm{t} \, W(x, y).$
- Complement: $\bar{R}(x, y) = 1 - R(x, y).$

Figure 4.4 illustrates the operations of union and intersection realized for some specific triangular norms.

Properties such as inclusion, dominance, and equality are defined in the same way as for fuzzy sets, namely

- Inclusion: $R \subset W$ $R(x, y) \leq W(x, y)$,
- Equality: $R = W$ $R(x, y) = W(x, y)$,

for all x in $\mathbf{X}$ and y in $\mathbf{Y}$.

The *transposition* operation produces from R a new fuzzy relation R^T, whose membership function is equal to

$$R^T(x, y) = R(y, x).$$

If R is represented in a matrix form, this operation rearranges the rows and columns. The following properties are evident:

$$(R^T)^T = R,$$

$$(\bar{R})^T = \overline{R^T}.$$

4.3   Compositions of Fuzzy Relations

Fuzzy relations can be composed with the aid of different set-theoretic operations. The most important (from a practical viewpoint) are summarized below. Assume that R, G, and W are fuzzy relations defined in the Cartesian products $\mathbf{X} \times \mathbf{Y}$, $\mathbf{X} \times \mathbf{Z}$, and $\mathbf{Z} \times \mathbf{Y}$, respectively.

- sup-t composition:

$$R = G \,\square\, W.$$

$$R(x, y) = \sup_{z \in \mathbf{Z}} [G(x, z) \, t \, W(z, y)].$$

The sup-min composition is a particular example of this family:

$$R(x, y) = \sup_{z \in \mathbf{Z}} [G(x, z) \wedge W(z, y)].$$

- inf-s composition:

$$R = G \,\blacklozenge\, W.$$

$$R(x, y) = \inf_{z \in \mathbf{Z}} [G(x, z) \, s \, W(z, y)].$$

The inf-max composition is viewed as a special case within this family:

$$R(x, y) = \inf_{z \in \mathbf{Z}} [G(x, z) \vee W(z, y)].$$

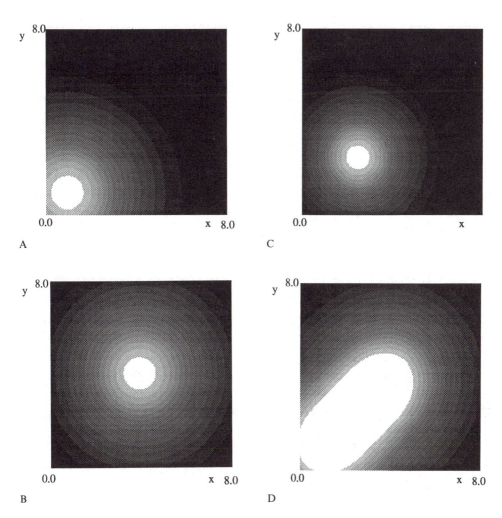

Figure 4.4
Union and intersection of two fuzzy relations R and W in $[0,8] \times [0,8]$ defined by Gaussian membership functions: (a) $R(x,y) = \exp(-(x-1)(x-1)/9)$; (b) $G(x,y) = \exp(-(x-3)(x-3)/9)$; (c) intersection t-norm: product; (d) union s-norm: Lukasiewicz OR connective.

The notations for the sup-min and inf-max compositions commonly used in the literature are $R = G \circ W$ and $R = G \bullet W$, respectively. We use different symbols to emphasize the class of t- or s-norms exploited in the composition operators.

The sup-t and inf-s composition operations possess a series of interesting and useful properties.

- Associativity:

$$R \,\square\, (P \,\square\, W) = (R \,\square\, P) \,\square\, W.$$

$$R \blacklozenge (P \blacklozenge W) = (R \blacklozenge P) \blacklozenge W.$$

- Distributivity over union (sup-t composition) and intersection (inf-s composition):

$$R \,\square\, (P \cup W) = (R \,\square\, P) \cup (P \,\square\, W).$$

$$R \blacklozenge (P \cap W) = (R \blacklozenge P) \cap (R \blacklozenge P).$$

- Weak distributivity over intersection (sup-t composition) and union (inf-s composition):

$$R \,\square\, (P \cap W) \subset (R \,\square\, P) \cap (R \,\square\, W).$$

$$R \blacklozenge (P \cup W) \supset (R \blacklozenge P) \cup (R \blacklozenge W).$$

- Monotonicity:

If $P \subset W$ then $R \,\square\, P \subset R \,\square\, W$ and $R \blacklozenge P \supset R \blacklozenge W$.

The composition operations and the transpose interact in the following way:

$$(R \,\square\, P)^T = P^T \,\square\, R^T$$

$$(R \cup P)^T = R^T \cup P^T$$

$$(R \cap P)^T = R^T \cap P^T$$

4.4   Projections and Cylindric Extensions of Fuzzy Relations

The operations of projection and extension are used to affect the size of the fuzzy relations. Let R be defined in $\mathbf{X} \times \mathbf{Y}$. The projection of R on $\mathbf{X}$ is defined as

$$R_{|\mathbf{X}}(x) = \mathrm{Proj}_{\mathbf{X}} R(x) = \sup_{y \in \mathbf{Y}} R(x, y),$$

$x \in \mathbf{X}$. This operation

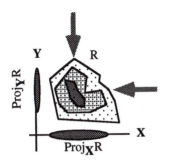

Figure 4.5
Projection of R on $\mathbf{X}$ and $\mathbf{Y}$ and its reconstruction

• reduces the dimension of the relation; in particular, for a two-dimensional relation, we arrive at a fuzzy set.

• leads to information compression and, quite often, implies its inevitable losses. This is caused by the supremum operation, which eliminates all but one maximal coordinate of the relation.

Figure 4.5 illustrates the projection effect. Analogously, one can define a projection of R on $\mathbf{Y}$:

$$R_{|\mathbf{Y}}(y) = \mathrm{Proj}_{\mathbf{Y}}\, R(y) = \sup_{x \in \mathbf{X}} R(x, y).$$

See also figure 4.5.

The projection operation can be interpreted in the framework of the sup-t composition, when the fuzzy set A in $\mathbf{X}$ assumes a value of 1 identically over the entire space. This yields

$$Y(y) = (A \,\square\, R)(y) = \sup_{x \in \mathbf{X}} [A(x) \,\mathrm{t}\, R(x, y)] = \sup_{x \in \mathbf{X}} R(x, y).$$

(In this sense the sup-t composition can be regarded as a selectively focused way of projecting fuzzy relations.) Note that the range and co-domain operations are examples of projections.

EXERCISE 4.3   For the fuzzy relation

$$R = \begin{bmatrix} 1.0 & 0.4 & 0.8 & 0.3 & 0.0 \\ 0.5 & 1.0 & 0.6 & 0.7 & 1.0 \\ 0.9 & 1.0 & 0.0 & 0.6 & 0.8 \\ 1.0 & 0.5 & 0.2 & 0.0 & 0.9 \\ 0.3 & 0.5 & 0.3 & 0.1 & 1.0 \end{bmatrix},$$

compute its projections and reconstruct the relation from these projections. Comment on the quality of this reconstruction

The cylindric extension elevates the dimension of an object on which it operates by expanding a fuzzy set into a fuzzy relation, a two-dimensional relation into its three-dimensional counterpart, and so forth. In this way, its behavior is complementary to the projection mechanism. The cylindric extension on $\mathbf{X} \times \mathbf{Y}$ of any A in $\mathbf{X}$ is a fuzzy relation, $(\mathrm{cyl}\, A)$, with the membership function given as

$$(\mathrm{cyl}\, A)(x, y) = A(x),$$

for all $y \in \mathbf{Y}$. If the fuzzy relation is viewed as a two-dimensional array, the operation of cylindric extension builds identical columns indexed by the successive values of y. The cylindric extension's main intent is to construct dimensionally compatible objects (sets, relations). For instance, let A be a fuzzy set in $\mathbf{X}$ and R a fuzzy relation in $\mathbf{X} \times \mathbf{Y}$. Consider an attempt at performing an intersection or union of A and R. Because the dimensionality of these two objects is not equal (X and R are dimensionally incompatible) these operations cannot be performed. By completing a cylindric extension of A, however, we arrive at the required compatibility. Hence the operations $(\mathrm{cyl}\, A) \cap R$, $(\mathrm{cyl}\, A) \cup R$, and the like do make sense.

The concepts of projection and cylindric extension generalize easily to multidimensional fuzzy relations. Here the definitions call for more attention, since many combinations of the coordinates of the relations can be affected. In general, let R be defined in $\mathbf{X} = \mathbf{X}_1 \times \mathbf{X}_2 \times \cdots \times \mathbf{X}_p$. Denote by $\mathbf{X}^\sim$ a certain collection of the coordinates of the above Cartesian product,

$$\mathbf{X}^\sim = \mathbf{X}_{i_1} \times \mathbf{X}_{i_2} \times \cdots \times \mathbf{X}_{i_s}$$

where $I = \{i_1, i_2 \ldots i_s\}$ denotes the indices of these coordinates. The projection of R on $\mathbf{X}^\sim$ retains the coordinates belonging to $\mathbf{X}^\sim$ and suppresses the rest of the coordinates of the original Cartesian product,

$$R_{|\mathbf{X}^\sim}(x_{i_1}, x_{i_2}, \ldots, x_{i_s}) = \sup_{\substack{x_j \in \mathbf{X}_j \\ j \notin I}} R(x_1, x_2 \ldots x_j \ldots x_p)$$

The cylindric extension involving $\mathbf{X}^\sim$ applies to any fuzzy relation G defined in $\prod_{i=1}^p \mathbf{X}_i \backslash \mathbf{X}^\sim$, and produces the relation $\mathrm{cyl}\, X$ defined in the entire Cartesian product of spaces,

$$\mathrm{cyl}\, X(x_1, x_2 \ldots x_p) = X(x_{i_1}, x_{i_2} \ldots x_{i_s}),$$

for all $x_j \in \mathbf{X}^\sim$.

The projection operations do not retain information conveyed by the original fuzzy relation, implying that one cannot "reconstruct" R from

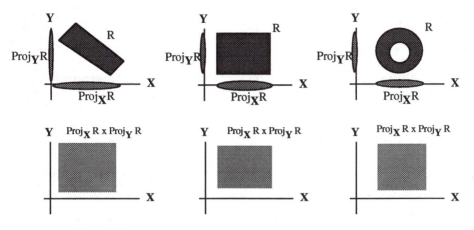

Figure 4.6
Fuzzy relations and their projections and reconstructions

its projections (shadows). $\text{Proj}_X R$ and $\text{Proj}_Y R$, the two projections of a fuzzy relation, do not necessarily lead to the original fuzzy relation. In general, we get

$$\text{Proj}_X R \times \text{Proj}_Y R \supseteq R.$$

If, however, the above relationship holds, we say that R is noninteractive. Figure 4.6 illustrates examples of interactive and noninteractive fuzzy relations. The following inclusion holds:

$$\text{Proj}_X R \times \text{Proj}_Y R \supset R.$$

The projections Proj_X and Proj_Y are sometimes viewed as marginal fuzzy restrictions.

4.5   Binary Fuzzy Relations

When talking about fuzzy relations, the notion "binary" concerns the universes of discourse in which these fuzzy relations are defined, rather than the membership values they assume. In short, R is a binary fuzzy relation if defined in $\mathbf{X} \times \mathbf{X}$,

$$R : \mathbf{X} \times \mathbf{X} \to [0, 1].$$

Some basic properties of the relations of this form are listed below.

• Reflexivity:

$$R(x, x) = 1,$$

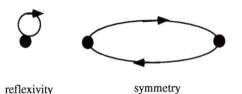

reflexivity   symmetry

Figure 4.7
Reflexivity and symmetry of fuzzy relations

which holds for each $x \in \mathbf{X}$. Or equivalently, we have $R \supset I$, where I is an identity Boolean relation, $I(x,y) = 1$ if $x = y$ and 0 otherwise. This property states that all diagonal elements of the relation are equal to 1. The concept of irreflexivity complements the above and leads to the expression

$$R(x,x) = 0.$$

• Symmetry:

$$R(x,y) = R(y,x),$$

satisfied for all x and $y \in \mathbf{X}$. The matrix representing this relation has its entries distributed symmetrically along the main diagonal. Making use of the previous notation, we note that if R is symmetric, its transpose equals the original relation, $R^T = R$. These two properties are easily represented in a graphic way, as depicted in figure 4.7. The reflexivity property can be relaxed by admitting a so-called ε-reflexivity stating

$$R(x,x) \geq \varepsilon,$$

where $\varepsilon \in [0,1]$.

Transitivity: The notion of transitivity is inherently associated with the composition operation applied to fuzzy relations. We say that R is sup-t transitive if it satisfies the inequality

$$R \,\square\, R \subset R,$$

namely,

$$\sup_{z \in \mathbf{X}} [R(x,z) \,\mathrm{t}\, R(z,y)] \leq R(x,y).$$

In particular, if the above holds for $\mathrm{t} = \min$, the relation is called sup-min transitive. Figure 4.8 clarifies the interpretation of the transitivity property. One can look at $R(x,z)$ and $R(z,y)$ as the levels (strengths) of association observed between the nodes (elements), that is, x and z and z and

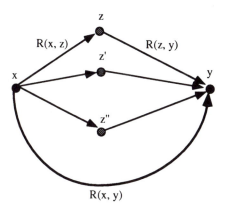

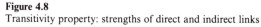

Figure 4.8
Transitivity property: strengths of direct and indirect links

y. The essence of transitivity is that the maximal strength among all possible "combined" links that are arranged in series $(R(x,z)$ and $R(z,y))$ does not exceed the strength of the direct link $(R(x,y))$ observed between x and y. The transitive closure of R, trans(R), is defined as

$$\text{trans}(R) = R \cup R^2 \cup \cdots \cup R^n,$$

where $n \ (= \text{card}(\mathbf{X}))$ is the number of elements of the universe of discourse. By definition,

$$R^2 = R \, \square \, R, \ldots, R^p = R \, \square \, R^{p-1}.$$

If R is reflexive and the t-norm is the minimum operation, the successive powers of R yield an increasing sequence of relations:

$$I \subset R \subset R^2 \subset \ldots R^{n-1} = R.$$

EXERCISE 4.4   Determine a transitive closure of the relation

$$R = \begin{bmatrix} 0.4 & 1.0 & 0.5 \\ 0.2 & 0.0 & 0.7 \\ 1.0 & 0.7 & 0.4 \end{bmatrix}$$

4.6   Some Classes of Fuzzy Relations

4.6.1   Equivalence and Similarity Relations

Equivalence relations are relations that are reflexive, symmetric and transitive. Consider that one of the arguments (say, x) of the relation is fixed. Then all elements related to x constitute a set known as an equivalence class of R with respect to x,

$$A_x = \{y \mid R(x,y) = 1\}.$$

The family of all equivalence classes defined by R and denoted by $\mathbf{X}/R$ is a partition of $\mathbf{X}$.

Similarity relations are fuzzy relations that are reflexive, symmetric and transitive. Like any fuzzy relation, a similarity relation can be represented as a nested family of its α-cuts. Each α-cut constitutes a fuzzy partition of R. These fuzzy partitions are nested, meaning that if $\alpha < \beta$, then $\mathbf{X}/R_\alpha$ is finer than the partition given by $\mathbf{X}/R_\beta$. For example, consider the similarity relation below, defined in $\mathbf{X} = \{a,b,c,d,e\}$.

	a	b	c	d	e
a	1.0	0.8	0.0	0.0	0.0
b	0.8	1.0	0.0	0.0	0.0
c	0.0	0.0	1.0	0.9	0.5
d	0.0	0.0	0.9	1.0	0.5
e	0.0	0.0	0.5	0.5	1.0

The levels of refinement of the similarity relation are represented in the form of a partition tree. The greater the value for α, the more refined the classes are, as figure 4.9 shows.

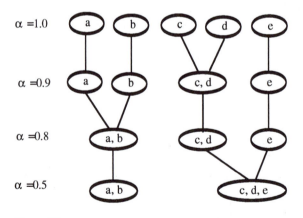

Figure 4.9
Partition tree induced by the fuzzy relation in the table above

4.6.2   Compatibility and Proximity Relations

Relations that are reflexive and symmetric are called compatibility (tolerance) relations. Fuzzy relations that feature reflexivity and symmetry are called proximity relations. Associated with any compatibility relation are sets named compatibility classes. A compatibility class is a subset A of $\mathbf{X}$ such that $R(x, y)$ for all x and y in A. For proximity relations one can introduce α-compatibility classes; A is an α-compatibility class of R if $R(x, y) \geq \alpha$ for $x, y \in A$.

Fuzzy relations possessing the properties of reflexivity, symmetry, and transitivity are called similarity relations (or fuzzy equivalence relations). Furthermore, fuzzy relations that are reflexive and symmetrical but not necessarily transitive are called resemblance (proximity or tolerance) relations. In light of the transitivity of the transitive closure of any fuzzy relation, we conclude that the transitive closure of any resemblance relation is a similarity relation.

If R is a similarity relation, then its complement,

$$\bar{R}(x, y) = 1 - R(x, y),$$

is called a dissimilarity relation. In this context the sup-min transitivity property develops an interesting interpretation. Let us denote by $d(x, y)$ the complement of R, $d(x, y) = 1 - R(x, y)$. Then

$$\max_{z \in \mathbf{X}} \left[\min(1 - d(x, z), 1 - d(z, y)) \right] \leq 1 - d(x, y).$$

Note that

$$\min(1 - d(x, z), 1 - d(z, y)) = 1 - \max(d(x, z), d(x, y)),$$

meaning that

$$\max_{z \in \mathbf{X}} \left[1 - \max(d(x, z), d(z, y)) \right] \leq 1 - d(x, y).$$

Because this inequality is satisfied for each $z \in \mathbf{X}$, we get

$$1 - \max(d(x, z), d(z, y)) \leq 1 - d(x, y),$$

$x, z, y, \in \mathbf{X}$. In other words,

$$d(x, y) \leq \max(d(x, z), d(z, y)).$$

This inequality is nothing but the ultrametric inequality. This relationship can be bounded as follows:

$$d(x, y) \leq \max(d(x, z), d(z, y)) \leq d(x, z) + d(z, y).$$

In general, the triangle inequality holds for any sup-t composition with a continuous t-norm. Furthermore $d(x,y)$ is a pseudometric[1], that is, a function satisfying the following properties:

1. If $x = y$, then $d(x,y) = 0$.
2. $d(x,y) = d(y,x)$.
3. $d(x,y) \leq d(x,z) + d(z,y)$.

4.7   Fuzzy-Relational Equations

4.7.1   Introductory Comments

The first research results on fuzzy relational equations emerged around 1976 (Sanchez 1976). In the form being introduced, they can be regarded as an essential generalization of Boolean equations. (See, for instance, Rudeanu 1974). In brief, this area developed along several clearly defined and important lines:

- exploiting new structures (topologies) of the equations, including their multilevel architectures
- discussing characterizations of families of solutions
- constructing approximate solutions to fuzzy-relational equations
- studying equations in more abstract spaces (e.g., complete lattices)

The areas of applications are numerous. Zadeh and Desoer (1963) first stressed the equivalence between the study of relations and general system theory.

In this section, we concentrate on fuzzy sets with membership values defined in [0, 1] (as opposed to other, more advanced structures, such as complete lattices), primarily in keeping with the current areas of applications. Furthermore, this restriction allows us to exploit the machinery of triangular norms as well as to take advantage of their translucent logical interpretation.

The agenda pursued here is threefold. First, we elaborate on some generic types of fuzzy-relational equations, proceeding with the interpretation of the composition operators used therein. Second, the analytical solutions are derived and discussed in depth. Third, semianalytic solutions and optimization schemes are provided and studied in the context of multilevel fuzzy-relational equations.

As a necessary prerequisite, we assume that the reader is familiar with triangular norms (t- and s-norms), discussed in chapter 2. To avoid getting bogged down in minute conceptual details, we confine ourselves to continuous triangular norms. We use two operators closely associated with the triangular norms (with $a, b \in [0, 1]$):

1. the φ-operator (implication) induced by t-norm:

$a \varphi b = a \rightarrow b = \sup\{c \in [0, 1] | a \, t \, c \leq b\}$.

2. the β-operator:

$a \beta b = \inf\{c \in [0, 1] | a \, s \, c \geq b\}$.

From this point on, we consider X and Y to be fuzzy sets in $\mathbf{X}$ and $\mathbf{Y}$, respectively, and R to be a fuzzy relation in $\mathbf{X} \times \mathbf{Y}$.

4.7.2   Interpretations of Composition Operators

It is worthwhile to interpret various set-relation rather than relation-relation compositions, as this can shed more light on the semantics of these operations. Such interpretation reveals essential links between some fundamental constructs of fuzzy computing and fuzzy relational equations.

4.7.2.1   sup-t (max-t) Composition

Among all set-relation operations encountered in the literature, the sup-t (max-t) composition is the most commonly encountered. The operation transforms X through R, yielding Y:

$Y = X \, \square \, R$.

$$Y(y) = \sup_{x \in \mathbf{X}} [X(x) \, t \, R(x, y)].$$

1. Possibility interpretation: Fixing $y \in \mathbf{Y}$, we rewrite the above equation accordingly,

$$Y(y) = \sup_{x \in \mathbf{X}} [X(x) \, t \, R_y(x)],$$

where $R_y(x)$ is a fuzzy set indexed by y (in other words, a slice of the fuzzy relation R whose location is specified by its second variable). Thus $Y(y)$ is the degree of possibility measure of X and R_y, $Y(y) = \mathrm{Poss}(X, R_y)$.

2. Logic interpretation: The supremum implements the existential quantifier (*there exists*), whereas the t-norm is viewed as the *and* connective. Then $Y(y)$ is a truth value of the compound statement "$X(x)$ and $R_y(x)$," that is,

$truth(\exists_x[X(x) \text{ and } R_y(x)]) = Y(y)$,

forming a conjunction of the two components.

3. The sup-t composition can be regarded as a specialized version of the projection operation. One can regard the max-t composition as a sort of focused (directed) projection of R, with the focus of attention specified by X. If X is equal to the entire space $\mathbf{X}$, then the composition formula reduces to the standard projection formula, namely,

$$Y(y) = \sup_{x \in \mathbf{X}} [X(x) \, t \, R(x, y)] = \sup_{x \in \mathbf{X}} [1 \, t \, R(x, y)] = \sup_{x \in \mathbf{X}} R(x, y).$$

4.7.2.2   inf-s Composition

The composition operator below exploits a tandem of infimum and s-norm (the maximum operation, in particular) producing

$$Y = X \blacklozenge R.$$

The membership of Y is equal to

$$Y(y) = \inf_{x \in \mathbf{X}} [X(x) \, s \, R_y(x, y)],$$

$y \in \mathbf{Y}$. Again assuming that an element of $\mathbf{Y}$ is fixed, we derive

$$Y(y) = \inf_{x \in \mathbf{X}} [X(x) \, s \, R_y(x)] = \inf_{x \in \mathbf{X}} [\overline{\overline{X}}(x) \, s \, R_y(x)] = \mathrm{Nec}(\overline{X}, R_y),$$

so $Y(y)$ becomes a necessity measure of $\overline{X}$ and R_y.

The logic-oriented interpretation of the composition operator employs the universal quantifier $\forall_x$ and the *or* type of connective. This yields

$$Y(y) = \forall_x [\overline{X} \text{ or } R_y].$$

4.7.2.3   Composition Operators Involving Implication (φ-operator)

Fix the second argument of R and denote the resulting fuzzy set by R_y. The φ composition of X and R_y reads now as

$$Y(y) = \inf_{x \in \mathbf{X}} [X(x) \, \varphi \, R_y(x)].$$

The interpretation can accordingly be expressed in the language of set theory. Since the implication operator models inclusion, we derive, for fixed y in $\mathbf{Y}$,

$$Y(y) = \inf_{x \in \mathbf{X}} [X(x) \subset R_y(x)].$$

The latter expression is nothing more than a pessimistic (minimal) degree of inclusion of X in the respective slice of the fuzzy relation R_y. In the logic-inclined setting, $Y(y)$ denotes a truth value of the statement

$$Y(y) = \forall_x [X(x) \rightarrow R_y(x)].$$

The optimistic version of the inclusion composition is formulated as

$$Y(y) = \sup_{x \in \mathbf{X}} [X(x) \subset R_y(x)],$$

or equivalently,

$$Y(y) = \exists_x [X(x) \rightarrow R_y(x)].$$

The dual form of the inclusion composition involves the arrangement of R and X in a reverse order. This composition operator returns a minimal or maximal degree to which X is included in R_y, which leads to the expressions

$$Y(y) = \inf_{x \in \mathbf{X}} [R(x, y) \ \varphi \ X(x)],$$

$$Y(y) = \sup_{x \in \mathbf{X}} [R(x, y) \ \varphi \ X(x)].$$

Making use of the logic quantifiers, one obtains

$$Y(y) = \forall_x [R_y(x) \subset X(x)],$$

$$Y(y) = \exists_x [R_y(x) \subset X(x)].$$

4.8   Estimation and Inverse Problem in Fuzzy-Relational Equations

Fuzzy relations (or fuzzy sets) are composed using various composition operators, which leads to the notion of relational (relation) equations. Consider the simplest set-relation composition,

$$Y = X \ \text{op} \ R,$$

where $X \in F(\mathbf{X})$, $R \in F(X \times Y)$, and subsequently $Y \in F(\mathbf{Y})$. The composition operator (op) is one among those studied in section 4.7. Two main problems are formulated:

1. Given X and Y, determine R.
2. Given Y and R, determine X.

The first task is usually referred to as an *estimation* (less precisely, identification) problem, and the second is regarded as an *inverse* problem. This terminology arises from system theory: The simplest single-input single-output form of the fuzzy relation models describe the dependency,

X and Y are *related*,

where X and Y are the input and output, respectively, whereas R quantifies the dependences between the elements of $\mathbf{X}$ and $\mathbf{Y}$. (See figure 4.10.)

Figure 4.10
Single input–single output fuzzy-relational model

Hence the estimation problem is concerned with determining the parameters of the relational model. The inverse problem is aimed at computing the input variable (X) for the given output Y.

To avoid any unnecessary mathematical ballast, we confine ourselves to continuous triangular norms. (The theory expounded herein holds for left-continuous norms.) The discussion that follows focuses on analytical solutions to the equations whose composition operator is specified as the sup-t (max-t) composition. Later, the discussion is extended to other composition operators.

4.9   Solving Fuzzy-Relational Equations with the sup-t Composition

4.9.1   Properties of the Implication Operator

The sup-t composition, as well as the sup-min and max-min, in particular, are all strongly linked with the implication $\rightarrow$ (φ-operator). This linkage will play a key role in constructing the solutions to equations formed from these compositions. One can show that the following properties hold $(a, b, c \in [0, 1])$:

1. $a\,\varphi\,\max(b, c) \geq \max(a\,\varphi\,b, a\,\varphi\,c)$
2. $a\,t\,(a\,\varphi\,b) \leq b$
3. $a\,\varphi\,(a\,t\,b) \geq b$

The proofs are quite brief. As an example, let us prove property 1. We get

$$a\,\varphi\,\max(b, c) = \sup\{d \in [0, 1]\,|\,a\,t\,d \leq \max(b, c)\}$$

$$\geq \sup\{d \in [0, 1]\,|\,a\,t\,d \leq b\} = a\,\varphi\,b.$$

Similarly,

$$a\,\varphi\,\max(b, c) \geq a\,\varphi\,c.$$

Combining these two inequalities, one gets

$$a\,\varphi\,\max(b, c) \geq \max(a\,\varphi\,b, a\,\varphi\,c).$$

We now define the composition operator induced by the given t-norm standing in the original sup-t composition:

- For any $X \in F(\mathbf{X})$ and $Y \in F(\mathbf{Y})$, $X \varphi Y$ is a fuzzy relation in $\mathbf{X} \times \mathbf{Y}$ whose membership function is specified pointwise as

$$\left(X \boxed{\varphi} Y \right)(x, y) = X(x) \varphi Y(y).$$

- For any $R \in F(\mathbf{X} \times \mathbf{Y})$ and $Y \in F(\mathbf{Y})$, $\hat{X}$ is a fuzzy set in $\mathbf{X}$ with a membership function equal to

$$\hat{X}(x) = \inf_{y \in \mathbf{Y}} \left[R(x, y) \varphi Y(y) \right].$$

Let us reveal more specific relationships between these two composition operators. The following lemmas establish the inclusion dependences between some fuzzy sets and fuzzy relations that are indispensable in tackling estimation problem.

LEMMA 4.1

$$\forall_{X \in F(\mathbf{X})} \forall_{R \in F(\mathbf{X} \times \mathbf{Y})} R \subset X \varphi (X \boxempty R)$$

LEMMA 4.2

$$\forall_{X \in F(\mathbf{X})} \forall_{Y \in F(\mathbf{Y})} X \boxempty (X \varphi Y) \subset Y$$

The proofs of these lemmas rely on the properties of the φ-operator. Starting with Lemma 4.1, one has

$$\left[X \boxed{\varphi} (X \boxempty R) \right](x, y) = X(x) \varphi \sup_{z \in \mathbf{X}} \left[X(z) \, \mathrm{t} \, R(z, y) \right].$$

We split the above expression into two components by distinguishing the case where $z = x$. This yields

$$X(x) \varphi \max \left\{ X(x) \, \mathrm{t} \, R(x, y), \sup_{z \in \mathbf{X}} \left[X(z) \, \mathrm{t} \, R(z, y) \right] \right\}.$$

Property 1 of the implication operator leads to the inequality

$$\left[X \boxed{\varphi} (x \boxempty R) \right](x, y) \geq X(x) \varphi (X(x) \, \mathrm{t} \, R(x, y)).$$

Making use of property 3, one derives

$$\left[X \boxed{\varphi} (x \boxempty R) \right](x, y) \geq R(x, y).$$

Lemma 4.2 is directly implied by property 2. Note that the relationship

$$X(x) \, \mathrm{t} \, (X(x) \, \varphi \, Y(y)) \le Y(y)$$

holds for any $x \in \mathbf{X}$.

Now we are in a position to solve the estimation problem. Denote by $\mathbb{R}$ a family of fuzzy relations satisfying $X \square R = Y$, namely,

$$\mathbb{R} = \{ R \in F(\mathbf{X} \times \mathbf{Y}) | \, X \square R = Y \}.$$

PROPOSITION 4.1

If $\mathbb{R} \ne \phi$, then its maximal element $\hat{R}$ (in terms of relation inclusion $R = \max \mathbb{R}$) is given as

$$\hat{R} = X \boxed{\varphi} Y.$$

Proof   To prove this proposition, we combine Lemmas 4.1 and 4.2 with the monotonicity property of the sup-t composition. Lemma 4.1 leads to the upper estimate of the solution, stating that

$$X \boxed{\varphi} (X \square R) = X \boxed{\varphi} Y = \hat{R} \supset R.$$

As the sup-t composition operation is monotonic, we obtain

$$X \square \hat{R} \supset X \square R = Y.$$

Moreover Lemma 4.2 implies

$$X \square \left(X \boxed{\varphi} Y \right) \subset Y,$$

that is

$$X \square \hat{R} \subset Y.$$

Since the two inequalities hold simultaneously,

$$X \square \hat{R} \supset Y$$

and

$$X \square \hat{R} \subset Y.$$

We therefore derive

$$X \square \hat{R} = Y.$$

We now proceed with the inverse problem. Denote by $\mathbb{X}$ a family of fuzzy sets in $\mathbf{X}$,

$\mathbb{X} = \{X \mid X \,\square\, R = Y\}.$

The following lemmas constitute the necessary prerequisites:

LEMMA 4.3

$$\forall_{X \in F(\mathbf{X})} \forall_{R \in F(\mathbf{X} \times \mathbf{Y})} \left(R \boxed{\varphi} Y \right) \,\square\, R \subset Y.$$

LEMMA 4.4

$$\forall_{X \in F(\mathbf{X})} \forall_{R \in F(\mathbf{X} \times \mathbf{Y})} X \subset R \boxed{\varphi} (X \,\square\, R).$$

As before, the proofs exploit the properties of the φ-operator. Lemma 4.3 exploits property 2, and proving lemma 4.4 requires property 3.

PROPOSITION 4.2
If $\mathbb{X} \neq \varnothing$ then the maximal element of $\mathbb{X}$ (in terms of set inclusion) is expressed as

$$\hat{X} = R \boxed{\varphi} Y$$

$$\hat{X} = \inf_{y \in \mathbf{Y}} \,[R(x, y) \,\varphi\, Y(y)]$$

The proof is completed along the same lines as that for proposition 4.1; now lemmas 4.3 and 4.4 play a central role.

EXAMPLE 4.4   Consider two fuzzy sets

$$X = [0.8 \quad 0.5 \quad 0.3], \quad Y = [0.4 \quad 0.2 \quad 0.0 \quad 0.7],$$

and specify $t = \min$. The greatest fuzzy relation $\hat{R}$ is computed as follows

$$\hat{R} = [0.8 \quad 0.5 \quad 0.3] \,\varphi\, \begin{bmatrix} 0.4 \\ 0.2 \\ 0.0 \\ 0.7 \end{bmatrix}$$

$$= \begin{bmatrix} 0.8 \,\varphi\, 0.4 & 0.8 \,\varphi\, 0.2 & 0.8 \,\varphi\, 0.0 & 0.8 \,\varphi\, 0.7 \\ 0.5 \,\varphi\, 0.4 & 0.5 \,\varphi\, 0.2 & 0.5 \,\varphi\, 0.0 & 0.5 \,\varphi\, 0.7 \\ 0.3 \,\varphi\, 0.4 & 0.3 \,\varphi\, 0.2 & 0.3 \,\varphi\, 0.0 & 0.3 \,\varphi\, 0.7 \end{bmatrix}$$

$$= \begin{bmatrix} 0.4 & 0.2 & 0.0 & 0.7 \\ 0.4 & 0.2 & 0.0 & 1.0 \\ 1.0 & 0.2 & 0.0 & 1.0 \end{bmatrix}$$

Let us recall that for that particular t-norm, the associated operator reads as

$$a \varphi b = \begin{cases} 1, & \text{if } a \le b \\ b, & \text{if } a > b. \end{cases}$$

Take, for instance, the first elements of both vectors. We easily derive

$$0.8 \varphi 0.4 = 0.4$$

One can easily verify that the relations

$$R' = \begin{bmatrix} 0.4 & 0.2 & 0.0 & 0.7 \\ 0.0 & 0.0 & 0.0 & 0.5 \\ 0.3 & 0.0 & 0.0 & 0.5 \end{bmatrix}$$

and

$$R'' = \begin{bmatrix} 0.0 & 0.0 & 0.0 & 0.7 \\ 0.4 & 0.0 & 0.0 & 0.2 \\ 0.6 & 0.2 & 0.0 & 1.0 \end{bmatrix}$$

are also elements of $\mathbb{R}$. By and large, these solutions are not comparable and cannot be ordered linearly by the inclusion relation. Thus the family of solutions can be visualized as in figure 4.11; clearly it involves a unique maximal solution and a series of lesser solutions.

Let us treat $\hat{R}$ and Y as provided and solve the inverse problem. Following proposition 4.2, we compute

$$\hat{X} = \min \left\{ \begin{bmatrix} 0.4 & 0.2 & 0.0 & 0.7 \\ 0.4 & 0.2 & 0.0 & 1.0 \\ 1.0 & 0.2 & 0.0 & 1.0 \end{bmatrix} \varphi [0.4 \quad 0.2 \quad 0.0 \quad 0.7] \right\}$$

where the minimum applies to each row separately. Subsequently

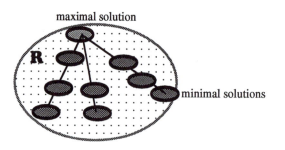

maximal solution

minimal solutions

Figure 4.11
Solutions to fuzzy-relational equations with sup-t composition

$$\hat{X} = \begin{bmatrix} \min(0.4 \; \varphi \; 0.4, \; 0.2 \; \varphi \; 0.2, \; 0 \; \varphi \; 0, \; 0.7 \; \varphi \; 0.7) \\ \min(0.4 \; \varphi \; 0.4, \; 0.2 \; \varphi \; 0.2, \; 0 \; \varphi \; 0, \; 1.0 \; \varphi \; 0.7) \\ \min(1.0 \; \varphi \; 0.4, \; 0.2 \; \varphi \; 0.2, \; 0 \; \varphi \; 0, \; 1.0 \; \varphi \; 0.7) \end{bmatrix} = \begin{bmatrix} 1.0 \\ 0.7 \\ 0.4 \end{bmatrix}$$

Interestingly, some other fuzzy sets are also solutions to this inverse problem; in particular $X' = [1 \; 0 \; 0]$, $X'' = [0 \; 0.7 \; 0]$, and so forth.

The estimation problem can be augmented by allowing for a series of relational constraints. The basic task is now reformulated as follows: Given pairs of fuzzy sets (relational constraints),

$$(X_1, Y_1), (X_2, Y_2), \ldots, (X_N, Y_N),$$

estimate R such that the system of equations (relational contraints),

$$X_k \; \square \; R = Y_k,$$

$k = 1, 2, \ldots, N$, is satisfied. The solution can be inferred by making some assumptions and noting that the solution $\hat{R}_k$ to the kth equation is the maximal one. First, suppose that each relational constraint (equation) is satisfied individually, namely,

$$\mathbb{R}_k = \{R \,|\, X_k \; \square \; R = Y_k\} \neq \phi.$$

Furthermore, let us assume that all the equations are solvable en block, that is,

$$\bigcap_{k=1}^{N} \mathbb{R}_k \neq \phi.$$

The maximal solution to the system of equations is then computed by intersecting the fuzzy relations $\hat{R}_k$,

$$\hat{R} = \bigcap_{k=1}^{N} \hat{R}_k.$$

The intersection of the partial results is legitimate in light of the maximality of the individual solutions.

EXERCISE 4.5 Solve the fuzzy relational equation with the max-min composition for the following input-output fuzzy sets:

$$X_1 = [1.0 \quad 0.7 \quad 0.1 \quad 0.0], \quad Y_1 = [0.4 \quad 0.5 \quad 0.6 \quad 0.7 \quad 0.1]$$
$$X_2 = [0.0 \quad 0.2 \quad 1.0 \quad 0.6], \quad Y_2 = [0.2 \quad 0.9 \quad 0.0 \quad 0.5 \quad 1.0]$$

Repeat the calculations for the min-max composition. Discuss the results.

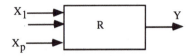

Figure 4.12
Multivariable fuzzy-relational equations

4.9.2   Extended Topologies of Fuzzy-Relational Equations

The single input–single output fuzzy-relational structure (equation) naturally extends to structures involving many variables, giving rise to multivariable fuzzy-relational equations. Consider

$$X_n \in F(\mathbf{X}_n), \quad Y \in F(\mathbf{Y}), \quad R \in F\left(\overset{p}{\underset{i=1}{\times}} \mathbf{X}_i \times \mathbf{Y} \right).$$

Then

$$Y = X_1 \,\square\, X_2 \,\square\, \cdots \,\square\, X_p \,\square\, R, \quad \text{and}$$

$$Y(y) = \sup[X_1(x_1) \,\mathrm{t}\, X_2(x_2) \,\mathrm{t}\dots\mathrm{t}\, X_p(x_p) \,\mathrm{t}\, R(x_1, x_2, \dots, x_p, y)],$$

with the supremum taken over all the spaces of input variables. Equations of this type describe relational models with many inputs, as depicted in figure 4.12.

The estimation and inverse problems are formulated in an analogous manner as done before:

Estimation problem: Given $X_1, X_2, \dots X_p$ and Y, estimate R.

Inverse problem: Here a diversity in the formulation of the problem is remarkable. First of all the problem can be solved with respect to a single variable. This implies n individual versions: For given $X_1, X_2, \dots X_{i-1}$, $X_{i+1} \dots X_p$ and R, determine X_i. The solution to this problem comes in the form of the maximal fuzzy set

$$\hat{X}_i = (X_1 \,\square\, X_2 \,\square\, \cdots \,\square\, X_{i-1} \,\square\, X_{i+1} \,\square\, \cdots \,\square\, X_n \,\square\, R) \,\varphi\, Y,$$

or

$$\hat{X}_i = R_i \,\boxed{\varphi}\, Y,$$

where R_i is a fuzzy relation defined over $\mathbf{X}_i \times \mathbf{Y}$:

$$R_i = (X_1 \,\square\, X_2 \,\square\, \cdots \,\square\, X_{i-1} \,\square\, X_{i+1} \,\square\, \cdots \,\square\, X_n \,\square\, R).$$

Finally,

$$\hat{X}_i(x_i) = \inf_{y \in \mathbf{Y}} [R_i(x_i, y) \,\varphi\, Y(y)].$$

If the above form of the inverse problem is assumed, one can enumerate p distinct versions, depending on which X_i needs to be computed. One can also anticipate another formulation of the inverse problem that involves an ensemble of fuzzy sets. Let X_I denote a fuzzy relation composed of some fuzzy sets

$$X_i = \underset{i \in I}{X} X_i,$$

where $I = \{i_1, i_2, \ldots i_s\}$ is a set of indices ranging from 1 to p. Let I' denote a collection of the remaining indices, $I' = \{i \mid i \notin I\}$. Here the Cartesian product is computed using the same t-norm as in the original equation. Let us rewrite the previous composition in a modified form by distinguishing the coordinates (fuzzy sets) of X_I,

$$Y = X_I \,\square\, X_{I'} \,\square\, R = X_I \,\square\, R_I,$$

where

$$R_I = X_I \,\square\, R.$$

Then

$$\hat{X}_I = R_I \,\varphi\, Y.$$

Any explicit computations of this fuzzy relations's coordinates call for its decomposability. If $\hat{X}_I$ is decomposable, then

$$\hat{X}_I = X_{i_1} \times X_{i_2} \times \cdots \times X_{i_s},$$

namely,

$$\hat{X}_I(x_{i_1}, x_{i_2}, \ldots, x_{i_s}) = \hat{X}_{i_1}(x_{i_1}) \,t\, \hat{X}_{i_2}(x_{i_2}) \,t \cdots t\, \hat{X}_{i_s}(x_{i_s}).$$

If this is not satisfied, the only final result comes in the form of the fuzzy relation $\hat{X}_I$.

4.9.3   Solvability Conditions

The theory developed so far is valid assuming that $\mathbb{R}$ (or $\mathbb{X}$) are nonempty, so that solutions to the equations exist. Although this may not always be satisfied, it is interesting to look at the conditions that guarantee the equations' solvability. These conditions are not always obvious and easy to articulate, but in some circumstances, nonempty solution sets are guaranteed under a quite mild condition.

For the estimation problem the condition is straightforward:

$$\mathrm{hgt}(X) \geq \mathrm{hgt}(Y).$$

In particular, if X is normal, one can always determine a fuzzy relation satisfying $X \square R = Y$. The proof is immediate. Note that the maximal fuzzy relation $\hat{R}$ is given as

$$\hat{R}(x,y) = X(x) \boxed{\varphi} Y(y).$$

Then for $y \in Y$ fixed, one derives

$$\sup_{x \in X} [X(x) \, t \, \hat{R}(x,y)] = \max \left\{ \sup_{x \in \Omega_+} [X(x) \, t \, (X(x) \, \varphi \, Y(y))], \, \sup_{x \in \Omega_-} [X(x) \, t \, 1] \right\}$$

$$\geq \sup_{x \in \Omega_+} [X(x) \, t \, (X(x) \, \varphi \, Y(y))],$$

with

$$\Omega_+ = \{x \in X \, | \, X(x) > Y(y)\},$$

$$\Omega_- = \{x \in X \, | \, X(x) \leq Y(y)\}.$$

On the other hand, we derive

$$\sup_{x \in \Omega_+} [X(x) \, t \, (X(x) \, \varphi \, Y(y))] \leq Y(y).$$

Especially if X is a singleton (a fuzzy set consisting of a single element),

$$X(x) = \delta(x - x') = \begin{cases} 1, & \text{if } x = x' \\ 0, & \text{if } x \neq x', \end{cases}$$

then the estimated fuzzy relation equals

$$\hat{R}(x,y) = X(x) \, \varphi \, Y(y) = \begin{bmatrix} 1, \text{if } x = x' \\ 0, \text{if } x \neq x' \end{bmatrix} \varphi \, Y(y) = \begin{cases} Y(y) & \text{if } x = x' \\ 1 & \text{if } x \neq x'. \end{cases}$$

This means that R consists of the entries equal to 1s for all xs except x'. For a finite case (finite universes of discourse), we obtain a fuzzy relation (matrix) whose row, indexed by xs, is simply formed by the membership function of Y,

$$\hat{R} = \begin{bmatrix} 1 \\ Y \\ 1 \end{bmatrix}.$$

The general solvability conditions for the system of equations are more demanding and cannot be stated explicitly. The only interesting exception comes in the form of $X_k s$ that are disjoint and "cover" the entire discrete space X, $\text{card}(X) = N$. Take, for example,

$$X_1 = [1 \quad 0 \quad 0 \dots 0]$$
$$X_2 = [0 \quad 1 \quad 0 \dots 0]$$
$$X_N = [0 \quad 0 \dots 0 \quad 1].$$

This system of the equations $X_k \,\square\, R = Y_k$ is always solvable. The fuzzy relation $\hat{R}$ is equal to

$$
\hat{R} = \begin{bmatrix} Y_1^T \\ \vdots \\ Y_i^T \\ \vdots \\ Y_N^T \end{bmatrix}.
$$

Observe that the successive rows of the relation are formed by the corresponding fuzzy sets that in turn form the collection of the relational constraints. The concise solvability conditions for the inverse problem are much more difficult to establish; see, for example, Gottwald 1984.

4.9.4 Relation-Relation Fuzzy Equations

Relation-relation equations express dependences among fuzzy relations. Consider the fuzzy relations

$$R \in F(\mathbf{X} \times \mathbf{Y}), \quad W \in F(\mathbf{X} \times \mathbf{Z}), \quad P \in F(\mathbf{Z} \times \mathbf{Y}).$$

In its generic form, the equation combines R, W, and P as follows:

$$R = W \,\square\, P,$$

or equivalently,

$$R(x, y) = \sup_{z \in \mathbf{Z}} [W(x, y) \,\mathrm{t}\, P(z, y)].$$

The taxonomy of estimation-inverse problems is not so evident as before, because no input-output variables are now clearly identified. Solutions are discussed for two cases:

• Given R and W, determine P.
• Given R and P, determine W.

We elaborate on the first case; the derivations for the second are very similar. It is convenient to convert the relation-relation equation into a system of fuzzy-relational equations as already studied. Denote by W_x and R_x two fuzzy sets indexed by the elements of $\mathbf{X}$,

$$W_x(z) = W(x, z),$$

$$R_x(y) = R(x, y),$$

$x \in \mathbf{X}$. In light of this tranformation, the original relation-relation composition unveils itself as the equivalent form of the system of set-relation compositions:

$$R_x(y) = \sup_{z \in \mathbf{Z}} [W_x(z) \, \mathrm{t} \, P(z, y)],$$

$x \in \mathbf{X}, y \in \mathbf{Y}$. For x fixed, the solution is straightforward:

$$\hat{P}_x = W_x \boxed{\varphi} R_x.$$

The combination of the partial solutions is completed through their intersection,

$$\hat{P} = \bigcap_{x \in \mathbf{X}} \hat{P}_x,$$

that is,

$$\hat{P}(z, y) = \inf_{x \in \mathbf{X}} [W_x(z) \, \varphi \, R_x(z)] = \inf_{x \in \mathbf{X}} [W(x, z) \, \varphi \, R(x, y)].$$

4.10   Solutions to Dual Fuzzy-Relational Equations

The inf-s composition operation describes dual fuzzy-relational equations. As before, the estimation and inverse problem take on the same format. The solutions to these equations are dual, entailing the following observations:

• The derived analytical solutions are the least fuzzy relations and fuzzy sets in the family of solutions. This property is meant in terms of inclusion of fuzzy sets.

• The solutions to the system of equations are constructed as unions of the solutions to the individual equations.

As before, $\mathbb{R}$ and $\mathbb{X}$ denote the solution sets to the estimation and inverse problems. The least solution to the estimation problem is expressed as

$$\tilde{R} = X \boxed{\beta} Y,$$

$$\tilde{R}(x, y) = X(x) \, \beta \, Y(y).$$

The least solution to the inverse problem is computed as

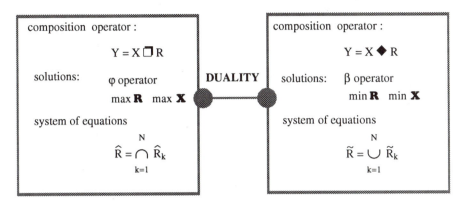

Figure 4.13
Duality of fuzzy-relational equations

$$\tilde{X} = X \boxed{\beta} Y,$$

$$\tilde{X}(x) = \sup_{y \in Y} [R(x, y) \, \beta \, Y(y)].$$

Figure 4.13 emphasizes the duality concept. (In this setting, the equations with the sup-t composition can be called direct.)

4.11   Adjoint Fuzzy-Relational Equations

Adjoint fuzzy-relational equations assume the form

$$Y = X \boxed{\varphi} R,$$

$$Y(y) = \inf_{x \in X} [X(x) \, \varphi \, R(x, y)],$$

$y \in Y$. In this way the equations exploit pessimistic version of the inclusion relationship.

The solution to the estimation problem can be given analytically. Denote by $\mathbb{R}$ a family of fuzzy relations satisfying the adjoint fuzzy-relational equation. The least fuzzy relation in $\mathbb{R}$ is defined as the Cartesian product of **X** and **Y** (whose combination is realized by the t-norm implied by the φ-operator occurring in the original equation to be solved):

$$\tilde{R} = X \times Y,$$

$$\tilde{R}(x, y) = X(x) \, \text{t} \, Y(y),$$

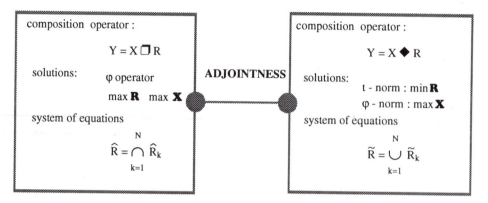

Figure 4.14
Adjointness of fuzzy-relational equations

$x \in \mathbf{X}, y \in \mathbf{Y}$. The maximal solution to the inverse problem is given as

$$\hat{X} = R \boxed{\varphi} Y,$$

$$\hat{X}(x) = \inf_{y \in \mathbf{Y}} [R(x, y) \; \varphi \; Y(y)].$$

Figure 4.14 emphasizes the property of adjointness of these two classes of equations. Observe that the composition operator in one type of the equation dictates the specific adjoint φ-operator in the solution to the second type of the equation.

The same form of adjointness can be established between the equations with the inf-s composition and those formed with the use of the β-operator. The latter assume the form

$$Y = X \boxed{\beta} R,$$

$$Y(y) = \sup_{x \in \mathbf{X}} [X(x) \; \beta \; R(x, y)].$$

4.12 Generalizations of Fuzzy-Relational Equations

4.12.1 Fuzzy-Relational Equations with an Equality Composition Operator

The degree of similarity between two membership grades a and b can be expressed as

$$ab = (a \; \varphi \; b) \wedge (b \; \varphi \; b).$$

This operator can be used as the composition operator in the equation,

$$Y = X \boxed{\&} R,$$

namely,

$$Y(y) = \sup_{x \in \mathbf{X}} [X(x)R(x,y)].$$

One can interpret $Y(y)$ as an optimistic degree of matching between the relational constraint R_y and fuzzy set X.

Let us concentrate on the estimation problem. The method exploits a slightly modified version of the φ-operator,

$$a \, \tilde{\varphi} \, b = \begin{cases} a \, \varphi \, b, & \text{if } b < 1 \\ a, & \text{if } b = 1. \end{cases}$$

Assuming that the solution set $\mathbb{R}$ is nonempty, the fuzzy relation

$$\tilde{R} = X \boxed{\varphi} Y$$

satisfies the equation under consideration. Unfortunately, this result is not as strong as those previously derived. In particular, nothing specific can be claimed about the maximal or minimal character of the solution obtained.

4.12.2   Multilevel Fuzzy-Relational Equations

Multilevel fuzzy-relational equations imply a series of such equations arranged into a cascade form. In general, the lth level cascade reads as

$$Z_1 = X \text{ op}_1 R_1$$
$$Z_2 = Z_1 \text{ op}_2 R_2$$
$$\vdots$$
$$Y = Z_1 \text{ op}_l R_l,$$

where $Z_i \in F(\mathbf{Z}_i)$, $X \in F(\mathbf{X})$, and $Y \in F(\mathbf{Y})$. The composition operations $(\text{op}_1, \text{op}_2, \dots, \text{op}_l)$ can also vary from equation to equation.

4.12.3   Fuzzy-Relational Equations with s-t and t-s Composition Operations

Figure 4.15 illustrates the panoply of composition operators used in fuzzy-relational equations. Note that what the theory covers well constitutes a small fraction of all possibilities. By and large, equations with the s-t (t-s) composition are not furnished with general solutions; analytical formulas can be derived only for some very specific forms of the equations.

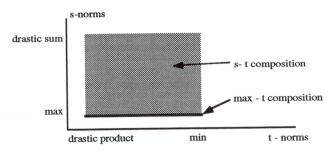

Figure 4.15
Fuzzy-relational equations with s-t and sup-t composition operations

4.13 Approximate Solutions to Fuzzy-Relational Equations

In general, analytical methods are primarily concerned with (and re-stricted to) determining extremal (minimal or maximal) solutions to the corresponding equations. Numerical algorithms come into the picture in all these situations when no exact solution sets exist or the structures of the equations are beyond analytic methods' reach. These latter cases emerge when we are concerned with multilevel fuzzy-relational equations or the equations with general s-t or t-s composition operations (Pedrycz 1993b), a very common, if not the dominant, scenario when processing real-world data. Under these circumstances, analytical methods have a quite limited applicability. The optimization tools used in fuzzy-relational equations attempt to determine approximate solutions, namely, those solutions that are the "best" according to some assumed performance index. Although the first publication in this area (Pedrycz 1983) treats the problem as a standard optimization task, some recent approaches involve fuzzy neurocomputations (cf. Pedrycz 1993b; 1995a; Pedrycz and Rocha 1993). Selected methods of approximate equations to be discussed include

- augmenting relational structures by new explanatory variables (Pedrycz 1985),
- clustering fuzzy data (relational constraints) (Pedrycz 1985),
- modifying fuzzy data through thresholding, and
- logical filtering.

Figure 4.16 outlines the general taxonomy of methods.

One general observation should be made: Optimization tools generate numerically efficient solutions (namely, they properly represent the fuzzy data within the process of a nonlinear mapping of the relational con-

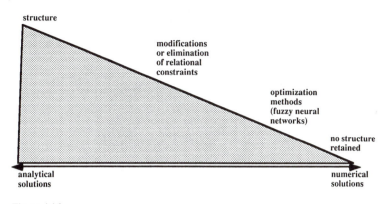

structure

modifications
or elimination
of relational
constraints

optimization
methods
(fuzzy neural
networks)

no structure
retained

analytical
solutions

numerical
solutions

Figure 4.16
Taxonomy of solutions to fuzzy relational equations

straints). Nevertheless it is not possible to interpret the objects (either
fuzzy relations or sets) derived in this way. In contrast, analytical solutions are easily interpretable but, as already emphasized, their mapping
capabilities become unacceptable. In the following discussion, we briefly
lay the foundations of selected approximation methods. To focus our
discussion, we concentrate on the sup-t (sup-min, in particular) type of
the fuzzy-relational equations. In principle, the approaches studied maintain validity for the remaining types of the equations; some modifications
could be required, however. The discussion concerns equations defined in
finite universes of discourse, $\text{card}(\mathbf{X}) = n$, $\text{card}(\mathbf{Y}) = m$. Because no exact
solutions exist, one must evaluate the performance of the approximation
method by looking at distances between corresponding fuzzy sets: the
smaller that distance, the greater the solvability of the equation(s).

4.13.1   Modifications of Relational Constraints via Thresholding

Using thresholding to modify relational constraints originates, in essence,
from an observation made in section 4.9.3. If the fuzzy sets X_k are disjoint (and their heights are not lower than those of the corresponding
Y_ks) then there exists a fuzzy relation satisfying this system of equations.
In fact, section 4.9.3 also envisioned the extreme scenario, in which all
fuzzy sets are singletons. The equations' solvability is increased by making
the X_ks as disjoint as possible. Introduce the transformation

$$X_k \rightarrow X_k^{\alpha},$$

with

$$X_k^{\alpha}(x) = \begin{cases} X_k(x), & \text{if } X_k(x) \geq \alpha \\ 0, & \text{otherwise.} \end{cases}$$

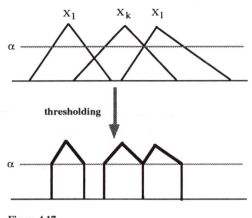

Figure 4.17
Thresholding of fuzzy data (sets) X_k

More descriptively, the transformation leads to a *sharpened* version of X_ks, as figure 4.17 shows. The higher the threshold level, the more radical modification is made to the fuzzy sets. If α is close to 1, a substantial amount of data is left unused, and the fuzzy relation obtained does not reveal most of essential dependences between fuzzy sets. As α approaches zero, the thresholding effect diminishes.

4.13.2   Preprocessing Fuzzy Data via Clustering

The mechanism of clustering pairs (X_k, Y_k) is exploited to eliminate some "noisy" data and concentrate primarily on solving the system of equations involving the prototypes (centers) of the clusters. Let $P_1, P_2, \ldots, P_c$ and $Q_1, Q_2, \ldots, Q_c$, respectively, denote the prototypes produced by a certain clustering procedure; c stands for the number of the prototypes, usually $c \ll N$. The solution to the system of equations

$$P_i \square R = Q_i,$$

$i = 1, 2, \ldots, c$, is easier to find, since P_i and Q_i tend to be more consistent than the original relational constraints we had to deal with.

We are interested in hierarchical clustering, and the clustering criterion involves the pairs of data (X_k, Y_k) and (X_l, Y_l). The rationale behind the criterion is this: If X_k and X_l are *similar*, whereas Y_k and Y_l are *different*, then these pairs of data should be allocated to two different clusters. On the other hand, if both X_k and X_l as well as Y_k and Y_l are *similar*, then they should be assigned to the same cluster. The performance index guiding the formation of the clusters assumes the form

$$V(k, l) = \text{sim}(X_k, X_l) \rightarrow \text{sim}(Y_k, Y_l),$$

where sim(.) is a normalized measure of similarity between the corresponding fuzzy sets. We can use here the equality index discussed in chapter 2. The clustering's goal is to maximize V. Pedrycz (1988) discussed the method of agglomerative clustering: Proceeding with N single-element clusters (original pairs of data X_k and Y_k), one combines the relational constraints and eliminates (averages) the most visible outliers.

4.13.3   Use of Auxiliary Variables

The extension algorithm adds some explanatory variables that make the relational constraints consistent (satisfiable by a fuzzy relation). Instead of the original relational equations

$$X_i \,\square\, R = Y_i,$$

$i = 1, 2, \ldots, N$, one considers their augmented version, assuming the form

$$X_i \,\square\, Z_j \,\square\, G = Y_i,$$

$j = 1, 2, \ldots, c$, where Z_j is a normal fuzzy set defined in an auxiliary space $\mathbf{Z}$; moreover G becomes a fuzzy relation defined in the Cartesian product of $\mathbf{X}$, $\mathbf{Z}$ and $\mathbf{Y}$. This method is closely linked with the clustering idea, because the auxiliary fuzzy sets are associated with the corresponding clusters of data. Observe that by keeping the Z_js pairwise disjoint, the system of extended equations can be rendered solvable.

4.13.4   Solving Fuzzy-Relational Equations via Logic Filtering

The standard aggregation of the solutions to the individual equations $R(k)$ can be generalized as

$$\mathscr{R} = \bigcap_{k=1}^{N} (R(k) \cup W(k)),$$

where, in general, W denotes a fuzzy selection relation (fuzzy selector) or, simply, a logical mask. Note that the role of this relation is to filter the components ($R(k)$) regarded as less relevant elements in the aggregation process. At the limit, the identity selection relation assigned to a certain (k_0) relational constraint makes $R(k_0)$ totally inactive, thus eliminating any contribution of the pair $\mathbf{x}_{k_0} \rightarrow \mathbf{y}_{k_0}$ to the construction of the fuzzy relation R. Note, however, that the above mechanism works in one direction by suppressing (but not enhancing) the contribution coming from the specific constraint: The term "masking" might sound more accurate. Describe $\mathscr{R}$ explicitly in terms of the corresponding membership functions,

$$\mathscr{R}(y_j, x_i) = \bigcap_{k=1}^{N} [R(k; y_j, x_i) \text{ s } W(k; y_j, x_i)],$$

where

$$R(k; y_j, x_i) = x_k(x_i) \; \varphi \; y_k(y_j).$$

In the former expression, the triangular conorm (s-norm) realizes a union of R and the mask relation W.

Depending on the spatial form of the filtering mask, $W(k)$, several options can be distinguished:

1. The mask $W(k)$ is space independent. This form constitutes the most condensed type of the filter:

$$\mathscr{R}(y_j, x_i) = \bigwedge_{k=1}^{N} [R(k; y_j, x_i) \text{ s } W(k)].$$

The filter's coefficients can be treated as weight factors (confidence levels) applied to the individual data points and determining their significance.

2. The filter $W(k)$ is partially space independent, implying its form as

$$\mathscr{R}(y_j, x_i) = \bigwedge_{k=1}^{N} [R(k; y_j, x_i) \text{ s } W(k, y_j)].$$

Thus, the coefficients discriminate among the data and the elements as well in the equation's output space ($[0, 1]^m$).

3. The fully distributed (space dependent) filter $W(k)$ depends upon y_j and x_i and is described as

$$\mathscr{R}(y_j, x_i) = \bigwedge_{k=1}^{N} [R(k; y_j, x_i) \text{ s } W(k, y_j, x_i)].$$

These three forms of the filters pose different memory requirements and can exhibit different levels of efficiency as well. In particular, the fully distributed version of the filter (3) could introduce a significant computational burden, whereas the space independent version (1) can be too restrictive. The partially distributed version (2) constitutes a viable compromise between these two extremes. Hereafter, we will confine ourselves to (1) and (2).

Undoubtedly, the computational model of the fuzzy relation based on the filtering mechanisms is more complex and computationally demanding than the original one (namely, (1)) calling for the constraint-free optimization of the fuzzy relation. To gain a better insight into the potential

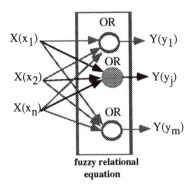

Figure 4.18
Fuzzy-relational equation as a collection of fuzzy neurons; jth neuron highlighted

differences in the learning mechanisms required, it is worth considering the corresponding, more detailed network interpretations.

4.13.5   Solving Fuzzy-Relational Equations as a Problem of Learning of Fuzzy Neurons

A fuzzy-relational equation can be considered as a collection of specialized processing units (fuzzy neurons), as seen in figure 4.18. These are, in fact, multiple-input single-output elements. In the case of the sup-t composition, they are usually referred to as OR neurons. Note that the connections of the jth neuron are arranged as the jth row of the fuzzy relation R. The estimation problem involves learning the network, whereas the inverse problem is concerned with determining the inputs of the network for fixed outputs. Chapter 11 is fully devoted to these constructs as well as their multilayer generalizations.

4.14   Chapter Summary

Fuzzy relations are straightforward yet highly powerful generalizations of relations. Because they are more general than functions, they allow dependences between various variables to be captured without necessarily committing to any particular directional characterization of these variables. This chapter introduced fuzzy relations and studied their main properties. It also analyzed a number of classes of fuzzy relations. The theory of fuzzy-relational equations emerged as a direct computational vehicle capable of capturing and processing relational constraints efficiently.

4.15 Problems

1. The patterns $p_1, p_2, \ldots, p_n$, $n = 5$, were assigned to c classes, $c = 3$, with the grades of membership in each class summarized in the $n \times c$ fuzzy relation given subsequently. How similar are the patterns? Collect the similarity results in an $n \times n$ fuzzy relation. Is the obtained relation reflexive, symmetric and transitive?

	class 1	class 2	class 3
pattern 1	0.3	0.1	0.6
pattern 2	0.1	0.8	0.1
pattern 3	0.1	0.0	0.9
pattern 4	1.0	0.0	0.0
pattern 5	0.3	0.4	0.3

2. Perform the max-t composition of the two fuzzy relations given below:

$$R = \begin{bmatrix} 0.5 & 1.0 & 0.7 & 0.9 \\ 0.4 & 1.0 & 0.2 & 0.1 \\ 0.6 & 0.9 & 1.0 & 0.4 \end{bmatrix}$$

and

$$G = \begin{bmatrix} 0.9 & 0.3 & 0.1 & 0.7 & 0.6 & 1.0 \\ 0.1 & 0.1 & 0.9 & 1.0 & 1.0 & 0.4 \\ 0.0 & 0.3 & 0.6 & 0.9 & 1.0 & 0.0 \\ 1.0 & 0.0 & 0.0 & 0.0 & 1.0 & 1.0 \end{bmatrix}.$$

When carrying out the composition of R and G, use several t-norms. Compare the results.

3. Establish a relationship between the max-t and min-s composition where these two triangular norms are dual (namely, $a\,s\,b = 1 - (1 - a)\,t\,(1 - b)$).

4. Input-output fuzzy sets (relational constraints) are given in the form

$\mathbf{x}(1) = [1.0\ 0.4\ 0.5\ 0.8\ 0.0], \quad \mathbf{y}(1) = [0.5\ 0.7\ 0.3\ 0.1]$

$\mathbf{x}(2) = [0.1\ 0.9\ 1.0\ 0.2\ 0.0], \quad \mathbf{y}(2) = [1.0\ 0.3\ 0.1\ 0.0]$

$\mathbf{x}(3) = [1.0\ 0.4\ 0.5\ 0.8\ 0.0], \quad \mathbf{y}(3) = [0.5\ 0.7\ 0.3\ 0.1].$

Solve the system of the resulting fuzzy-relational equations considering several types of composition operator (max-t, min-s, adjoint, etc.).

5. An image can be conveniently represented as a fuzzy relation. (See figure 4.19).

(a) In the context of image processing, interpret basic operations on fuzzy relations, such as contrast intensification, dilation, and the like.

(b) The fuzzy relation can also be stored as a finite collection of input-output fuzzy sets $\{\mathbf{x}(k), \mathbf{y}(k)\}$, $k = 1, 2, \ldots, N$. (See figure 4.20). This mechanism could lead to a sort of compression of the original image. Elaborate on the quality of reconstruction based on the number of fuzzy sets used therein (N) as well as their membership functions. (Hint: The quality of reconstruction is greatly affected by the form of $\mathbf{x}(k)$s and their distribution.)

Figure 4.19
Image as a two-dimensional fuzzy relation

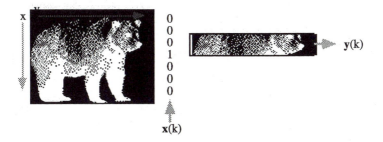

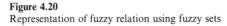

Figure 4.20
Representation of fuzzy relation using fuzzy sets

6. Analyze the boundary conditions for different types of fuzzy-relational equations

    $\mathbf{x}$ op $R = \mathbf{y}$.

    Namely, consider $\mathbf{x} = \varnothing$ and $\mathbf{x} = \mathbf{X}$.

7. A fuzzy-relational system (figure 4.21) is described by two fuzzy-relational equations with the max-min composition. Solve the estimation problem: Determine R and G for A, B, C, and D provided.

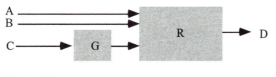

Figure 4.21
Fuzzy-relational system

8. This problem concerns an issue of reducing linguistic dimensionality. A fuzzy relation R is defined in $\mathbf{X} \times \mathbf{Y}$, both of which spaces involve a significant number of elements, $\text{card}(\mathbf{X}) = n$, $\text{card}(\mathbf{Y}) = m$. Our objective is to build a linguistic summarization of this relation. To accomplish that, let us define a number of fuzzy sets—a frame of cognition for $\mathbf{X}$ and $\mathbf{Y}$. Denote these families of fuzzy sets by $\mathscr{A}$ and $\mathscr{B}$; $\mathscr{A} = \{A_1, A_2, \ldots, A_c\}$, $\mathscr{B} = \{B_1, B_2, \ldots, B_p\}$. Obviously, $\text{card}(\mathscr{A}) \ll \text{card}(\mathbf{X})$; similarly the same relationship holds for $\mathscr{B}$ and $\mathbf{Y}$. Linguistic dimension reduction aims to produce a fuzzy relation G defined in $\mathscr{A} \times \mathscr{B}$ (see figure 4.22) that captures the essence of the original fuzzy relation. Obviously, some details could be missing. What constitutes a reduction in the size of the problem? Discuss an eventual inverse problem, that is, a problem in the reconstruction of R from G provided.

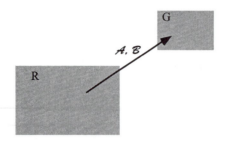

Figure 4.22
From fuzzy relation R to G—a concept of linguistic dimensionality reduction

9.   Elaborate on the following two-argument relations:

   • country-currency

   • computer industry: education-job

   • real estate market: price-quality of house

   • digital equipment: power consumption-speed

   Are these relations Boolean (two-valued) or fuzzy? Justify your answer. Come up with some relevant examples by enumerating the corresponding universes of discourse and representing the above relations in a matrix form.

10.   In software engineering, a number of design factors are considered. Very often the relationships between these factors can be viewed as fuzzy relations. Consider two fuzzy relations: The first describes the relationship between software complexity and its reliability, the second relates software reliability to the development costs. Both are fuzzy relations defined in finite universes of discourse. Propose fuzzy relations capturing the above dependences and discuss how to construct a fuzzy relation of complexity and development costs.

Note

1. The stronger notion of metric requires (2) and (3) whereas (1) needs to be reformulated as an iff statement: $(1')$ $d(x,y) = 0$ iff $x = y$. We cannot guarantee that the iff requirement holds, as one could have $R(x,y) = 1$ (i.e., $d(x,y) = 0$) even for x different from y.

References

Gottwald, S. 1984. On the existence of solutions of systems of fuzzy equations. *Fuzzy Sets and Systems* 12:301–2.

Pedrycz, W. 1983. Numerical and applicational aspects of fuzzy relational equations. *Fuzzy Sets and Systems* 11:1–18.

Pedrycz, W. 1985. Applications of fuzzy relational equations for methods of reasoning in presence of fuzzy data. *Fuzzy Sets and Systems* 16:163–75.

Pedrycz, W. 1988. Approximate solutions of fuzzy relational equations. *Fuzzy Sets and Systems* 26:183–202.

Pedrycz, W. 1991. Neurocomputations in relational systems. 1991. *IEEE Trans. on Pattern Analysis and Machine Intelligence* 13:289–96.

Pedrycz, W. 1993a. Fuzzy neural networks and neurocomputations. *Fuzzy Sets and Systems* 56:1–28.

Pedrycz, W. 1993b. *Fuzzy Control and Fuzzy Systems* (2d ed.). Taunton, NY: Research Studies Press/J. Wiley.

Pedrycz, W. 1995a. Relational neural structures. In *Progress in Neural Networks* vol. III, ed. O. M. Omidvar, 177–210. Norwood, NJ: Ablex Publishing Corp.

Pedrycz, W. 1995b. *Fuzzy Sets Engineering*. Boca Raton, FL: CRC Press.

Pedrycz, W. and A. F. Rocha. 1993. Fuzzy-set based models of neurons and knowledge-based networks. *IEEE Trans. on Fuzzy Systems* 1:254–66.

Rudeanu, S. 1974. *Boolean Functions and Equations*. Amsterdam: North Holland.

Sanchez, E. 1976. Resolution of composite fuzzy relation equations. *Information and Control* 34:38–48.

Zadeh, L. A., and C. A. Desoer. 1963. *Linear System Theory*. New York: Mc Graw Hill.

5 Fuzzy Numbers

Fuzzy numbers model imprecise quantities, which tend to be omnipresent when describing complex systems. They could be used to model approximate quantities (numbers) such as *about* five, *below* 100, and the like. In this chapter, we start with basic notions and definitions and contrast fuzzy numbers with the concepts of interval mathematics, their direct predecessors. The chapter includes detailed studies on computing with fuzzy numbers, including several categories of membership functions. In addition, we discuss several essential phenomena characteristic to fuzzy numbers, such as accumulation of fuzziness during iterative computing and inverse operations on fuzzy numbers.

5.1 Defining Fuzzy Numbers

Let us begin with a basic description of numbers (Dubois and Prade 1979a,b, 1980; Dijkman, van Haeringen, and De Lange 1983; Kaufmann and Gupta 1988). Several definitions exist. The one exhibiting a pragmatic flavor regards fuzzy numbers as mappings from the real line **R** to the unit interval that satisfy a series of properties such as normality, unimodality, continuity, and boundness of support. These requirements can be relaxed, if necessary. For instance, on can replace the property of continuity by upper semicontinuity. Using the terminology of interval analysis, some of the above conditions can be translated into set-oriented requirements. For instance, unimodality and boundness of support translate into the following format: α-cuts are convex, and α-cuts are closed intervals of **R**, $\alpha \in (0, 1]$.

In general, one considers a family of so-called LR fuzzy numbers. As introduced by Dubois and Prade (1980), the L(R) membership functions satisfy several properties necessary to model approximate quantities:

- $L(x) = L(-x)$ [symmetry]
- $L(0) = 1$ [normality]
- $L(x)$ is nonincreasing on $[0, \infty)$

The family of such functions is denoted by L. A fuzzy number is said to be an LR fuzzy number if its membership function is constructed with the aid of some L and R membership functions. More precisely, a two-parameter modification of an L type of membership function applies to all $x \leq m$, whereas the R membership function contributes to the definition of A for $x > m$. This yields

$$A(x) = \begin{cases} L\left(\dfrac{m-x}{\alpha}\right), & \text{if } x \le m, \ \alpha > 0 \\[2ex] R\left(\dfrac{x-m}{\beta}\right), & \text{if } x > m, \ \beta > 0, \end{cases} \tag{5.1}$$

where $L, R \in \mathsf{L}$. To explicitly articulate the parameters of the membership functions, we use an extended notation to identify them; for instance $A(x; \alpha, m, \beta)_{LR}$. In this notation m is normally referred to as the modal value of A whereas α and β form the spreads of the number. Several selected examples of the family of functions L are listed below. (See also figure 5.1.)

$$L(x) = \begin{cases} 1, & \text{if } x \in [-1, 1] \\ 0, & \text{otherwise.} \end{cases} \tag{5.2}$$

$$L(x) = \max(0, 1 - |x|^p). \tag{5.3}$$

$$L(x) = \frac{1}{1 + |x|^p}. \tag{5.4}$$

$$L(x) = e^{-|x|^p}, \quad p \ge 0. \tag{5.5}$$

Consider now (5.4) and (5.5) as the membership functions from L selected to model an imprecise quantity. This entails the following expression:

$$A(x) = \begin{cases} e_1^{-|(x-m)/\alpha|^p}, & \text{if } x \le m \\[2ex] \dfrac{1}{1 + |(m-x)/\beta|^p}, & \text{if } x > m. \end{cases} \tag{5.6}$$

See also figure 5.2, which illustrates this fuzzy number for some parameters of the membership function.

EXERCISE 5.1   Using the LR representation of fuzzy numbers, propose suitable elements of L to describe the fuzzy interval shown in figure 5.3.

EXERCISE 5.2   Figure 5.4 illustrates the response of a simple resistor-capacitor (RC) circuit. Model U (voltage on the capacitor) as a fuzzy number defined in $\mathbf{R}_+$.

5.2   Interval Analysis and Fuzzy Numbers

The roots of computing with fuzzy numbers originate from interval analysis (Moore 1966), a branch of mathematics developed to deal with the

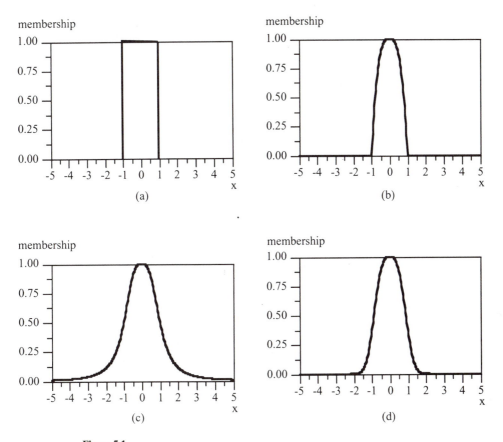

Figure 5.1
Functions of L for $p = 3.0$: (a) given by (5.2); (b) given by (5.3); (c) given by (5.4); (d) given by (5.5)

calculus of tolerances. In this setting, the basic objects in which one is interested are just intervals in **R**, such as [4, 6], $[a, b]$, $[-1.5, 3.2]$, and so forth. The resulting formulas describe the basic arithmetic operations: addition, subtraction, multiplication, and division. We have

$$[a, b] \oplus [c, d] = [a + c, b + d],$$

$$[a, b] - [c, d] = [a - d, b - c],$$

$$[a, b] \otimes [c, d] = [\min(ac, ad, bc, bd), \max(ac, ad, bc, bd)],$$

$$[a, b]/[c, d] = \left[\min\left(\frac{a}{d}, \frac{a}{c}, \frac{b}{c}, \frac{b}{d}\right), \max\left(\frac{a}{d}, \frac{a}{c}, \frac{b}{c}, \frac{b}{d}\right)\right].$$

EXERCISE 5.3   Perform addition, subtraction, multiplication and division of [2,7] and $[-5, 1]$.

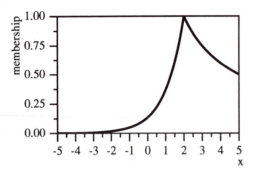

Figure 5.2
Membership function of A for $m = 2.0$, $\alpha = 1.0$, $\beta = 2.0$ and $p = 1.0$

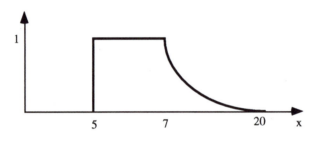

Figure 5.3
An example fuzzy interval

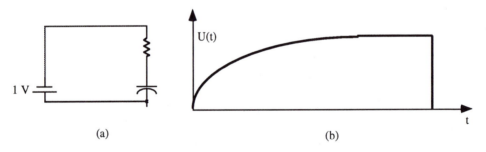

(a)   (b)

Figure 5.4
(a) RC circuit and (b) its impulse response

EXERCISE 5.4   Consider an interval $[a, b]$ and show that its inversion $[a, b]^{-1}$ can be expressed in the form

$$[a, b]^{-1} = \left[\min \left(\frac{1}{a}, \frac{1}{b} \right), \max \left(\frac{1}{a}, \frac{1}{b} \right) \right].$$

EXERCISE 5.5   In an electric circuit, a resistance R whose nominal value is 200 Ω has 10 percent tolerance, meaning that it could take on any value from 180 to 220 Ω. Consider that a current source is 5A with a 20 percent tolerance. Compute the voltage that could be measured on this resistance.

5.3   Computing with Fuzzy Numbers

Calculations on fuzzy numbers rely on the extension principle (Mizumoto and Tanaka 1976). Consider a function F transforming two fuzzy numbers A and B and producing a third number C such that

$$C = F(A, B).$$

The extension principle determines the membership function of C to be

$$C(z) = \sup_{x, y \in \mathbf{R}: z = f(x, y)} [A(x) \wedge B(y)],$$

$z \in \mathbf{R}$; $f : \mathbf{R}^2 \to \mathbf{R}$ is a real function that is pointwise consistent with F, meaning that

$$F(\{x\}, \{y\}) = f(x, y).$$

As usual, $\{x\}$ describes a single-element set. Let us also assume that there are two two-variable functions $g, h : \mathbf{R}^2 \to \mathbf{R}$ such that $f(x, g(x, y)) = y$ and $f(h(x, y), y) = x$ for any x and y in $\mathbf{R}$. The fundamental question is whether C obtained in this way constitutes a fuzzy number. The following statement expresses the conditions that need to be met (these requirements can be relaxed even further, as the pertinent literature discusses; see Di Nola, Pedrycz, and Sessa 1985):

If f is continuous monotone function, then C is a fuzzy number.

Because A and B are fuzzy numbers, they are, by definition, normal. This means that we get two arguments a and b in $\mathbf{R}$ such that $A(a) = B(b) = 1$. Then

$$C(f(a, b)) = \sup_{x, y \in \mathbf{R}: f(a, b) = f(x, y)} [A(x) \wedge B(y)] = A(a) \wedge B(b) = 1.$$

Hence C is normal. Clearly, C is continuous and has a bounded support.

Let us decompose A by looking at its increasing and decreasing parts separately. This yields

$$A(x) = \max[A^+(x), A^-(x)].$$

The same decomposition is performed for B:

$$B(x) = \max[B^+(x), B^-(x)],$$

with the split point located at $x = b$. By virtue of distributivity of the minimum over the maximum we derive

$$C(z) = \sup_{x,y \in \mathbf{R}:z=f(x,y)} [A(x) \wedge B(y)]$$

$$= \sup_{x,y \in \mathbf{R}:z=f(x,y)} \{[\max(A^-(x), A^+(x)] \wedge [\max(B^-(y), B^+(y)]\}$$

$$= \sup_{x,y \in \mathbf{R}:z=f(x,y)} \{\max[A^+(x) \wedge B^+(y), A^+(x) \wedge B^-(y), A^-(x) \wedge B^+(y),$$

$$A^-(x) \wedge B^-(y)]\}$$

$$= \max \left\{ \sup_{x,y \in \mathbf{R}:z=f(x,y)} [A^+(x) \wedge B^+(y)], \sup_{x,y \in \mathbf{R}:z=f(x,y)} [A^+(x) \wedge B^-(y)], \right.$$

$$\left. \sup_{x,y \in \mathbf{R}:z=f(x,y)} [A^-(x) \wedge B^+(y)], \sup_{x,y \in \mathbf{R}:z=f(x,y)} [A^-(x) \wedge B^-(y)] \right\}.$$

Thus

$$C = \max[F(A^+, B^+), F(A^+, B^-), F(A^-, B^+), F(A^-, B^-)].$$

The monotone function $F(A^+, B^+)$ is nondecreasing in $(-\infty, f(a,b))$ and $F(A^-, B^-)$ is nonincreasing in $[f(a,b), +\infty)$. More specifically, one derives

$$F(A^+, B^-)(z) = \begin{cases} A^+(h(z,b)), & \text{if } z \in (-\infty, f(a,b)) \\ B^-(g(a,z)), & \text{if } z \in [f(a,b), +\infty, \end{cases}$$

as well as

$$F(A^-, B^+)(z) = \begin{cases} A^-(h(z,b)), & \text{if } z \in [f(a,b), +\infty) \\ B^+(g(a,z)), & \text{if } z \in (-\infty, f(a,b)). \end{cases}$$

This implies that $F(A, B)$ is nondecreasing in $[f(a,b), +\infty)$ and nonincreasing in $(-\infty, f(a,b))$. Summarizing, C is unimodal, which concludes the proof of the statement.

5.4   Triangular Fuzzy Numbers and Basic Operations

In this section we concentrate on a family of triangular fuzzy numbers. These are the simplest models of uncertain numerical quantities. The analysis focusing on this class allows us to reveal the most visible properties of fuzzy arithmetic. Consider two triangular numbers $A = (x; a, m, b)$ and $B = (x; c, n, d)$. More specifically, the membership functions of these numbers are defined through the following piecewise linear relationships:

$$A(x; a, m, b) = \begin{cases} \dfrac{x-a}{m-a}, & \text{if } x \in [a, m) \\[2mm] \dfrac{b-x}{b-m}, & \text{if } x \in [m, b] \\[2mm] 0, & \text{otherwise;} \end{cases}$$

$$B(x; c, n, d) = \begin{cases} \dfrac{x-c}{n-c}, & \text{if } x \in [c, n) \\[2mm] \dfrac{d-x}{d-n}, & \text{if } x \in [n, d] \\[2mm] 0, & \text{otherwise.} \end{cases}$$

The modal values (m and n) identify a dominant (typical) value of the corresponding quantity whereas the lower and upper bounds (a or c and b or d) reflect the spread of the concept. Furthermore, to simplify computations, for the time being we consider fuzzy numbers with positive lower bounds.

5.4.1   Addition

The extension principle applied to A and B yields

$$C(z) = \sup_{x, y: z = x + y} [A(x) \wedge B(y)].$$

The resulting fuzzy number is normal; that is, $C(z) = 1$ for $z = m + n$. Two cases capturing computations of the spreads of C are dealt with separately.

Consider first that $z < m + n$. In this situation, the calculations involve the increasing parts of the membership functions of A and B. There exist values x and y such that $x < m$ and $y < n$ and satisfying the relationship

$$A(x) = B(y) = \omega, \quad \omega \in [0, 1].$$

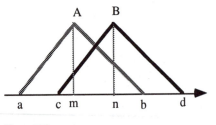

Figure 5.5
Addition of triangular fuzzy numbers

The linearly increasing sections of the membership functions A and B (see figure 5.5) embrace the interval $[a, n]$. Based on that we derive

$$\frac{x - a}{m - a} = \omega,$$

along with

$$\frac{y - c}{n - c} = \omega,$$

$x \in [a, m]$ and $y \in [c, n]$. Expressing x and y as functions of ω, one derives

$$x = a + (m - a)\omega,$$

$$y = c + (n - c)\omega.$$

Furthermore, since $z = x + y$, this produces

$$z = x + y = a + (m - a)\omega + c + (n - c)\omega = a + c + (m + n - a - c)\omega.$$

Now we derive similar relationships for the decreasing portions of the membership functions. The formulas below apply to $x \in [m, b]$ and y situated in $[n, d]$:

$$1 - \frac{x - m}{b - m} = \omega,$$

$$1 - \frac{y - n}{d - n} = \omega,$$

or equivalently

$$x = m + (1 - \omega)(b - m),$$

$$y = n + (1 - \omega)(d - n).$$

Adding x and y gives

$$z = x + y = m + (1 - \omega)(b - m) + n + (1 - \omega)(d - n)$$
$$= m + n + (1 - \omega)(b + d - m - n).$$

Summarizing, the membership function of $A \oplus B$ equals

$$C(z) = \begin{cases} \dfrac{z - (a + c)}{(m + n) - (a + c)}, & \text{if } z < m + n \\ 1, & \text{if } z = m + n \\ 1 - \dfrac{z - (m + n)}{(b + d) - (m + n)}, & \text{if } z > m + n. \end{cases}$$

Interestingly, C also has a triangular membership function. To emphasize that, we use a concise notation:

$$C = (x; a + c, m + n, b + d).$$

In general, when triangular fuzzy numbers are added, a fuzzy set of the same form results. More precisely, if $A_i = (x; a_i, m_i, b_i)$, $i = 1, 2, \ldots, n$, then the sum of these numbers equals

$$A = \sum_{i=1}^{n} A_i, \quad A(x; m, b) = A\left(x; \sum_{i=1}^{n} a_i, \sum_{i=1}^{n} m_i, \sum_{i=1}^{n} b_i\right).$$

One can observe that the spreads of the result start accumulating. For instance, let $m_i = 1$, $a_i = 0.95$, $b_i = 1.05$ for all $i = 1, 2, \ldots, n$. Iterating, one gets a sequence of fuzzy numbers

$$A_1 \oplus A_2 = (1.90, 2.00, 2.10),$$

$$A_1 \oplus A_2 \oplus A_3 = (2.85, 3.00, 3.15),$$

$$A_1 \oplus A_2 \oplus A_3 \oplus A_4 = (3.80, 4.00, 4.20).$$

EXERCISE 5.6   Consider the sup-product in the extension principle and perform the addition of two triangular fuzzy numbers. What is the result of this operation? Could you still retain a triangular fuzzy number as the outcome of this operation? Dubois and Prade (1981) have studied this generalized version of addition of fuzzy numbers.

Owing to the fundamental representation theorem, each fuzzy number can be regarded as a family of nested α-cuts. Subsequently, these α-cuts can be used to reconstruct the resulting fuzzy number. For instance, we get

$$(A \oplus B)_\alpha = A_\alpha \oplus B_\alpha.$$

Generally, the use of α-cuts results in a sort of a "brute force" method of carrying out computing with fuzzy quantities.

5.4.2 Multiplication

As with addition, we look first at the increasing parts of the membership functions, expressed as

$$x = \omega(m - a) + a,$$

$$y = \omega(n - c) + c.$$

Their product then becomes

$$z = xy = [\omega(m - a) + a][\omega(n - c) + c]$$

$$= ac + \omega x(m - a) + \omega a(n - c) + \omega^2(n - a)(n - c) = F_1(\omega).$$

If $ac \leq z \leq mn$, then the membership function of C is an inverse of F_1:

$$(A \otimes B)(z) = F_1^{-1}(z). \tag{5.7}$$

Similarly, consider the decreasing parts of A and B, meaning that $mn \leq z \leq bd$:

$$z = xy = [m + (1 - \omega)(b - m)][n + (1 - \omega)(d - n)]$$

$$= mn + (1 - \omega)[b - m + d - n] + (1 - \omega)^2(b - m)(d - n) = F_2(\omega).$$

As before, for any z in $[mn, bd]$, we derive

$$(A \otimes B)(z) = F_2^{-1}(z). \tag{5.8}$$

Evidently, multiplication does not return a piecewise linear membership function. Instead, as clearly indicated by (5.7) and (5.8), it produces a quadratic form of the resulting fuzzy number. The linearization of this quadratic form can be treated as an approximation of the previously derived membership function. We require that this linearization consist of two linear functions coinciding with Cz at $= mn, ac,$ and bd. The quality of this linearization depends very much on the spreads of the fuzzy numbers. To quantify the resulting linearization error, let us consider that $m = n = 1$ and $a = 1 - \delta, c = 1 - \delta, b = 1 + \delta, d = 1 + \delta$. The integral of error

$$I(\delta) = \int_{1-\delta}^{1+\delta} [C(z) - C^*(z)]^2 \, dz$$

is of interest here, where C^* is a piecewise linear approximation of the original fuzzy number C. Assuming that δ tends to zero, one can accept the linear approximation of the form

$$C(z) = (A \otimes B)(z) \approx C^*(z; ac, mn, bd).$$

5.4.3   Division

Like multiplication, the operation of division does not preserve the linearity of the arguments' membership functions. For the increasing parts of A and B, the following expression holds for $z \in [a/c, m/n]$:

$$z = \frac{x}{y} = \frac{\omega(m - a) + a}{\omega(n - c) + c} = G_1(\omega),$$

so that

$$(A/B)(z) = G_1^{-1}(z).$$

Analogously, for $z \in [m/n, b/d]$, we obtain

$$z = \frac{x}{y} = \frac{m + (1 - \omega)(b - m)}{n + (1 - \omega)(d - n)} = G_2(\omega),$$

which leads to

$$(A/B)(z) = G_2^{-1}(z).$$

Note that the membership function of A/B is a rational function of z.

In the following sections, we derive detailed formulas for selected single-argument functions operating on triangular fuzzy numbers. In general, the result is a fuzzy number with a nonlinear membership function.

5.4.4   Inverse

Let

$$z = \frac{1}{x}, \quad x \neq 0.$$

Then

$$A^{-1}(z) = A\left(\frac{1}{z}\right).$$

For the increasing part of the membership function the computations are carried out in the following way. The obtained result is valid for $x \in [1/m, 1/a]$. Similarly, the second range of arguments is spread between $1/b$ and $1/m$.

Let us start with the first range of the arguments, that is $[1/m, 1/a]$ and carry out detailed computations. As before

$$\frac{x - a}{m - a} = \omega,$$

$\omega \in [0, 1]$.

From this we calculate

$$x = a + \omega(m - a).$$

As $z = 1/x$, this yields

$$z = \frac{1}{a + \omega(m - a)}.$$

Expressing ω as a function of z one derives

$$z[a + \omega(m - a)] = 1,$$

and after some algebra

$$\omega = \frac{1}{(m - a)z} - \frac{a}{m - a}.$$

Hence

$$A^{-1}(z) = \frac{1}{(m - a)z} - \frac{a}{m - a}.$$

We stress that this formula holds for z in $[1/m, 1/a]$. Similarly, the computations are computed for the decreasing part of the membership function. The ensuing formula holds for the range $[1/b, 1/m]$. First note that

$$\frac{x - b}{m - b} = \omega$$

and

$$x = b + \omega(m - b).$$

In the sequel

$$z = \frac{1}{b + \omega(m - b)}.$$

Thus

$$\omega = \frac{1}{z(m - b)} - \frac{b}{m - b}.$$

and

$$A^{-1}(z) = \omega.$$

Summarizing, one gets

$$A^{-1}(z) = \begin{cases} \dfrac{1}{(m-b)z} - \dfrac{b}{m-b}, & \text{if } z \in \left[\dfrac{1}{b}, \dfrac{1}{m}\right] \\[2ex] \dfrac{1}{(m-a)z} - \dfrac{a}{m-a}, & \text{if } z \in \left[\dfrac{1}{m}, \dfrac{1}{a}\right] \\[2ex] 0, & \text{otherwise.} \end{cases}$$

EXERCISE 5.7   Derive a membership function of $\exp(A)$ where A is triangular fuzzy number. Hint: Observe that $\exp(A)(z) = A(\ln z)$. Plot the obtained membership function for $A(x; 1, 2, 4)$.

5.4.5 Fuzzy Minimum

The extension principle implies the following expression:

$$C(z) = (\min[A, B])(z) = \sup_{x, y \in \mathbf{R}: z = \min(x, y)} [A(x) \wedge B(y)],$$

$z \in \mathbf{R}$. Let us restrict our analysis first to $x \leq m$ and $y \leq n$, meaning that $z \leq \min(m, n)$. We are thus working with the two linearly increasing parts of the membership functions. This subsequently leads to the following expressions:

$$x = a + (m - a)\omega,$$

$$y = c + (n - c)\omega,$$

$\omega \in [0, 1]$. Thus,

$$z = \min(x, y) = \min(a + (m - a)\omega, c + (n - c)\omega].$$

Two possible cases result, each requiring separate consideration.

Case A: Consider a situation in which $a < c$ and $m < n$, or $c < a$ and $n < m$ (figure 5.6a). This implies that there is no ω for which the equation

$$a + (m - a)\omega = c + (n - c)\omega \tag{5.9}$$

holds, and the membership function for any $z \in [\min(a, c), \min(m, n)]$ equals the membership function of A or B, whichever is lower. Assume, for instance, that $a < c$ and $m < n$. Then the membership function in the above range is expressed as

$$C(z) = \frac{z - a}{m - a}.$$

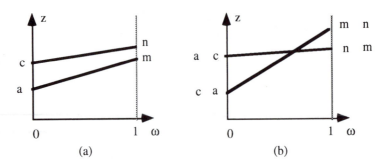

Figure 5.6
Computation of a fuzzy minimum in two separate cases, (a) and (b)

Analogously, if $c < a$ and $n < m$, we obtain

$$C(z) = \frac{z - c}{n - c}.$$

Case B: Consider a situation in which $a < c$ and $n < m$ or $c < a$ and $m < n$ (figure 5.6b). Then there exists ω_0 satisfying (5.9) such that

$$\omega_0 = \frac{a - c}{m + n - (a + c)}.$$

The corresponding z_0 is then given in the form

$$z_0 = a + \frac{(m - a)(a - c)}{m + n - (a + c)}.$$

Depending on the position of the two lines, figure 5.6 (b), we distinguish two situations:

1. If $a < c$ and $n < m$, then the membership function of C defined for $z \in [a, n]$ is expressed as

$$C(z) = \begin{cases} \dfrac{z - a}{m - a}, & \text{if } z \in [a, z_0] \\[2ex] \dfrac{z - c}{n - c}, & \text{if } z \in [z_0, n]. \end{cases}$$

2. If $c < a$ and $m < n$, we have

$$C(z) = \begin{cases} \dfrac{z - c}{n - c}, & \text{if } z \in [c, z_0] \\[2ex] \dfrac{z - a}{m - a}, & \text{if } z \in [z_0, m]. \end{cases}$$

The calculations to obtain the membership function of C for the decreasing parts of A and B are performed in the same way as those just completed for the increasing portions.

EXERCISE 5.8   Assuming that A and B are triangular fuzzy number, determine the result of their maximum, $C = \max(A, B)$, that is

$$C(z) = (\max[A, B])(z) = \sup_{x, y \in \mathbf{R}: z = \max(x, y)} [A(x) \wedge B(y)], \quad z \in \mathbf{R}.$$

5.5   General Formulas for LR Fuzzy Numbers

Before we examine computational formulas for fuzzy numbers, let us consider a few definitions. A fuzzy number A is said to be positive if its membership function assumes nonzero membership values for positive arguments; $A(x) = 0$ for all $x < 0$. Similarly, a fuzzy number A is negative if $A(x) = 0$ for $x > 0$. Moreover, we use the notation $M = (m, \alpha, \beta)_{LR}$ and $N = (n, \gamma, \delta)_{LR}$.

The general computational formulas for arguments that are LR-type fuzzy numbers are summarized below. (For more details, refer to Dubois and Prade 1980).

1. Addition: $(m, \alpha, \beta)_{LR} \oplus (n, \gamma, \delta)_{LR} = (m + n, \alpha + \gamma, \beta + \delta)_{LR}.$

2. Opposite: $-(m, \alpha, \beta)_{LR} = (-m, \beta, \alpha)_{RL}.$

(Note here that the reference membership function (L and R functions) have been exchanged.)

3. Subtraction: $(m, \alpha, \beta)_{LR} - (n, \gamma, \delta)_{RL} = (m - n, \alpha + \delta, \beta + \gamma)_{LR}.$

4. Multiplication: The multiplication formulas are approximate and hold under the assumption that the spreads of the arguments are small in comparison to the modal values of the fuzzy numbers.

- If $M > 0, N > 0$:

$(m, \alpha, \beta)_{LR} \otimes (n, \gamma, \delta)_{LR} \approx (mn, m\gamma + n\alpha, m\delta + n\beta)_{LR}.$

- If $M > 0, N < 0$:

$(m, \alpha, \beta)_{RL} \otimes (n, \gamma, \delta)_{RL} \approx (mn, m\alpha - n\delta, m\beta - n\gamma)_{RL}.$

- If $M < 0, N > 0$:

$(m, \alpha, \beta)_{RL} \otimes (n, \gamma, \delta)_{RL} \approx (mn, n\alpha - m\delta, n\beta - m\gamma)_{RL}.$

- If $M < 0, N < 0$:

$(m, \alpha, \beta)_{LR} \otimes (n, \gamma, \delta)_{LR} \approx (mn, -n\beta - m\delta, -n\alpha - m\gamma)_{RL}.$

In particular, for scalar multiplication, we obtain:

- If a is positive: $a(m, \alpha, \beta)_{LR} = (am, a\alpha, a\beta)_{LR}$.
- If a is negative: $a(m, \alpha, \beta)_{LR} = (am, -a\beta, -a\alpha)_{RL}$.

5. Division: Again, this is an approximate formula assuming that the spreads are small in comparison with the modal values.

$$(m, \alpha, \beta)_{LR} / (n, \gamma, \delta)_{RL} \approx \left(\frac{m}{n}, \frac{\delta m + \alpha n}{n^2}, \frac{\gamma m + \beta n}{n^2} \right)_{LR}$$

5.6   Accumulation of Fuzziness in Computing with Fuzzy Numbers

We need to mention here a phenomenon of accumulation of fuzziness, in manifestation quite similar to that encountered in error accumulation in numerical computations, for example, when performing some iterative calculations. To illustrate this effect, consider an iterative process of adding a fuzzy number with triangular membership function, $A = (0, 1, 2)$, to a single numerical quantity $(1, 1, 1)$. Following the addition formula for triangular fuzzy numbers, we get

$B = A \oplus 1 = (1, 2, 3),$

$C = A \oplus B = (0, 1, 2) \oplus (1, 2, 3) = (1, 3, 5),$

$D = A \oplus C = (0, 1, 2) \oplus (1, 3, 5) = (1, 4, 7),$

$E = A \oplus D = (0, 1, 2) \oplus (1, 4, 7) = (1, 5, 9) \ldots.$

We note that the supports of the successive fuzzy numbers expand very quickly; the lower bound remains fixed, but the upper increases from step to step, skewing the resulting membership function. Clearly, one should proceed with caution when computing with fuzzy numbers: the length of the chain of iterative computations should eventually be reduced to keep the obtained results meaningful.

EXERCISE 5.9   Consider the problem of an iterative addition of fuzzy numbers following the recurrence formula

$A_{k+1} = A_k \oplus A_{k-1}, \quad k = 1, 2, \ldots.$

(a) Assuming that $A_0 = (0.95, 1, 1.10)$ and $A_1 = (3.45, 3.50, 3.55)$, determine how many iterations are permissible if result whose support is broader than 2.00 is not acceptable.

(b) Consider a reverse problem: For the same admissible size of support (that is, 2.00) what should be the size of support of the first two fuzzy numbers, assuming that we want to iterate 50 times? That is, compute A_{51}.

(c) What is the energy measure of fuzziness of A_k viewed as a function of k? Determine also its entropy measure of fuzziness and express it as a function of k.

5.7   Inverse Problem in Computation with Fuzzy Numbers

The overall discussion in this chapter has been exclusively devoted to computing operations whose arguments are fuzzy numbers. An interesting problem, which one can refer to as an inverse problem in fuzzy arithmetic, concerns inverse operations. To illustrate the nature of the problem, let us again consider an addition operation. Obviously, for real numbers, this task is trivial. Given a and b such that

$$a + x = b,$$

one immediately derives

$$x = b - a.$$

This relationship is not valid for fuzzy numbers, meaning that for a given two fuzzy numbers A and B,

$$A \oplus X = B, \tag{5.10}$$

one cannot determine X simply by subtracting A from B,

$$X = B - A,$$

where the subtraction operation is considered in its extended version. The simple explanation of this fact is that there is no group structure as far as operations on fuzzy numbers are concerned. Namely, the expression

$$A \oplus (B - A) = B$$

does not return a real zero (real number). There is, however, a way to solve this equation with respect to X. Note that (5.10) is a fuzzy-relational equation with the sup-min composition,

$$B = X \circ R,$$

with the relation R defined as

$$R(z, x) = A(z - x).$$

Assuming that the equation is solvable, the maximal fuzzy set X for which the equation holds is determined as

$$X(x) = \inf_{z \in \mathbf{R}} [R(x,z) \rightarrow B(z)] = \inf_{z \in \mathbf{R}} \begin{cases} 1, & \text{if } R(x,z) \leq B(z) \\ B(z), & \text{if } R(x,z) > B(z). \end{cases}$$

EXERCISE 5.10   Formulate the inverse problem for (a) multiplication and (b) division of fuzzy numbers.

5.8   Fuzzy Numbers and Approximate Operations

The main feature of all computations with fuzzy numbers is that the objects to be manipulated are nonpointwise (fuzzy sets), but the operations themselves are defined in a very standard and precise way. Together, these two features can (and, in fact, do) produce some semantic deficiencies.

Consider, for instance, the addition of fuzzy numbers and real numbers in two cases: *about* $1 \oplus 100$, and *about* $100 \oplus 1$. In both cases the result implied by this addition gets too precise (*about* 101) and hence becomes somewhat semantically doubtful. Note that in both cases we obtain fuzzy numbers whose modal values are situated exactly at 101, even though the addition of these quantities is slightly different and the semantics of the results may dictate notions such as (exactly) 100 (here we do not care about fuzzy 1) in the first example and *about* 100 in the second. The extension principle governing fuzzy arithmetic is noticeably in need of more refinement.

Let us start with an equivalent formulation of the addition operation by looking at some relational constructs. We introduce a three-argument Boolean relation,

$$R(x,y,z) = \begin{cases} 1, & \text{if } z = x + y \\ 0, & \text{otherwise,} \end{cases}$$

$z, y, z \in \mathbf{R}$. The essence of the semantics-driven approach (or, so to speak, a customization of the operations on fuzzy numbers) is to make this relation fuzzy by accommodating some essential semantic constraints. Intuitively we accept that if $|x| \gg |y|$, then $z = x$, no matter what the specific value of x is. Similarly, the same argument can be used for the second inequality: If $|x| \ll |y|$, then $z = y$. Looking at these two requirements, we can propose a relation of the type

$$R(x,y,z) = \begin{cases} 1, & \text{if } z = x + y \text{ and } |x| \approx |y| \\ 1, & \text{if } z = x \text{ and } |x| \gg |y| \\ 1, & \text{if } z = y \text{ and } |x| \gg |x|. \end{cases}$$

The notions *approximately* equal and *much greater* (*smaller*) in the above definition are modeled through the corresponding binary (two-argument)

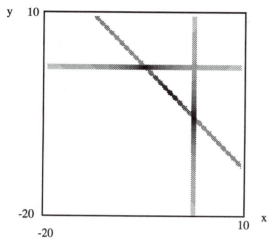

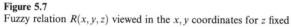

Figure 5.7
Fuzzy relation $R(x, y, z)$ viewed in the x, y coordinates for z fixed

fuzzy relations. For instance, we may consider the models

approximately equal $(x, y) = \exp(-\rho|x - y|)$,

much greater $(x, y) = \dfrac{|x|}{|x| + \kappa|y|}$,

where ρ and κ are two positive scaling factors. Figure 5.7 shows an example of such a fuzzy relation for $\rho = 0.12$ and $\kappa = 4$, here z is fixed and equal to 10. The computations of the fuzzy number C are now governed by the formula

$$C(z) = \sup_{x, y \in \mathbf{R}} [A(x) \wedge B(y) \wedge R(x, y, z)]$$

that can be regarded as a fuzzy-relational equation with two input variables, A and B, respectively.

5.9   Chapter Summary

This chapter has covered the main topics concerning fuzzy numbers, addressing both their conceptual background and computational aspects. The calculus of fuzzy numbers relies on the extension principle. Assuming some classes of membership functions that are usually sufficient in many applications, one can derive concise computation formulas that help reduce computational overhead. Any computing with fuzzy numbers primarily dealing with the families of elements should be carried out with caution—longer chains of operations on fuzzy numbers can lead to accumulation of fuzziness. This phenomenon should be monitored closely; otherwise, the results obtained could be quite meaningless. We also discussed the inverse problem and examined the idea of more semantically enhanced operations on fuzzy numbers (fuzzy operations on fuzzy numbers as opposed to operations on fuzzy numbers).

5.10 Problems

1.   The standard transformation from polar (r, ϕ) to Cartesian (x, y) coordinates reads as

$$x = r \cos \phi,$$

$$y = r \sin \phi.$$

Consider now a fuzzy point defined in the polar coordinates whose values are given as

$$R(r) = \exp[-(r-1)^2],$$

$$\Phi(\phi) = \exp[-(\phi - 60)^2].$$

Find the point's Cartesian coordinates.

2.   Let us consider the simple relationship associating distance with velocity and time travelled,

$$d = vt.$$

Assuming that the velocity is fixed and given as a triangular fuzzy number $V(v; 10, 20, 40)$, compute the distance covered in two hours. What happens with the result if the time of this travel is *around* two hours, modeled by the triangular fuzzy number $T(t; 1.5, 2, 2.5)$?

3.   Calculate the arithmetic mean of two triangular fuzzy numbers $T(x; 1, 3, 5)$ and $R(x; 2, 4, 6)$. Repeat the calculations for the expression

$$(T_1 + T_2 + \cdots + T_n)/n,$$

where all arguments are the same: $T_1 = T_2 = \cdots = T_n = T(x; 1, 3, 5)$. Discuss the results obtained when the number of arguments (n) increases.

4.   Let A and B be two fuzzy numbers with Gaussian membership functions $A(x; 1, 2, 3)$ and $B(x; 2, 5, 8)$. Determine X such that

$$A \oplus X = B.$$

5.   Discuss how to solve a system of equations involving fuzzy numbers:

$$(A_{11} \otimes X) \oplus (A_{12} \otimes Y) = B_1,$$

$$(A_{21} \otimes X) \oplus (A_{22} \oplus Y) = B_2,$$

where $A_{11}, A_{12}, A_{21}, A_{22}, B_1$, and B_2 are specified fuzzy numbers.

6.   A signal is described as follows:

$$X(t) = \sin(\Omega t + \Phi),$$

where Ω and Φ are Gaussian fuzzy numbers with the membership functions

$$\Omega(\omega) = \exp[-(\omega - 60)^2/5],$$

$$\Phi(\phi) = \exp[-\phi^2].$$

Determine $X(t)$. Plot its membership function for selected time instants (t). Study the same problem assuming that Ω and Φ are intervals defined as follows:

$$\Omega = [50, 70],$$

$$\Phi = [-1, 1].$$

Compare the results of computing using fuzzy numbers and intervals.

7.   Consider a fuzzy model governed by the expression

$$Y = (A \otimes X) \oplus B,$$

where A and B are triangular fuzzy numbers with the membership functions $A = (0.5, 1, 2)$ and $B = (0.9, 1, 1.1)$.

(a) Assume that X is a singleton. Plot Y as a function of x.

(b) Treat the above expression as an equation for A, B and Y given. Solve it with respect to X.

8.   Consider the addition of two images (fuzzy relations A and B),

$$C = A \oplus B.$$

(See figure 5.8.) Propose the pertinent computational formulas. Illustrate your findings for Gaussian membership functions.

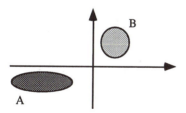

Figure 5.8
Addition of two images

9.   Consider an iterative process of subtraction a triangular fuzzy number $A = (1, 0.8, 1.2)$ from a numeric value $B = 10$,

$$B - A, (B - A) - A, ((B - A) - A) - A, \dots.$$

Analyze the results of successive iterations and discuss the effect of accumulation of fuzziness. Repeat the same analysis for an iterative multiplication of A and B.

10.   The classic quadratic map is described as

$$y = 4x(1 - x).$$

Assuming that A is a triangular fuzzy number, $A = (0.5, 0.45, 0.55)$, analyze the results of iteration of A through this nonlinear mapping.

References

Dijkman, J., H. van Haeringen, and S. I. De Lange. 1983. Fuzzy numbers. *J. Math. Anal. and Applications* 92:301–41.

Di Nola, A., W. Pedrycz, and S. Sessa. 1985. Processing of fuzzy numbers by fuzzy relation equations. *Kybernetes* 15:43–7.

Dubois, D., and H. Prade. 1979a. Operations on fuzzy numbers. *Int. J. Systems Science* 9:613–26.

Dubois, D., and H. Prade. 1976b. Fuzzy real algebra—some results. *Fuzzy Sets and Systems* 2:327–48.

Dubois, D., and H. Prade. 1980. Fuzzy Sets and Systems. New York: Academic Press.

Dubois, D., and H. Prade. 1981. Additions of interactive fuzzy numbers. *IEEE Trans. on Automatic Control* 26:926–36.

Kaufmann, A., and M. M. Gupta. 1988. Fuzzy Mathematical Models in Engineering and Management Science. Amsterdam: North Holland.

Moore, R. 1966. Interval Analysis. Englewood Cliffs, NJ: Prentice Hall.

Mizumoto, M., and K. Tanaka. 1976. The four operations of arithmetic on fuzzy numbers. *Systems, Computers, Controls* 7:73–81.

This chapter deals with the description of uncertainty from the point of view of fuzzy sets and probability. It aims to distinguish these two ideas for modeling uncertainty and to introduce the concepts of probability of fuzzy events, linguistic probabilities, and fuzzy random variables. It shows that, since no single universal approach exists to handle uncertainty, fuzzy sets and probability are complementary. As part of the theory of fuzzy sets, however, linguistic probabilities have a unique character that enables them to describe many real-world situations that probability theory cannot otherwise describe.

6.1   Introduction

Probability and fuzzy sets (Dubois and Prade 1994) are usually contrasted as two distinct conceptual and computational vehicles aimed at representing and processing uncertainty.[1] Uncertainty is not directly associated with any real-world system but is primarily linked with the prudently selected process of describing the system itself—hence, the way in which it manifests itself and the method by which it can be captured properly depend on the observer (figure 6.1). Subsequently, no single universal vehicle exists to cope with uncertainty. Some other formal frameworks may well be developed (and likely will be developed) to address specific issues of uncertainty. In this sense, fuzzy sets and probability seem to be complementary rather than antagonistic ideas. In fact, one can contemplate several hybrid models embracing fuzziness and probability that have already been carefully investigated in the literature. For more details on probability-fuzziness, the reader may refer to the special issue of *IEEE Transactions on Fuzzy Systems 2* (1994).

6.2   Probability and Fuzzy Sets

The distinction between probability and fuzzy sets is self-evident. Probability is concerned with *occurrence* of *well-defined* events (being regarded as sets). Tossing a coin, drawing a ball from an urn, accomplishing a space mission—these are examples of events described by sets. Probabilities are or can be associated with each of them. For instance, the probability of picking randomly a red ball from an urn containing 5 red and 7 black balls is $5/(5+7) = 5/12$, assuming that each ball can be picked up with equal probability, $1/12$. Probabilities are assessed or estimated based on a repetition of a certain series of experiment carried out in a stationary environment. Then, to make probabilities meaningful, one has to propose a way to determine them. Engineering relies heavily on probabilities treated as limit frequencies and determined based on a finite

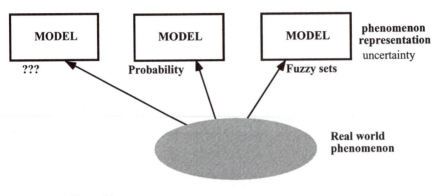

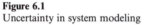

Figure 6.1
Uncertainty in system modeling

number of experiments. Thus the probability of event A is estimated as

$$p(A) = \lim_{n \to \infty} \frac{n_A}{n} \tag{6.1}$$

where n_A is the number of experiments in which event A has occurred and n is the total number of experiments in the series. Hence, the occurrence of an event becomes a central notion of probability: The calculus of probability deals with *occurrence* of events.

On the other hand, fuzzy sets deal with *graduality* of concepts and describe their boundaries. They therefore have nothing to do with frequencies (repetition) of an event. This is the dominant difference between probabilities and fuzzy sets.

To make the difference even clearer, consider an experiment whose outcome (A) can eventually occur. Only before the experiment can one think of the probability of A, $P(A)$. Once the experiment is over, the probabilistic facet of uncertainty vanishes. The outcome is unambiguously known: A has happened or not. In contrast, let A be a fuzzy set; after the experiment, the idea is still valid and fully intact.

The conceptual difference between these two notions of uncertainty makes the mathematical frameworks of fuzzy sets and probability also very distinct. The former relies on set theory and logic, whereas the latter hinges on the concepts of (additive) measure theory.

6.3   Hybrid Fuzzy-Probabilistic Models of Uncertainty

6.3.1   Probability of Fuzzy Events

The previous section strongly emphasized that, as mechanism for coping with occurrence and graduality, probability and fuzziness are highly

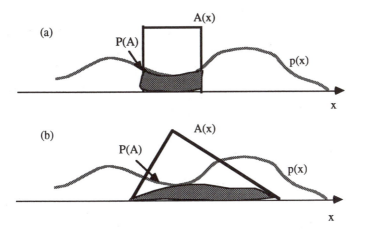

Figure 6.2
Probabilities (a) of an event and (b) of a fuzzy event

orthogonal. In spite of that, some interesting situations lead to a useful symbiosis of fuzzy sets and probability.

Let us consider a statement,

$\text{Prob}(X \text{ is } A)$,

where A is a fuzzy set defined in $\mathbf{X} \subset \mathbf{R}$. Because $\mathbf{X}$ is equipped with some probability density function, the probability of this fuzzy event is quantified as (Zadeh 1968)

$$\text{Prob}(X \text{ is } A) = \int_{\mathbf{R}} A(x)p(x)\,dx. \tag{6.2}$$

Formula (6.2) provides what is known as a probability of a fuzzy event (assuming that the integral makes sense). Figure 6.2b illustrates the underlying idea. The probability density function $p(x)$ is integrated over the support of A, whereas $A(x)$ plays the role of a weighting function. Illustrated for comparison is a situation where A is a set (figure 6.2a). The shadowed areas denote the probability of the corresponding events.

If $A \subset B$, then the probabilities entailed satisfy the inequality

$\text{Prob}(X \text{ is } A) \le \text{Prob}\,(X \text{ is } B)$,

implying an increase in the resulting probability. Similarly, considering a certain level of probability (λ),

$\text{Prob}(X \text{ is } A) = \lambda$,

we can determine a fuzzy event A satisfying this relationship. (The choice of A is definitely not unique). These expressions, similar to those given above, exemplify probabilistically qualified linguistic statements. The fundamental requirements for these statements can be formulated by looking at the following constraints:

• Specificity constraint: The statement (liguistic label) A should be specific enough.

• Probability constraint: The statement should be meaningful enough to make the probability associated with it sufficiently high.

These requirements translate into the formal expressions

$$\frac{1}{\text{Card}(\mathbf{X})} \int_x A(x)\,dx \leq \mu, \tag{6.3}$$

$$\int_{\mathbf{R}} A(x)p(x)\,dx \geq \lambda, \tag{6.4}$$

where $\mu,\ \lambda \in [0,1]$. The first constraint restricts the normalized energy measure of fuzziness, as it should not exceed μ (maintaining sufficient specificity). A similar formulation of this requirement can be found in Pardo (1985). At the same time, the probability of the statement needs to be meaningful (greater than λ); this requirement is captured via the second condition. Figure 6.3 shows the region of feasible linguistically qualified statements. These two requirements are in conflict: Less specific statements (fuzzier As) lead to higher probabilities, but their semantics are weakened. On the other hand, too specific statements are not sufficiently supported by visible probabilistic evidence.

Let us now consider that A is a set. The previous formulation is read as

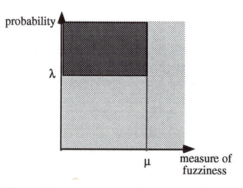

Figure 6.3
(μ, λ) region of feasible linguistically quantified statements

$$\text{Prob}(X \text{ is } A) = \int_{x_0-\varepsilon}^{x_0+\varepsilon} p(x)\,dx \geq \lambda \tag{6.5}$$

The range $[x_0 - \varepsilon, x_0 + \varepsilon]$ plays the role of a confidence interval and is commonly employed in statistical analysis. The higher the probability (specific values are usually set at 0.95, 0.99, and 0.999; in fact, the tradition is to use their complements, that is, 0.05, 0.01, and 0.001), the broader the confidence interval. What occurs at higher confidence levels, particularly at the highest probability, 0.999, is that the associated confidence intervals get too broad and tend to become almost meaningless.

In addition to fuzzy probabilities, one can easily envision some other generalizations (Zadeh 1968) of the standard probabilistic notions, such as

- Expected (mean) value of A: $m_A = \int_{\mathbf{R}} xA(x)p(x)\,dx$. $\tag{6.6}$
- Variance of A: $\sigma_A^2 = \int_{\mathbf{R}}[A(x) - m_A]^2 p(x)\,dx$. $\tag{6.7}$

EXERCISE 6.1   Compute the probabilities of events or fuzzy events for the situations listed below. In all cases, assume that the probability density function $p(x)$ is uniform and defined over $[-10, 10]$.

(a) $A(x) = \exp(-x^2/5)$ for $x > 0$, and $A(x) = 0$ otherwise.

(b) $A(x) = \begin{cases} 1, & \text{if } x \in [-5, 5] \\ 0, & \text{otherwise.} \end{cases}$

(c) $A(x) = \dfrac{|x|}{|x| + 10}$,

(d) A is triangular, with the modal value at 3.5 and the bounds equal to 1 and 5.

EXERCISE 6.2   Assume a triangular probability density function,

$$p(x) = \begin{cases} \dfrac{2}{7}\dfrac{x+3}{3}, & \text{if } x \in (-3, 0) \\[2mm] \dfrac{2}{7}\dfrac{4-x}{4}, & \text{if } x \in (0, 4) \\[2mm] 0, & \text{otherwise,} \end{cases}$$

and a fuzzy set characterized by the Gaussian membership function

$$A(x) = \exp[-(x - 1)^2/2].$$

Determine the probability of this fuzzy event, its expected value and its variance.

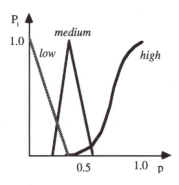

Figure 6.4
Example membership functions of linguistic probabilities P_i

6.3.2 Linguistic Probabilities

Any text on probability calculus and statistics employs numeric probabilities, but in many practical situations we are confronted with probabilities that are linguistic rather then numeric. For instance, we talk about a *high* probability of rain, *very probable* failure of the system, *low* probability of recession, *medium* probability of completing the project on time, and so forth. All these statements contain notions that can be called linguistic probabilities. Figure 6.4 illustrates the intuitively appealing membership functions of some linguistic probabilities.

Because the linguistic probabilities P_i are fuzzy sets, the operations on them need to be carried out accordingly; the results become fuzzy sets as well. To illustrate the basic idea, let us perform detailed computations of the expected value of the random variable

$$z = \sum_{i=1}^{n} a_i p_i,$$

$a_i \in \mathbf{R}$ where a_i, occurs with probability p_i. The extension of this expression to incorporate fuzzy probabilities reads as

$$Z = \sum_{i=1}^{n} a_i P_i. \tag{6.8}$$

The problem is solved through the extension principle, resulting in an optimization problem of the form

$$Z(z) = \max[\min(P_1(p_1), P_2(p_2), \dots, P_n(p_n))],$$

subject to

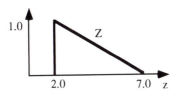

Figure 6.5
Membership function of $Z = 2$ likely $+ 7$ unlikely

$$z = \sum_{i=1}^{n} a_i p_i,$$

and

$$\sum_{i=1}^{n} p_i = 1.$$

The second restriction reflects the probabilistic nature of the constraints, which need to sum to 1.

As an illustrative example, let us compute the membership function of the expression

$Z = a_1$ likely $+ a_2$ unlikely,

where both terms (*likely* and *unlikely*) are linguistic probabilities; moreover, we assume that their membership functions satisfy the antonym relationship,

likely $(p) = $ *unlikely* $(1 - p)$,

$p \in [0, 1]$. We then obtain

$$Z(z) = likely\left(\frac{z - a_2}{a_1 - a_2}\right).$$

More specifically, let us consider that $a_1 = 2$ and $a_2 = 7$. Assume a linear membership function for the fuzzy probability, $likely(u) = u$, for all u in $[0,1]$. The expected value Z is a fuzzy set with the membership function

$$Z(z) = likely\left(\frac{7 - z}{5}\right).$$

See also figure 6.5.

6.4   Probability-Possibility Transformations

Although no strict identity mapping exists between membership functions and probability functions (or probability density functions), the notion of possibility entails a relationship known as the consistency principle (Zadeh 1978): "what is *possible* may not be *probable* and what is improbable need not to be impossible." In other words, the degree of possibility (membership) of a certain element is greater than or equal to its probability.

Prompted by the inequality between a numeric quantification of probability and fuzziness, a bijective transformation was proposed for completing mapping between these two. Consider a finite universe $\mathbf{X}$ where $p_1, p_2, \ldots, p_n$ denote discrete probabilities defined for the individual elements of this space. Furthermore, let us assume that these probabilities are ordered in a nonincreasing sequence,

$$p_1 \geq p_2 \geq \cdots \geq p_n.$$

As usual, we require that

$$\sum_{i=1}^{n} p_i = 1. \tag{6.9}$$

The respective membership values μ_i are determined as (Dubois and Prade 1983)

$$\mu_i = 1 - \sum_{j=1}^{i-1} (p_i - p_j), \quad i = 1, \ldots, n. \tag{6.10}$$

Let us enumerate the membership values for $n = 3$:

$$\mu_1 = 1 - \sum_{j=1}^{0} (p_i - p_j) = 1 \quad \text{(no sum, since the lower index exceeds the upper),}$$

$$\mu_2 = 1 - \sum_{j=1}^{1} (p_2 - p_j) = 1 - (p_2 - p_1) = 1 - p_2 + p_1,$$

$$\mu_3 = 1 - \sum_{j=1}^{3} (p_3 - p_j) = 1 - [(p_3 - p_1) + (p_3 - p_2)]$$

$$= 1 - (p_3 - p_1) - (p_3 - p_2) = 1 + (p_1 - p_3) + (p_2 - p_3).$$

The induced fuzzy set is normal because $\mu_1 = 1$. Note that one also has

$$\mu_i = 2 - i p_i - \sum_{j=i+1}^{n} p_j.$$

Without making any particular assumption regarding any arrangement of the probabilities, the possibility-probability expression is reformulated as

$$\mu_i = \sum_{j=1}^{n} \min(p_i, p_j), \quad i = 1, 2, \ldots, n. \tag{6.11}$$

The transformation leading to the probability function is inferred from (6.10). Treating it as a system of equations for given μ_i and adding an additional normalization condition (6.9), on derives

$$p_i = \sum_{j=i}^{n} \frac{1}{j} (\mu_j - \mu_{j+1}), \tag{6.12}$$

$i = 1, 2, \ldots, n$; we acknowledge that $\mu_{n+1} = 0$. This possibility-necessity transformation is not unique.

EXERCISE 6.3   Given the probabilities below, determine the induced membership values.

Element	a	b	c	d	e
Probability	0.6	0.05	0.20	0.10	0.05

EXERCISE 6.4   Provided is a discrete membership function. Calculate the corresponding probability function.

Element	a	b	c	d	e
Membership	0.6	0.5	0.9	1.0	0.2

6.5   Probabilistic Sets and Fuzzy Random Variables

Probabilistic sets and fuzzy random variables are two concepts synthesizing the symbiosis of fuzzy sets and the elements of randomness. Both constructs regard randomness as an operational component that helps express linguistic terms in an algorithm. The notions are conceptually straightforward; the formal underlying apparatus, however, looks far more complicated, because it alludes either to random fields (probabilistic sets) or random sets (fuzzy random variables). In brief, fuzzy sets are no longer considered pointwise structures (single membership values); rather, the factor of randomness reflected in their description (characterization) gives rise to their more setlike manifestation.

6.5.1   Probabilistic Sets

When dealing with probabilistic sets (Hirota 1981), we assume that for each element of **X** the membership grade is essentially a truncated random variable A_x whose probability density function indexed by x is denoted by $p(x, u)$, $u \in [0, 1]$.[2] The mean value of the random variable A_x is a membership value

$$\bar{A}(x) = \int_0^1 A_x(u) p(x, u) \, du.$$

Thus, when only mean values of the probabilistic sets are considered, they reduce to fuzzy sets. Higher moments of A_x characterize the spread of the A_xs. In particular, the variance of A_x (called the vagueness function) is of special interest:

$$V(x) = \int_0^1 [A_x(u) - \bar{A}(x)]^2 p(x, u) \, du.$$

Higher moments (called monitor functions) are introduced in the integral form, namely,

$$M_r(x) = \int_0^1 [A_x(u) - \bar{A}(x)]^r p(x, u) \, du,$$

$r = 3, 4, \dots$. The values of the monitors vanish quite quickly:

If $r > r'$, then $M_r(x) < M_{r'}(x)$.

In general, only $\bar{A}(x)$ and $V(x)$ convey significant amounts of information about any probabilistic set (Hirota 1981). From an experimental point of view, probabilistic sets are directly linked with the horizontal method of membership estimation (see chapter 1).

6.5.2   Fuzzy Random Variables

The idea of combining fuzziness and randomness, introduced by Kwakernaak (1978), extends the concept of random variables by acknowledging their values as fuzzy numbers (fuzzy sets in **R**). Let us recall that a random variable x defined in the probability space $(\Omega, \mathbf{B}, P)$ is a mapping

$$x : \Omega \to \mathbf{R}.$$

A fuzzy random variable X generalizes the previous mapping, assuming now the form

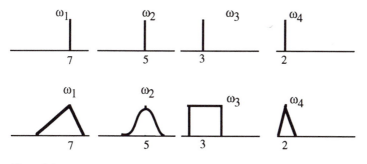

Figure 6.6
Example of a random variable and its counterpart fuzzy random variable

$$X : \Omega \to \mathbf{S},$$

where $\mathbf{S}$ is a family of membership functions of fuzzy numbers. (More strictly, these mappings need to be restricted to satisfy certain mild measurability requirements.) Figure 6.6 illustrates a discrete random variable and its generalization.

Because the values of the fuzzy random variables are nonpointwise, any operation on them is carried out by working with a finite sequence of the corresponding fuzzy sets. Following figure 6.6, let us compute the expected value of the random variable and its fuzzy-set generalization. To complete calculations, also given are probabilities $p(\omega_1) = 0.3, p(\omega_2) = 0.4$, $p(\omega_3) = 0.2, p(\omega_4) = 0.1$. The expected value of the random variable equals

$$EX = \sum_{i=1}^{4} p_i x_i = \sum_{i=1}^{4} p(\omega_i)\omega_i = 4.9.$$

The fuzzy random variable assumes nonpointwise values X_i, implying that EX becomes a fuzzy set (number). Although the extension principle can be readily invoked, one can convert the resulting optimization problem into a series of nested computations employing certain α-cuts of X_is. Finally, EX is reconstructed by combining the series of previously computed α-cuts. The algorithm involves a single iteration, indexed by the threshold level of the membership function, as follows:

Start with $\alpha = 0$ and repeat the body of the loop (1–2), increasing the value of α by $\Delta\alpha$ until it attains 1:

1. Determine the bounds of α-cut of X_i, x_i^l and x_i^u, using the formulas

$$x_i^l = \inf[x \in \mathbf{R} | X_i(x) \geq \alpha],$$

$$x_i^u = \sup[x \in \mathbf{R} | X_i(x) \geq \alpha].$$

2. Aggregate the bounds, producing successive α-cuts of the fuzzy expected value:

$$EX_\alpha = \left[\sum_{i=1}^{n} p_i x_i^l, \sum_{i=1}^{n} p_i x_i^u \right].$$

As usual, EX is approximated by taking the union of EX_α:

$$EX = \bigvee_\alpha \alpha EX_\alpha.$$

6.6   Chapter Summary

This chapter began with a general discussion about the roots of uncertainty, then proceeded to introduce the main issues concerning fuzziness and probability, exploring in particular their complementary aspects. The central point has been that uncertainty modeling becomes more amenable with the contributions of both fuzzy sets and probability. In addition, the ability to handle linguistic probabilities via fuzzy sets increases our strategies for handling real-world systems and situations.

6.7   Problems

1. Consider the probability density function $p(x)$ and the fuzzy set A defined in the form

$$p(x) = \begin{cases} \dfrac{1}{c}, & \text{if } x \in [0, c] \\ 0, & \text{otherwise,} \end{cases}$$

$$A(x) = \begin{cases} 1, & \text{if } x \in [a, b] \\ 0, & \text{otherwise .} \end{cases}$$

Furthermore, assume that $0 < a < c$ and $c < b$. Compute the probability of A, its expected value and its variance. Next, swap $p(x)$ and A, meaning that

$$p(x) = \begin{cases} \dfrac{1}{b-a}, & \text{if } x \in [a, b] \\ 0, & \text{otherwise,} \end{cases}$$

$$A(x) = \begin{cases} 1, & \text{if } x \in [0, c] \\ 0, & \text{otherwise.} \end{cases}$$

Repeat the computations. Interpret the results obtained.

2. An error distribution of a certain sensor can be modeled by a triangular probability density function, as shown in figure 6.7(a). A certain reading of the sensor comes in the form of the membership function visualized in figure 6.7(b). Assuming that noise affects the measurement additively, determine the probability of this fuzzy event.

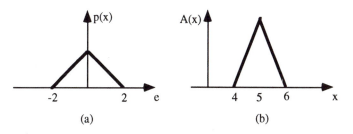

Figure 6.7
Probability density function and membership function

3.   Table 6.1 summarizes the outcomes of a combustion process; see also figure 6.8. Propose a collection of fuzzy sets to model meaningful linguistic categories (labels) A_1, A_2, A_3, and A_4.

(a) Determine the probability, expected value, and variance for each of the linguistic labels.

(b) Choose the label with the lowest probability value. Suggest how you could modify this membership function to increase the associated probability.

(c) By analyzing the histogram of the data set, adjust A_ks in such a way that their probabilities can be made almost equal.

Table 6.1
Experimental data in successive discrete time moments

53.8	53.6	53.5	53.5	53.4	53.1	52.7	52.4	52.2	52	52	52.4
53	54	54.9	56	56.8	56.8	56.4	55.7	55	54.3	53.2	52.3
51.6	51.2	50.8	50.5	50	49.2	48.4	47.9	47.6	47.5	47.5	47.6
48.1	49	50	51.1								

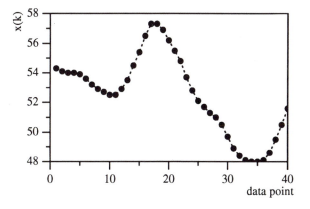

Figure 6.8
Experimental data

Notes

1. Here we use the term "uncertainty" in a broad sense. When being more precise, one should distinguish between uncertainty and vagueness. The term "vagueness" is primarily reserved for describing objects without clearly defined boundaries. (In this sense, fuzzy sets deal with vagueness.) Uncertainty concerns a state of knowledge of an agent (observer) about the state of the system under discussion. Let us emphasize that the notion of vagueness pertains to the nature of entities of the system rather than that of status of knowledge possessed by the observer.

2. This is a somewhat simplified version of the idea introduced in Hirota 1981. Usually the elements in **X** are not independent; one should therefore analyze them as a collection of random variables—this gives rise to the notion of a random field. As an ensemble (family) of random variables with well-defined interaction, random fields exhibit the continuity and smoothness necessary to capture the key feature of gradual changes in membership values.

References

Dubois, D., and H. Prade. 1994. Fuzzy sets—A convenient fiction for modeling vagueness and possibility. *IEEE Trans. on Fuzzy Systems* 2:6–21.

Dubois, D., and H. Prade. 1983. Unfair coins and necessity measures: Toward a possibilistic interpretation of histograms. *Fuzzy Sets and Systems* 10:15–20.

Hirota, K. 1981. Concepts of probabilitic set. *Fuzzy sets and Systems* 5:31–46.

Kwakernaak, H. 1978. Fuzzy random variables I—Definitions and theorems. *Information Science* 15:1–29.

Pardo, L. 1985. Information energy of a fuzzy event and a partition of fuzzy events. *IEEE Trans. on Systems, Man, and Cybernetics*, SMC-15:139–44.

Zadeh, L. A. 1968. Probability measures of fuzzy events. *J. Math. Analysis and Applications* 22:421–7.

Zadeh, L. A. 1978. Fuzzy sets as a basis for a theory of possibility. *Fuzzy Sets and Systems* 1:3–28.

7 Linguistic Variables

The concept of linguistic variables is fundamental within fuzzy set theory. Informally, a linguistic variable is a variable whose values are words or sentences rather than numbers. This chapter defines this notion and introduces related concepts such as linguistic hedges, linguistic approximation, and linguistic quantifiers as well. Several mechanisms for computing with linguistic variables are also discussed, and we show that they are crucial in many applications of fuzzy set theory.

7.1 Introduction

Very often we are inclined to describe observations about a certain phenomenon by characterizing its states, which we naturally translate in terms of an idea called a variable. For instance, when we refer to environmental conditions, we may express our observations by statements like *warm* place or, *clean and green* place or, *very wild and quite cute* place, and so on. The state of being *warm* could be translated by the variable *temperature*, with values in a set such as the interval 0–50 °C. Alternatively, temperature could be quantified (coded) using labels such as *cold*, *warm*, *hot*. Clearly, a percise numerical value such as 25 °C seems simpler than the ill-defined term *warm*. But the linguistic label *warm* is a choice of one out of three possible values, whereas 25 °C is a choice of one out of many, perhaps, in the entire 0–50 °C range. Linguistic characterizations are, in general, less specific than numerical, but it would certainly be much safer, unless one actually knew the exact temperature, to state that an environment temperature is *warm* than that it is 25 °C. The statement could be strengthened if the underlying meaning of *warm* is conceived as around 25 °C. In this setting, whereas the numerical value 25 can be visualized as a point in a set, the linguistic value *warm* can be viewed as a collection of objects (temperatures) within a bounded region whose center is at 25. The situation with the state of being *clean and green* or *very wild and quite cute* is more complex, because the scale involved in their quantification is quite subjective, and it is not natural to translate them into numerical values. But they do convey useful information.

What is worth noting in this simple example are the essential motivations for using linguistic variables. First, linguistic variables may be regarded as a form of information compression called granulation (Zadeh 1994a,b). Second, they serve as a means of approximate characterization of phenomena that are either too ill-defined or too complex, or both, to permit a description in sharp terms (Zadeh 1975). Third, they provide a means for translating linguistic descriptions into numerical, computable ones. Therefore, the duality between symbolic and numerical processing becomes natural instead of antagonistic.

Also important, especially from the practical point of view, is that through the use of the extension principle, many of the existing tools for system analysis and design can be extended to handle linguistic variables. Therefore mechanisms for computing with linguistic variables become feasible and useful in a wide range of application domains.

7.2   Linguistic Variables: Formalization

In contrast to the idea of a variable, as used in its common meaning, whose values are numbers, the notion of linguistic variable may be regarded as a variable whose values are fuzzy numbers. Linguistic variables can be envisaged in a wider context, however, since they may assume values consisting of words or sentences in a language. More specifically, they are defined as follows (Zadeh 1975):

DEFINITION 7.1   A linguistic variable is characterized by a quintuple denoted by

$$\langle X, T(X), \mathbf{X}, G, M \rangle$$

in which

X is the name of the variable,

$T(X)$ is the term set of X whose elements are labels of linguistic values of X,

G is generally a grammar for generating the names of X,

M is a semantic rule for associating with each label $L \in T(X)$ its meaning $M(L)$, which is a fuzzy set on the universe $\mathbf{X}$ whose base variable is x.

For example, consider a linguistic variable named *temperature*, that is, $X = temperature$, with $\mathbf{T} = [0, 50]$ and base variable $t \in \mathbf{T}$. The term set associated with *temperature* could be $T(temperature) = \{very\ low,\ low,\ medium,\ high,\ not\ low\ and\ not\ very\ high,\ very\ high \ldots.\}$ where each term in $T(temperature)$ is a label of a linguistic value of *temperature*. The meaning $M(T)$ of a label $T \in T(temperature)$ is defined to be the constraint $T(t)$ on the base variable t imposed by the name of T. Therefore $M(T)$ is a fuzzy set of $\mathbf{T}$ whose membership function $T(t)$ conveys the semantics of name T. Figure 7.1 illustrates the concept.

In general terms, it should be clear that the semantics of a linguistic variable yield a mapping,

$$M : T(X) \rightarrow \mathscr{F}(\mathbf{X}),$$

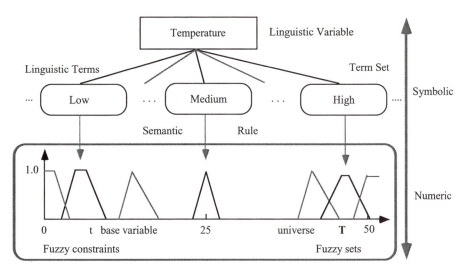

Figure 7.1
An example of a linguistic variable

that assigns to each term of $T(X)$ a corresponding fuzzy set in **X**; remember that $\mathscr{F}(\mathbf{X})$ denotes a family of fuzzy sets defined in **X**.

As stated in the definition, the elements of the term set are, in general, generated by a context-free grammar $G = \langle V, \Sigma, P, S \rangle$, in which V is the set of terminal symbols, Σ is the set of nonterminal symbols, S is the starting symbol, and P is a production system. Recall that in this setting, P means a set of syntactic rules to construct well-formed sentences in G. For instance, in the example above, the elements *not low and not very high* and *very high* of T (*temperature*) can be generated as follows:

Let $V = \{low, high, medium, very, not, and, \dots\}$, $\Sigma = \{S, A, B, C, D, E,$ $F, \dots\}$, and P be

$S \rightarrow A$

$C \rightarrow E$

$A \rightarrow B$

$A \rightarrow A$ *and* B

$B \rightarrow C$

$B \rightarrow$ *not* C

$C \rightarrow D$

$C \rightarrow F$

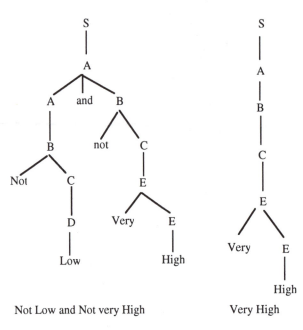

Figure 7.2
An example of term generation

$D \to very \; D$

$E \to very \; E$

$D \to low$

$E \to high$

$F \to medium$

Figure 7.2 displays the corresponding syntactic trees.

Typically, a linguistic variable involves a finite number of primary terms (*low, medium, high,* ...), a finite number of hedges (*very, more or less, quite,* ...), the connectives *and* and *or*, and the negation *not*. Hedges, connectives and negation are referred to as modifiers. Thus, the syntax of a linguistic variable given by a grammar $G = \langle V, \Sigma, P, S \rangle$ is such that the set of terminal symbols V consists of primary terms T and modifiers M, whereas the set of nonterminal symbols Σ contains all symbols used in the production rules (productions). That is,

$$V = \{T_1, T_2, \ldots, T_n, M_1, M_2, \ldots, M_m\},$$

$$\Sigma = \{S, \langle expression \rangle, \langle simple\_expression \rangle\},$$

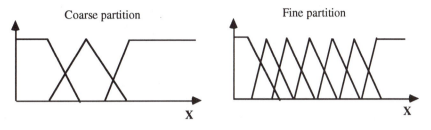

Figure 7.3
Examples of fuzzy partitions of different granularity

The set of productions P as represented in Backus-Naur form (BNF) notations is written as

$S ::= \langle \text{expression} \rangle$

$\langle \text{expression} \rangle ::= \langle \text{simple\_expression} \rangle | \langle \text{expression} \rangle$

$\langle \text{simple\_expression} \rangle ::= T_i \, | \, M_j T_i, \; i = 1, \ldots, n \text{ and } j = 1, \ldots, m.$

The fuzzy set attached to any value of a linguistic variable can be derived from the membership functions of the primary terms, as we shall see next. Before proceeding with more detailed studies, it should be noted that most applications of fuzzy set theory exploit a simpler approach to dealing with linguistic variables. Often the membership functions are assumed to have a known shape (triangular, trapezoidal, sigmoidal, etc.), and a small number of linguistic values is used. The number of linguistic values defines the granulation, and therefore the fuzzy partition, of the corresponding universe. A small the number of values induces a coarse partition, and a bigger number a finer partition, as figure 7.3 illustrates.

EXERCISE 7.1   Assume you are driving down a highway with a speed limit of 100 km/h. How would you characterize descriptions such as *slow, medium, fast* in terms of a linguistic variable? What about *very fast, not slow, not slow and not fast*?

7.3 Computing with Linguistic Variables: Hedges, Connectives, and Negation

The values of a linguistic variable are labels of fuzzy sets on a universe $\mathbf{X}$ that have the form of sentences in a language. In general, a value of a linguistic variable is a composite term $T = L_1 L_2 \ldots L_n$, which is a concatenation of atomic terms $L_1, L_2, \ldots, L_n$, as in *very high* and *not low and not very high*, for example. These atomic terms may be grouped into three main classes:

1. Primary terms: labels of fuzzy sets in **X** associated with their corresponding meaning, as *high* and *low* in the preceding example.

2. Hedges: act as linguistic modifiers of the primary terms, such as *very, much, more or less,*

3. Connectives and negation: the connectives *and* and *or* and the negation *not* act as linguistic modifiers.

An important problem arises in this context: Given the fuzzy sets representing the meaning of each primary term and the meaning of the connectives and negation as well, compute the meaning of a composite term (Zadeh 1972). A preliminary step in solving the problem is to consider first the problem involving composite terms of the form $T = HL$, where H is a hedge and L is a term with specified meaning such as $T = $ *very high*, with $H = $ *very* and $L = $ *high*. The point in finding the meaning of T relies on viewing a hedge as an *operator* H that transforms the fuzzy set $L(x)$, attached to the meaning of L, into the fuzzy set $T(x) = HL(x)$. For this purpose, we may employ some of the basic operators introduced in section 1.4, especially concentration, dilation, and fuzzification. In the following, we show, by a series of examples using typical hedges, how computations can be done (MacVickar-Whelan 1978; Zadeh 1975).

1. If $T = $ *very L*, then $T(x) = $ Con_$L(x)$. Thus *very* acts as an intensifier. For example considering the linguistic variable *temperature*, we may have *temperature = very high*, and if the meaning of *high* is specified by the fuzzy set

$$High(x) = \begin{cases} 0, & \text{if } 0 \le t < 38 \\ \sqrt{0.2(t - 38)}, & \text{if } 38 \le t < 43 \\ 1, & \text{if } 43 \le t, \end{cases}$$

then the meaning of *very high* can be found as *Very High*$(x) = $ Con_*High*(x), and for $p = 2$, *Very High*$(x) = [High(x)]^2$. (This effect is displayed in figure 7.4.)

$$Very\ High(x) = \begin{cases} 0, & \text{if } 0 \le t < 38 \\ 0.2(t - 38), & \text{if } 38 \le t < 43 \\ 1 & \text{if } 43 \le t. \end{cases}$$

EXERCISE 7.2   Compute the meaning of *very very high*. Hint: Regarded as an operator, *very* can be composed with itself. Assume for simplicity that $p = 2$.

2. If $T = $ *plus L*, then $T(x) = $ Con_$L(x)$, with, for example, $p = 1.5$.

3. If $T = $ *minus L*, then $T(x) = $ Dil_$L(x)$, with, for example, $r = 0.75$.

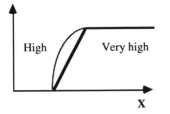

Figure 7.4
Effect of hedge *very*

The modifiers *plus* and *minus* can be used to define some hedges whose meaning differs only slightly from others. For instance, *highly* could be defined as *highly = plus very* or possibly as *highly = minus very very*.

4. If $T = much \ L$, then $T(x) = \text{Fuzz\_}L(x)$.

5. If $T = more \ or \ less \ L$, then $T(x) = \text{Dil\_}L(x)$ with, for example, $r = 0.5$.

After we know how to compute the meaning of a composite term such as $T = HL$, finding the meaning of more complex terms becomes easier. This includes the case where connectives and negation appear in addition to terms involving only hedges. If we identify *and* with *intersection*, *or* with *union*, and *not* with *complementation* (or, more generally, with *negation*) the meaning of a composite term can be computed straightforwardly, most of the time by inspection. For example, reconsidering the linguistic variable *temperature* and assuming a value such as *temperature = not low and not very high*, we have

$$not \ low \ and \ not \ very \ high = \overline{low} \cap \overline{very \ high}.$$

Thus, considering standard complementation and viewing intersection as a t-norm, we get:

Not Low And Not Very High(x)

$$= [1 - Low(x)] \ t \ [1 - \text{Con\_}High(x)], \quad \forall x \in \mathsf{T}.$$

Similarly, for *temperature = very low or high*, figure 7.5, looking at the union as a s-norm:

$$very \ low \ or \ high = very \ low \cup high,$$

Very Low Or High$(x) = [\text{Con\_}Low(x)] \ s \ [High(x)], \quad \forall x \in \mathsf{T}.$

We may certainly define hedges based on the ideas just discussed. For instance, *slightly* might be defined using one of the following expressions (Zadeh 1972):

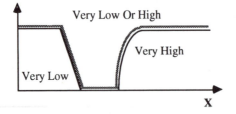

Figure 7.5
Linguistic expression *very low or high* realized with the maximum s-norm

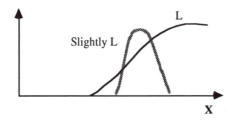

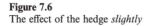

Figure 7.6
The effect of the hedge *slightly*

Slightly L(*x*) = Norm_*L And Not Very L*(*x*),

Slightly L(*x*) = Int_Norm_*Plus L And Not Very L*(*x*),

Slightly L(*x*) = Int_Norm_*Plus L And Not Plus Very L*(*x*),

These expressions describe possible approximations of an operator that transforms a fuzzy set *L* into a fuzzy set *Slightly L*, as visualized in figure 7.6.

EXERCISE 7.3   Why do we need a normalization operation when defining the hedge *slightly* as above? Elaborate your answer, considering $L = 4/0.1 + 5/0.2 + 6/0.4 + 7/0.5 + 8/0.7 + 9/0.8 + 10/1.0 + 11/1.0 + 12/1.0$ in $X = \{1, 2, 3, 4, 5, 6, 7, 8, 9, 10, 11, 12\}$, and computing *Slightly L* according to the given expressions.

7.4   Linguistic Approximation

Once the primary terms and modifiers are defined, the meaning of a composite term can be computed, resulting in a fuzzy set that may not correspond to any of the terms in the term set of a linguistic variable. Thus, if we wish to assign a linguistic label to the resulting fuzzy set, an approximation to the exact label must be admitted. Such an approxi-

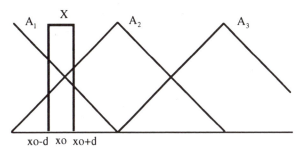

Figure 7.7
A collection of linguistic terms

mation is referred to as a linguistic approximation (Zadeh 1975; Pedrycz 1993).

In practice, the notion of linguistic approximation is used to denote a process of matching a given fuzzy set against a collection of primitive fuzzy sets associated with the terms available for a linguistic variable. Assuming that we have at our disposal a collection of fuzzy sets $T_1, T_2, \ldots, T_n$ and modifiers $M_1, M_2, \ldots, M_m$, approximating a fuzzy set A means expressing it in terms of T_is and M_js. A two-step procedure is as follows:

Step 1. Approximate A by one of the T_is. The selection is completed using the values obtained by computing the equality index for A and each T_i. Thus A is approximated by A^*, where $A^* = \max_i [(A \equiv T_i)_{av}]$, $i = 1, \ldots, n$.

Step 2. Improve approximation, selecting a modifier M^* such that

$$M^* = \max_j [(A \equiv M_j A^*)_{av}], \quad j = 1, \ldots, m.$$

As a result of this approximation, fuzzy set A is represented as one among the properly modified primitive fuzzy sets. Alternative similarity measures can be used in the procedure.

It is instructive to gain better insight into the performance of linguistic approximation by completing more detailed numerical experiments. Let us consider a collection of generic linguistic terms characterized by triangular membership functions defined in $\mathbf{X} = [0 \ldots 100]$. The terms are distributed uniformly in the space and overlap at the membership level of $1/2$, as shown in figure 7.7.

The datum to be (linguistically) approximated is a numerical interval (X) of width $2d$ that slides along the universe of discourse. The experiments are carried out for several levels of specificity of X (quantified by *d*) and some selected number of the linguistic terms. The syntax of the grammar is defined as follows:

$S ::= \langle \text{expression} \rangle$

$\langle \text{expression} \rangle ::= \langle \text{simple\_expression} \rangle | \langle \text{simple\_expression} \rangle$ or $\langle \text{expression} \rangle$

$\langle \text{simple\_expression} \rangle ::= \text{basic\_term} \,|\, \text{modifier (basic\_term)}$

More specifically, the grammar has five primary terms and two modifiers; hence the set of terminal symbols is given as

$V = \{A_1, A_2, A_3, A_4, A_5, \text{ very, more or less}\}.$

$A_i, i = 1, 2, \ldots, 5$, are triangular fuzzy numbers specified as

$A_1 = (0, 0, 25), \quad A_2 = (0, 25, 50), \quad A_3 = (25, 50, 75), \quad A_4 = (50, 75, 100),$

$A_5 = (75, 100, 100).$

The two modifiers (hedges) used in the experiment are defined as before

$Very \; A(x) = A^2(x) \quad More \; Or \; Less \; A(x) = A^{0.5}(x)$

Figures 7.8 and 7.9 plot the similarity index for X sliding along the universe of discourse. The results are reported for two term sets (codebooks) consisting of three and seven linguistic terms, respectively. There is a clear and justifiable regularity: as X increases in granularity, the results of

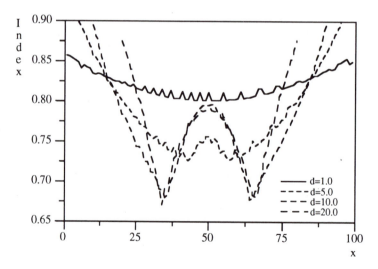

Figure 7.8
Similarity index as a function of x for three-element term set

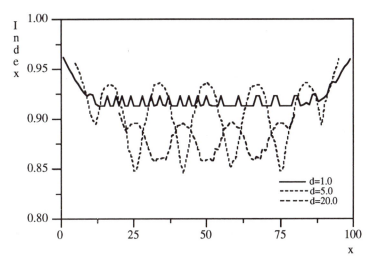

Figure 7.9
Similarity index as a function of x for seven-element term set

approximation lose their uniqueness as several terms are regarded as the optimal approximators of X.

EXERCISE 7.4   Assume the fuzzy sets $Low = 1.0/1.0 + 0.8/2 + 0.5/3$ and $High = 0.7/8 + 1.0/9 + 1.0/10$, defined on $\mathbf{T} = \{1, 2, \ldots, 10\}$, Further, assume the two sets are the primitives representing the meaning of the terms *low* and *high* of the linguistic variable *temperature*, with the term set $T(temperature) = \{low, high, more\ or\ less\ high, almost\ high\}$. Compute the meaning of *more or less high* and of *almost high* using the dilation and fuzzification operators, respectively. How would you assign a linguistic label to the fuzzy set $A = 0.84/8 + 1.0/9 + 1.0/10$? Can you guess the exact meaning of A?

7.5   Linguistic Quantifiers

Most, if not all, formal definitions are categorical; they commonly use universal ($\forall$, *for all*) and existential ($\exists$, *there exists*) quantifiers. To illustrate the point, let us resort to the definition of identity relationship between fuzzy sets:

$A = B$ iff $A(x) = B(x)$ $\forall x \in \mathbf{X}$.

Although formally meaningful, this definition seems to be very much in the spirit of defining very precise objects. Obviously, when using fuzzy sets, we are concerned with representing linguistic notions, and this level of precision might not look very realistic. For example, if A and B have the same membership function except a single isolated point, are they equal or not equal? We are very much tempted to regard A and B as equal. Or, if most membership values of A and B are similar, are these fuzzy sets equal? These questions lend themselves to a framework of computations with linguistic quantifiers, for we are concerned with determining truth values considering the conceptual approaches as envisioned below:

$Truth(A = B) = Truth(A$ and B have the same membership values over *most* elements of **X**), or

$Truth(A = B) = Truth(A$ and B have *similar* membership over *most* elements of **X**).

The first model is a special case of the latter one, in which we admit similarity rather than equality of membership values. The second model exploits the concept of gradual equality and introduces linguistic quantifiers as well.

Linguistically quantified statements such as "Most trucks are heavy" can be put in the form

Qxs are B,

where Q is a linguistic quantifier, x is a generic element of a set **X**, and B is the label of a fuzzy set of **X** describing a property of $x \in$ **X**. In the example, we identify Q with *most*, **X** with the set of trucks, and *heavy* with a property of trucks. We may also wish to assign a particular characteristic to the members of **X** as in "Most *big* trucks are heavy." A characteristic may be added to provide the general form of quantified statements (Yager 1983; Zadeh 1983):

QAxs are B,

where A is the label of a fuzzy set of **X** describing the meaning of the added characteristic.

Formally speaking, a linguistic quantifier Q is assumed to be represented by a fuzzy set of the unit interval (Kacprzyk, Fredizzi, and Nurmi 1992). For instance, *most* could be given as

$$Most(x) = \begin{cases} 0, & \text{if } 0 \leq x < 0.5 \\ 2x - 1, & \text{if } 0.5 \leq x \leq 1. \end{cases}$$

(See figure 7.10.)

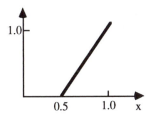

Figure 7.10
An example of the linguistic quantifier *most*

Assuming a finite universe $\mathbf{X} = \{x_1, x_2, \ldots, x_n\}$, and $Truth(x$ is $B) = B(x)$, $x \in \mathbf{X}$, the truth value of Qxs are B is computed by the following two-step procedure (Zadeh 1983):

Step 1. Determine $r = \dfrac{Card(B)}{Card(\mathbf{X})} = \dfrac{1}{n} \sum_{i=1}^{n} B(x_i)$.

Step 2. Set $Truth(Qxs$ are $B) = Q(r)$.

Essentially, the ratio r determines some mean proportion of elements of $\mathbf{X}$ satisfying the property B under consideration, whereas $Truth(Qxs$ are $B)$ determines the degree to which this proportion is compatible with the meaning of the quantifier Q.

Similarly, the truth value of a quantified statement in the form $QAxs$ are B can be found, once we rewrite it as $Q(A$ and $B)xs$ are B and interpret the connective *and* as an intersection. Thus, the two-step procedure becomes:

Step 1. Determine $r = \dfrac{Card(A \cap B)}{Card(B)}$.

Step 2. Set $Truth(QAxs$ are $B) = Q(r)$.

For example, if the intersection operation is carried out via the min t-norm, the value of r is computed as

$$r = \frac{\sum_{i=1}^{n} \min[A(x_i), B(x_i)]}{\sum_{i=1}^{n} B(x_i)}.$$

As a simple numerical example, consider $\mathbf{X} = \{x_1, x_2, x_3\}$, a set of trucks, $Big = 0.4/x_1 + 0.6/x_2 + 0.8/x_3$; $Heavy = 0.2/x_1 + 0.5/x_2 + 0.9/x_3$; and *Most* as defined in figure 7.10. Thus the truth of "*Most big* trucks are *heavy*" is found as follows:

$$r = \frac{\sum_{i=1}^{3} \min[Big(x_i), Heavy(x_i)]}{\sum_{i=1}^{3} Heavy(x_i)} = 0.9375 \text{ and } Q(r) = 2r - 1 = 0.875.$$

Therefore $Truth$ (*Most big* trucks are *heavy*) $= 0.875$.

7.6   Applications of Linguistic Variables

The concept of linguistic variable plays a major role in many applications of fuzzy set theory. Linguistic variables are an essential ingredient in approximate reasoning as a result of, for instance, treating truth as a linguistic variable whose truth values form a term set, or treating words as a form of granulation in describing dependencies and commands. This is a key point in fuzzy logic and the calculi of fuzzy rules and fuzzy graphs, respectively (Zadeh 1994a, b).

Treating truth as a linguistic variable leads to fuzzy logic, which provides a basis for computing with unsharp propositions through an approximate mode of reasoning. Typically, the term set of the linguistic variable *Truth* is assumed to be (Zadeh 1975)

$$T(Truth) = \{true, not\ true, very\ true, \ldots, false, not\ false, very\ false, \ldots\},$$

where *true is* a primary term, and *false* is defined as the mirror image of *true* with respect to the point 0.5 in [0,1]. Therefore, using the rules of fuzzy logic, we can compute statements of the form "(*Temperature* is *high*) is *true*," with the meaning of *high* being specified by a fuzzy set in **X**, and the meaning of *true* by a fuzzy set in [0,1].

The calculi of fuzzy rules provides mechanisms for computing with statements in the form of fuzzy if-then rules. A fuzzy rule relates n antecedent variables $X_1, X_2, \ldots, X_n$ to m consequent variables $Y_1, Y_2, \ldots, Y_m$ and has the pattern

If X_1 is A_1 and $\ldots$ and X_n is A_n, then Y_1 is B_1 and $\ldots$ and Y_m is B_m,

where the X_is and Y_js are linguistic variables and the A_is and B_js their respective linguistic values. For example, "If *temperature* is *high* and *humidity* is *low* then *thermal comfort* is *medium*." In this setting, the fuzzy rule can be interpreted as a fuzzy point or granule in an appropriate space. A collection of fuzzy granules comprises a fuzzy graph and, similar to the way we compute with ordinary graphs of functions, we can perform inferences using the framework of the calculi of fuzzy rules (see chapter 10).

Another important area of application for the notion of a linguistic variable lies in the realm of probability theory. If probability is viewed as a linguistic variable, typically its term set would be something like

$$T(Probability) = \{likely, very\ likely, \ldots, probable, improbable,$$
$$very\ probable, \ldots\}.$$

Thus, questions such as "What is the probability that the environment temperature will be higher tomorrow?" could be answered by *very prob-*

able. The method for computing linguistic probabilities generally applies also to solving nonlinear programming problems (Zadeh 1976).

7.7   Chapter Summary

This chapter dealt with linguistic variables, a key notion within fuzzy set theory and its applications. A linguistic variable is a variable whose values are words or sentences in a language, interpreted as labels of fuzzy sets. Generally speaking, they may be regarded as a language with context-free grammar and rules for attributing semantics to the terms generated. Linguistic modifiers on linguistic variables were presented as basic operations on the fuzzy sets defining the meaning of the primary terms. Applications of linguistic variables highlighted their relevance in fuzzy computing and information processing.

7.8   Problems

1. Consider the grammar G as delineated in section 7.2. Construct the syntactic trees that generate the terms *not very high and not very low*, *very very high*, and *not high*. What modifications in G are necessary to generate terms such as *not very low or not very high*?

2. Translate the productions of the example in section 7.2 into the BNF notation. Repeat the exercise 7.1 using the BNF to generate the syntactic tree.

3. Assume *big* as a primary term of a linguistic variable whose meaning is given by the fuzzy set $Big(x) = \int_0^\infty [1 + (x/a)^{-2}]^{-1}/x$, where a is the crossover point of $Big(x)$. Compute the meaning of *slightly big*, assuming *slightly* is defined by the normalization of *big and not very big*.

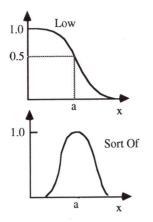

Figure 7.11
A model of the linguistic modifier *sort of*

4. Provide a definition, in terms of basic operators, for the hedge *sort of*. Hint: See figure 7.11, where $Low(x) = \int_0^\infty [1 + (x/a)^2]^{-1}/x$, and a is the crossover point.

5. Using linguistic modifiers as well as ideas of the degree of inclusion (dominance), propose a linguistic model for inclusion of fuzzy sets.

6. Assume that x takes a sequence of values $x_1, x_2, \ldots, x_n$ in a universe **X**. Then, the quantifier *usually*, in its unconditioned sense, may be defined by *Usually* $(x$ is $B) \cong$ *Most* xs are U, where U denotes a usual value of x, a fuzzy set in **X**. How would you compute the truth of a quantified statement of the form *Usually* $(x$ is *about_a*), where $a \in$ **X**?

7. Suggest a procedure for computing the square of *usually*, that is, *usually*^{2}. Hint: Use the extension principle. Is your procedure applicable for performing similar operations with linguistic hedges?

8. Consider the example of section 7.5. Compute *Truth* (*most* trucks are *heavy*).

References

Kacprzyk, J., M. Fredizzi, and H. Nurmi. 1992. Fuzzy logic with linguistic quantifiers in group decision making. In *An Introduction to Fuzzy Logic Applications in Intelligent Systems*, ed. R. Yager and L. Zadeh, pp. 263–80. Norwell, MA: Kluwer Academic Publishers.

MacVickar-Whelan, P. 1978. Fuzzy sets, the concept of height and the hedge very, *IEEE Trans. on Systems, Man and Cybernetics* 8:507–11.

Pedrycz, W. 1993. *Fuzzy Control and Fuzzy Systems*, 2 ed. Somerset, England: RSP Press.

Yager, R. R. 1983. Quantifiers in the formulation of multiple objective decision functions. *Information Sciences* 31:107–39.

Zadeh, L. A. 1972. A fuzzy-set-theoretic interpretation of linguistic hedges. *J. of Cybernetics* 2(2):4–34.

Zadeh, L. A. 1975. The concept of linguistic variable and its application to approximate reasoning. *Information Sciences* 8:199–249 (part I), 8:301–57 (part II).

Zadeh, L. A. 1976. The concept of linguistic variable and its application to approximate reasoning. *Information Sciences* 9:43–80 (part III).

Zadeh, L. A. 1983. A computational approach to fuzzy quantifiers in natural languages. *Computers and Mathematics with Applications* 9:149–84.

Zadeh, L. A. 1994a. Fuzzy logic, neural networks, and soft computing. *Communications of the ACM* 3(3):77–84.

Zadeh, L. A. 1994b. Soft computing and fuzzy logic. *IEEE Software* 11(6):48–56.

8 Fuzzy Logic

Logic concerns the formal principles that support the laws of thought and mirrors the foundations of many engineering and computer science areas. Two-valued logic systems consider statements whose values are either true or false. In many-valued logic systems a statement may be true or false or can assume intermediate truth values. In contrast, fuzzy logic goes far beyond the limits of many-valued logics and admits truth values that are fuzzy sets of the unit interval. Truth values may be regarded as a linguistic characterization of numerical truth values. Thus, fuzzy logic concerns the principles of approximate reasoning. This chapter exposes the reader to the fundamental notions, operations, and computational aspects of fuzzy logic and aims to provide the underlying principles, concepts, and methods.

8.1 Introduction

Most engineering tasks involve problem-solving or decision-making capabilities that involve reasoning as an essential part of the effort. Informally speaking, reasoning is the ability to infer information about some unknown facet of a problem based on the available information about domain knowledge. For instance, reasoning is performed when we attempt to infer some information about a system fault based on the observable symptoms of its subsystems.

A reasoning task depends on the nature and type of the available information. For example, when sentences of a language are used to represent domain knowledge, we may be able to attach truth values to some of them. In this case, the reasoning task is specified as: Given the known information in terms of truth values of the respective sentences and the domain knowledge in the form of sentences of a language, determine the truth value of some other sentences of interest.

A statement of the form

X is A,

where X is a name of an object and A its attribute (property), is called a proposition, or alternatively, an atomic proposition. Given a set of logical connectives (logical symbols) composed of negation (not), disjunction (or), conjunction (and) and implication (if-then), we may construct more complex propositions. The construction process should follow certain syntactic rules to deliver well-formed propositions. For instance, we may state that if X_1 is A_1 and X_2 is A_2 are propositions, then "X_1 is A_1 and X_2 is A_2," "X_1 is A_1 or X_2 is A_2," and "If X_1 is A_1, then X_2 is A_2" are also propositions. The set of propositions constructed in this way is called a (formal) language, the language of logic. This language can be enriched assuming the attributes as *n*-ary predicates. In this case logical symbols called quantifiers can be added to refer to the *n*-ary terms that appear at the predicates.

Although the syntactic structure of many logical systems can be viewed in this form, the issue is quite diverse when semantics, that is, the meaning of the propositions, is of interest. The meaning of a proposition is not fixed a priori. It depends on an assignment of truth values to the atomic propositions or predicates of the associated language. In addition, truth values can be any element of a finite set or countable infinite sets. This leads to two-, three-, ..., and many-valued logics. For instance, when there is no room for opinion, question or uncertainty in particular domain, propositions are either true or false. Therefore an overall formal model collapses to the level of the classic two-valued logic. Otherwise we are within the world of may-valued logics. Additionally, if propositions involve precise attributes, we resort to the scope of traditional logic systems. Propositions with imprecise attributes may lead to the framework of multivalued or fuzzy logic. Before we discuss these three main classes of logics, it is instructive to make a clear distinction between propositional and predicate calculus.

8.2 Propositional Calculus

The simplest logical system concerns propositions. The syntax of propositions governs the combination of basic building blocks such as propositions and logical connectives. Propositions are built upon elementary atomic statements called atomic propositions. For instance, the statement

Car is red

is an atomic proposition. More complex propositions are formed from atomic propositions using the logical connectives such as *not, and, or, if ... then*, and *if and only if*. For example, the statement

Car is red *and* sky is blue

is again a proposition constructed with the use of the *and* connective. In two-valued logic, propositions may be either true or false and cannot take on any other value. Traditionally, the truth is denoted by 1, whereas 0 is used to express the idea that the proposition is false. Several auxiliary examples of propositions follow:

P: 5 is integer.

Q: 13 is an odd number.

R: Two parallel lines intersect.

Hereafter capital letters denote propositions, and the following symbols are used for logical connectives: $-$ (*not*), $\wedge$ (*and*), $\vee$ (*or*), $\rightarrow$ (*if . . . then*), $\leftrightarrow$ (*if and only if*). The truth of a proposition P is denoted by p; thus p may have the value 1 if P is true or 0 if P is false, that is, $p \in \{0, 1\}$. Accordingly, the examples above read $p = 1$, $q = 1$, $r = 0$. In other words, we may state that

5 is integer is *true*.

13 is an odd number is *true*.

Two parallel lines intersect is *false*.

We define the syntax of propositional languages in a recursive fashion:

Atomic propositions are propositions.

If P and Q are propositions, then $\bar{P}$, $P \wedge Q$, $P \vee Q$, $P \rightarrow Q$ and $P \leftrightarrow Q$ are propositions.

All propositions are generated from a finite number of the previous constructs.

EXERCISE 8.1   Which of the following statements are meaningful propositions, that is, syntactically correct propositions in a propositional language:

(a) $(P \leftrightarrow Q) \wedge (P \vee Q) \vee \bar{Q}$?

(b) $W \vee \bar{P} \vee QP \rightarrow Q$?

(c) $(P \vee P \vee Q) \wedge (P \rightarrow Q)(Q \rightarrow R)\bar{P}$?

(d) $(Q \rightarrow \bar{P}) \vee (Q \wedge S)$?

The meaning or semantics of a proposition is simply the value, true or false. In other words, semantics deals with an assignment of a truth value to the proposition. An interpretation for a proposition or a group of propositions comes as an assignment of a truth value to each atomic proposition. We may, in one interpretation, assign 1 to P and 0 to Q in the proposition $P \vee Q$; in another we may assign 0 to P and 1 to Q. Clearly, four distinct interpretations exist for this proposition. In general, if we have a proposition with n distinct atomic propositions (or, equivalently, n distinct propositional symbols), then there are 2^n different

Table 8.1
Semantic rules for propositional logic

$P \wedge Q$	$P \vee Q$	$\bar{P}$
$\min(p, q)$	$\max(p, q)$	$1 - p$

interpretations. Once an interpretation has been assigned to a proposition, its truth value can be computed by a successive application of semantic rules to larger and larger parts of the proposition until a single truth value is determined. Computing the meaning of propositions for all interpretations gives rise to truth tables, a well-known mechanism to assign meaning to the propositions. Table 8.1 summarizes the basic semantic rules. Note that the semantics table provides the for the negation, disjunction and conjunction of the propositions, because it can be readily verified that the proposition $P \to Q$ is equivalent (in the sense that the propositions have the same truth values for all corresponding interpretations) to $\bar{P} \vee Q$. Likewise,

$$P \leftrightarrow Q \equiv (P \to Q) \wedge (Q \to P),$$

where the symbol $\equiv$ denotes an equivalence of propositions.

We can easily determine the meaning of the proposition

$$R \equiv (Q \to \bar{P}) \vee (Q \wedge S),$$

given an interpretation of its components. Assume that $p = 1$, $q = 0$ and $s = 1$. First, we note that $(Q \to \bar{P}) \equiv (\bar{Q} \vee \bar{P})$. From table 8.1, we read

$$r = \max[\max(1, 0), \min(0, 1)] = \max[1, 0] = 1$$

and conclude that the proposition R is true. A truth table can be constructed for proposition R that considers all of its 2^3 interpretations. This is left as an exercise.

Note that evaluating the truth value of a proposition P in $\{0, 1\}$ imposes a binary decision. Dichotomy, an idea discussed in chapter 1, arises again as the cornerstone of two-valued logic. This is not a mere coincidence but rather results from the isomorphism of set theory and two-valued logic, which explains why basic logic operations such as conjunction, disjunction, and negation are directly associated with intersection, union, and complement in set theory. In other words, set theory and two-valued logic are isomorphic mathematical structures in the sense that all the properties in one system have their counterparts in the other.

Dichotomy is also present when we interpret propositions from the set-theoretic point of view. Consider a proposition P:

The temperature is between 20 and 30 is *true*.

Symbolically,

$$P: T \text{ is } A \text{ is } v, \tag{8.1}$$

where T is a generic temperature value in **T**, A denotes the concept "between 20 and 30" and v stands for true. Therefore A can be regarded as a set whose characteristic function assumes 1 between 20 and 30. This means

$$A(T) = \begin{cases} 1, & \text{if } T \in A \\ 0, & \text{if } T \notin A. \end{cases}$$

It is intuitively clear that the truth value of P should be $p = 1$ if $T \in [20, 30]$, that is, if $A(T) = 1$; and $p = 0$ if $T \notin [20, 30]$, that is, if $A(T) = 0$. In addition, if the following proposition holds,

The temperature is between 20 and 30 is *true*,

then the next proposition,

The temperature is other than that between 20 and 30 is *false*,

is valid as well. Let Q be the statement,

The temperature is other than that between 20 and 30 is *false*,

Thus we may write

$$Q : T \text{ is } \bar{A} \text{ is } f, \tag{8.2}$$

where $\bar{A}$ is the complement of A and f stands for false. Similarly, as before, the truth value of Q should be $q = 0$, if $T \in [20, 30]$ (or, equivalently, if $A(T) = 1$), and $q = 1$ otherwise. Figure 8.1 summarizes these observations graphically.

Generally speaking, the truth value of a proposition can be provided by a truth function τ defined as a mapping,

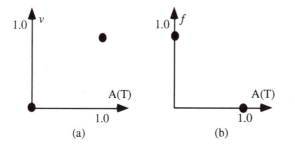

Figure 8.1
Truth functions: (a) $v = \tau_{true}$ and (b) $f = \tau_{false}$

$$\tau : \{0, 1\} \rightarrow \{0, 1\}, \tag{8.3}$$

such that the truth functions for *true* and *false* are specified accordingly:

$\tau_{true} : \{0, 1\} \rightarrow \{0, 1\}$ with $\tau_{true}(0) = 0$ and $\tau_{true}(1) = 1$.

$\tau_{false} : \{0, 1\} \rightarrow \{0, 1\}$ with $\tau_{false}(0) = 1$ and $\tau_{false}(1) = 0$.

Problem solving and decision making involve reasoning about a collection of information, represented in an appropriate way, to infer new information. Logic provides a formal framework for reasoning. Many concepts that can be described via a natural language can be translated into symbolic structures that closely represent their meaning. These symbolic structures can then be manipulated to *deduce* various facts, that is, to carry out a form of reasoning. In propositional logic, the symbolic structures are propositions that form the basic building blocks for information (knowledge) representation. Once symbolic structures have been created to represent basic facts, procedures or other types of knowledge, reasoning can be applied to compare, combine, and transform these structures (known information) into new, deduced structures (new information). The reasoning is performed through inference rules.

The inference rules of propositional logic provide an important means of performing reasoning or deduction in a propositional language framework. Essentially, reasoning involves the following problem: Given a set of propositions $P = \{P_1, P_2, \ldots, P_n\}$ (premises), find the truth of a proposition Q (conclusion). We summarize a few fundamental inference rules.

1. *Modus ponens*: From P and $P \rightarrow Q$, infer Q. We write it concisely as

$$P$$

$$\frac{P \rightarrow Q}{} \tag{8.4}$$

$$Q$$

2. Substitution: If P is a valid proposition, that is, a proposition that is true for every interpretation, then the proposition derived from P by consistent substitution of the symbols in P is also valid.

3. Conjunction: From P and from Q, infer $P \wedge Q$.

4. Projection: From $P \wedge Q$, infer P.

8.3   Predicate Logic

Expressiveness is one of the requirements for any useful knowledge representation scheme, which should be able to represent accurately most, if not all, concepts that can be expressed in a natural language. Propositional logic does not fulfill this requirement in some important aspects. In particular, it lacks mechanisms to express relations among objects. It also does not allow generalizations about categories of similar objects. These are serious limitations when dealing with real-world objects.

Predicate logic is a framework for enhancing propositional logic's expressiveness. It is a generalization that permits one to represent knowledge as well as to reason about objects as relational entities and classes and subclasses of objects. This generalization starts from the introduction of atomic predicates in place of atomic propositions, the use of functions and the use of variables together with variable quantifiers. In brief, predicate calculus involves statements containing variables. It is important to stress that the statements are neither true nor false unless the arguments are specified. Since predicate logic is a formal language, we must define its syntax and semantics to represent knowledge properly. To reason about a given amount of knowledge, inference rules must also be provided.

The symbols and rules of combination that make sense within the framework of predicate logic include the same connectives as in propositional logic together with constants, variables, and two variable quantifiers, denoted by $\forall$ (universal quantification) and $\exists$ (existential quantification). In addition, functional symbols denote functions defined on a domain D. They map n elements, $n \geq 0$, to a single element of the domain $f : D \rightarrow D$. An n-place (n-ary) function is written as $f(t_1, t_2, \ldots, t_n)$ where the t_i are terms (constants, variables, or functions) defined in the same domain. Predicates denote relations in a domain and map elements of the domain to the values true or false, $P : D \rightarrow \{0, 1\}$. Capital letters are used to represent predicates, and like functions, predicates may have $n \geq 0$ terms for arguments, written as $P(t_1, t_2, \ldots, t_n)$. Note that a 0-ary predicate is a proposition, that is, a constant predicate. For instance, $E(x, y)$ may be regarded as a predicate representing the fact that two real numbers x and y are equal. A predicate representing a basic unit of knowledge is called atomic formula. Following syntactic rules govern the writing of well-formed statements (well-formed formulas or sentences) in the language of predicate logic.

An atomic formula is a well-formed formula (wff). If P and Q are well-formed formulas, then $\bar{P}$, $P \wedge Q$, $P \vee Q$, $P \rightarrow Q$, $P \leftrightarrow Q$, $\forall x P(x)$, and $\exists x P(x)$ are also well-formed formulas. All well-formed formulas are formed only by applying these rules a finite number of times. The expression

$$\forall x \, \exists y (P(x) \vee Q(y)) \rightarrow (R(a, y) \wedge Q(b))$$

is a well-formed formula in which P, Q are unary predicates, R is a binary predicate, x and y are variables, and a and b are constants.

Again, the semantics of a sentence in the predicate logic language is just a true (1) or false (0) value; that is predicate logic is also a two-valued logical system because truth values lay in the set $\{1,0\}$. When values are assigned to each term and to each predicate in a sentence, we say an interpretation is given to the sentence. Therefore, we may compute the truth value of a sentence using the same semantic rules as those in table 8.1. When determining the truth value of a sentence, however, we must be careful in evaluating predicates that have variables in their arguments, since they evaluate as true only if they are true for the appropriate value(s) of the variables. For instance, the sentence $\forall x P(x)$ is true only if the predicate $P(x)$ is true for every value of x of the domain. Similarly, $\exists x P(x)$ is true only if there is at least one value of x in the domain for which $P(x)$ is true.

EXERCISE 8.2   Consider the sentence $\forall x\{[P(a, x) \vee Q(f(x))] \wedge R(x)\} \rightarrow S(x)$. Let the domain be $D = \{1, 2\}$. An interpretation allocates the following truth values: $a = 2$; $f(1) = 2$ and $f(2) = 1$; $P(2, 1) = 1$ and $P(2, 2) = 0$; $Q(1) = 1$ and $Q(2) = 0$; $R(1) = 1$ and $R(2) = 0$; $S(1) = 0$ and $S(2) = 1$. Does the sentence evaluate as true or false?

The inference rules for predicate logic are essentially the same as those for propositional logic, except that care should be taken when looking at variables. For instance, a key inference rule in predicate logic is modus ponens. In general, if a has a property P and all objects that have property P also have property Q, we conclude that a has property Q. Symbolically:

$$P(a)$$

$$\frac{\forall x P(x) \rightarrow Q(x)}{} \tag{8.5}$$

$$Q(a).$$

Observe that in concluding $Q(a)$, a substitution of a for x was necessary. This was possible because the implication $P(x) \rightarrow Q(x)$ is assumed true for all x, in particular, for $x = a$.

8.4 Many-Valued Logic

Predicate logic is a step toward augmenting propositional languages' expressiveness. It is still a two-valued system, however. It seems natural to extend predicate logic's ideas one step further by keeping all of its syntactic constructs but allowing truth to be any value in an appropriate truth set, say, the unit interval [0,1]. These values are then viewed as degrees of truth and, if they take values on [0,1], there are infinitely many. This leads to many-valued logical systems, that is, logics that admit an arbitrary number, $n \geq 2$, of truth values. For example, if truth values are taken as true (1), false (0), and indeterminate (1/2), we develop a three-valued system.

One of the pioneers in developing n-valued logic systems was a Polish mathematician, J. Lukasiewicz (1920). Actually, in Lukasiewicz's work, a chain $L_2, L_3, \ldots, L_\infty$ of logical systems emerged in which L_2 and L_∞ are the extreme cases two-valued and infinite-valued logic, respectively. The motivation behind this form of generalization of logic stems from the fundamental difficulty in assigning truth values to propositions expressing future events; the truth value at the time of evaluation is unknown. Lukasiewicz suggested using 1/2 (neutral, indeterminate) to quantify this situation. This gives rise to the three-valued truth set $T = \{0, 1/2, 1\}$. The basic operations are defined as

$$p \vee q = \max(p, q),$$

$$p \wedge q = \min(p, q),$$

$$p \rightarrow q = \min(1, 1 - p + q).$$

Clearly, this definition is equivalent to that for L_2, the two-valued case, p, $q \in \{0, 1\}$. More generally, L_n are n-valued systems in which the truth values are distributed uniformly across the unit interval:

$$T = \left\{ 0, \frac{1}{n-1}, \frac{2}{n-1}, \ldots, \frac{n-2}{n-1}, 1 \right\}.$$

Infinitely valued Lukasiewicz propositional logic with a denumerable set of logical values assumes that those values are rational numbers in the interval [0, 1]. In the case of an uncountable set of truth values, they are assumed to be real numbers in [0, 1] (Rescher 1969). Uncountably-valued Lukasiewicz logic L_∞ is considered in the foundations of fuzzy set theory as a basic logic for fuzzy reasoning (Rasiowa 1992).

In many-valued logic, reasoning is performed according to the same inference rules as discussed in section 8.2 except for the multivalued nature of the truth values and quantifiers. See Rescher 1969 for more details. The reader may also consult Epstein 1993 and Muzio and Wesselkamper 1986.

EXERCISE 8.3 Assume a propositional language with truth values in the set $\{0, 1/2, 1\}$ and the Lukasiewicz semantic rules. Construct the truth tables for the statements $P \wedge Q$, $P \vee Q$, $P \rightarrow Q$, $P \leftrightarrow Q$.

The basic ideas behind the logics we have reviewed are just a few among many alternatives. Modal, intuitionistic, and paraconsistent logics are examples of logic systems whose bases constitute significant departures from the systems we have discussed. For a broad overview of these and alternative approaches, see Rasiowa 1992.

8.5 Fuzzy Logic

Fuzzy logic arises as a generalization of multivalued logic that permits reasoning about unsharply defined world objects as relational entities. The fact that fuzzy logic deals with approximate rather than precise objects implies modes of reasoning that are also approximate. In fuzzy logic everything, including truth, is a matter of degree (Zadeh 1988). A truth value is allowed to be either a point in [0, 1], in the case of a numerical truth value, or a linguistic label such as *true*, *very true*, ..., *more or less false*, *false*, are the like, in the case of linguistic truth values. In the latter case, truth is viewed as a linguistic variable for which *true* and *false* are just two of the primary terms in its term set, instead of a pair of extreme points in the set of truth values (Zadeh 1975).

We may introduce fuzzy logic in an abstract fashion similar to that employed in introducing traditional logic systems: in other words, viewing fuzzy logic as a first-order language characterized by its syntax and semantics (Novak 1992). Algebraic approaches are a possible alternative (Rasiowa and Cat Ho 1992). Nevertheless, our intent is to discuss fuzzy logic on more intuitive and algorithmic grounds, since the aim is to address its main constructs in knowledge representation and reasoning. Furthermore the focus is on the capabilities of fuzzy logic as a tool for computations in problem solving and decision making. Thus, for our purposes, the lines originally suggested by Zadeh (1975, 1979), Tsukamoto (1979), and Pedrycz (1995) are adopted. The term "fuzzy logic" used in this chapter carries a sound logic connotation and constitutes a generalization of constructs of many-valued logics. The truth values in fuzzy logic are fuzzy sets, not single numeric truth values. The computations lead to fuzzy truth values and are again governed by logic inferences.

8.6 Computing with Fuzzy Logic

In this section we elaborate on the principles of fuzzy logic regarded primarily as a vehicle for computing with linguistic truth values.

8.6.1   Truth Space Methods

8.6.1.1   Fuzzy Truth Values and Fuzzy Truth Qualification

Atomic propositions denote basic pieces of knowledge of the form X is A, where X is the name of an object and A is the name of a fuzzy set in a universe **X**. Such a proposition is associated with two fuzzy sets; the meaning of A, a fuzzy set in **X** named A, and the truth value τ of X is A, a fuzzy set in [0,1]. When the truth value is a point in [0,1], it is a numeric truth value. Otherwise, the truth value is a linguistic variable in [0,1]. Some examples of truth values are

Presure is small is *very true*.

Temperature is low is *false*.

Compex propositions may be assembled combining atomic propositions using logical connectives, that is, disjunctions, conjunctions, negations, and implications. Thus statements of the form "X is A and Y is B," "X is A or Y is B," "X is not A" and "If X is A then Y is B" are also propositions. In particular, if-then statements are conditional propositions, frequently referred to as fuzzy rules.

To address the basic notions and operations for fuzzy logic and approximate reasoning, it is necessary to generalize the meaning of logical connectives involving linguistic rather than simply numerical truth values. For example, if the linguistic truth value of the proposition X is A is τ_A and of the proposition Y is B is τ_B, we should have a mechanism to compute the truth value of any compound proposition such as "X is A and Y is B." Note that such a mechanism requires the truth values τ_A and τ_B to be given. Before getting into this construct, we first proceed with a notion of fuzzy truth qualification.

The essence of truth qualification is to derive a fuzzy truth value for the relationship between two items viewed as fuzzy sets and defined in the same universe. Consider the following statement, including a piece of evidence A_i and a linguistic truth value associated with it:

$$(\text{Input datum is } A_i) \text{ is } \tau_i. \tag{8.6}$$

See also figure 8.2(a), with linguistic truth value $\tau_i : [0, 1] \to [0, 1]$. For instance, we may have

(Input datum is large) is *true*.

(Input datum is large) is *false*.

We are interested in determining a fuzzy set A such that

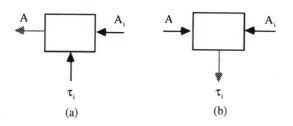

Figure 8.2
Truth qualification: (a) Direct and (b) inverse

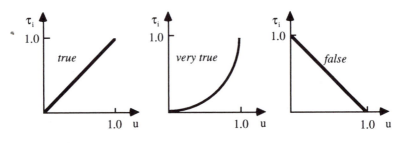

Figure 8.3
Selected models of fuzzy truth values

(Input datum is A_i) is τ_i = Input datum is A.

Truth qualification (Zadeh 1975) ensures that τ_i acts as an elastic constraint on A_i in the following way:

$$A(x) = \tau_i(A_i(x)), \qquad \forall x \in \mathbf{X}.$$

The form of A depends on the linguistic format of the truth τ_i. Figure 8.3 depicts several characteristic models of the truth values. (A justification for some of the models will be provided later.)

Because fuzzy truth values are one-to-one mappings, the truth computations are straightforward. We derive

$$A(x) = true(A_i(x)) = A_i(x),$$

$$A(x) = very\ true(A_i(x)) = A^2_i(x),$$

$$A(x) = more\ or\ less\ true(A_i(x)) = A^{0.5}_i(x).$$

The results are highly appealing. If the statement is *true*, then the result A becomes an original datum (no extra linguistic qualification). Any concentration-like hedges (*very*) lead to a more specific format for the resulting fuzzy set, say $A(x) < A_i(x)$. The dilation hedges (*more or less*, etc.) imply an inverse effect, $A(x) > A_i(x)$. If the truth value is *false* and represented as

$$false(u) = 1 - u$$

over all u in the unit interval, we get

$$A(x) = 1 - A_i(x).$$

The last expression generates an interesting interpretation. Consider A_i to be *small* and represented as

$$A_i(x) = \exp(-x)$$

for $x > 0$. Then the statement

(Input datum is small) is *false*

yields

$$A(x) = false(\exp(-x))$$

that in turn can be interpreted as

(Input datum is small) is *false* = Input datum is large.

If we confine ourselves to two-valued logic, the resulting truth values are just limit (boundary) truth values of fuzzy logic; let us call them *absolutely true* and *absolutely false*, respectively:

$$\tau_{absolutely\ true}(V) = \begin{cases} 1, & \text{if } v = 1 \\ 0, & \text{if } v \neq 1 \end{cases} = \delta(v - 1),$$

$$\tau_{absolutely\ false}(V) = \begin{cases} 1, & \text{if } v = 0 \\ 0, & \text{if } v \neq 0 \end{cases} = \delta(v).$$

A is then a set of x in $\mathbf{X}$ for which $A_i(x)$ assumes a value of 1. Interestingly, one can regard truth qualification as acting as a logic filter eliminating all membership grades but one (figure 8.4). The linguistic truth value of *absolutely false* acts in a complementary way, leaving out all nonzero membership values.

EXERCISE 8.4   Consider the linguistic truth value defined as the piecewise linear membership function shown in figure 8.5. What is the result of this truth evaluation applied to the Gaussian-like membership function $A_i(x) = \exp(-(x - m)^2)$?

8.6.1.2   Inverse Truth Qualification

The mechanism of inverse truth qualification arises in situations in which A_i and A are given but the fuzzy truth value τ_i must be determined, as in figure 8.2(b). The underlying formula (which is just the extension principle) reads

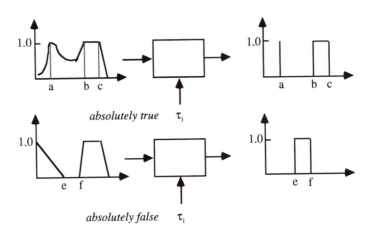

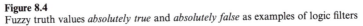

Figure 8.4
Fuzzy truth values *absolutely true* and *absolutely false* as examples of logic filters

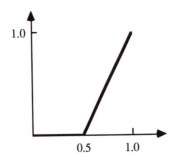

Figure 8.5
Example linguistic truth value

$$\tau_i(v) = \sup_{x:A_i(x)=v} A(x). \tag{8.7}$$

In the simplest case, set $A_i = A$. Then

$$\tau_i(v) = \sup_{x:A_i(x)=v} A(x) = v.$$

The last equation justifies the form of the generic linguistic truth value *true* proposed in figure 8.3. By careful inspection one finds that (8.7) is a compatibility measure of A taken with respect to A_i. In another specific case, when A is pointwise centered at a single point x_0 (that is, $A(x) = \delta(x - x_0)$), the fuzzy truth value becomes

$$\tau_i(v) = \begin{cases} 1, & \text{if } v = A_i(x_0) \\ 0, & \text{otherwise.} \end{cases}$$

8.6.1.3   Operations in Fuzzy Logic

The basic logic operations in fuzzy logic are defined with the aid of the extension principle. Let the fuzzy truth values of the statements A and B be given as τ_A and τ_B, respectively. We then derive

$$\tau_{A \text{ and } B}(v) = \sup_{w,z \in [0,1]: v = w\, t\, z} [\tau_A(w) \wedge \tau_B(z)]; \qquad (8.8)$$

$$\tau_{A \text{ or } B}(v) = \sup_{w,z \in [0,1]: v = w\, s\, z} [\tau_A(w) \wedge \tau_B(z)]; \qquad (8.9)$$

$$\tau_{\text{not } A}(v) = \sup_{u \in [0,1]: v = 1 - u} [\tau_A(u)] = \tau_A(1 - u). \qquad (8.10)$$

where the logic operations are defined by some t-norm and s-norm. The underlying multivalued *and*, *or*, and negation operations *induce* the linguistic truth values. Similarly, we introduce fuzzy implication starting with a certain form of the multivalued implication

$$\tau_{A \to B}(v) = \sup_{w,z \in [0,1]: v = w \to z} [\tau_A(w) \wedge \tau_B(z)]. \qquad (8.11)$$

8.6.1.4   Reasoning in the Framework of Truth Space

In the following, we concentrate on the basic reasoning scheme and study its fuzzy logic–based generalization. The generic inference model reads as

A

$$\frac{A_i \to B_i}{B}$$

where $A_i \to B_i$ denotes a rule (viewed as an implication), A is an input datum, and B stands for a conclusion. When, for a given $\to$, $A = A_i$ produces $B = B_i$, we have *modus ponens*. For the time being we consider a single rule only.

The first method we consider for fuzzy inference in the truth space (Tsukamoto 1979; Pedrycz 1995) assumes the ith rule linguistically qualified (e.g., *true*, *very true*, etc.), that is, it comes with a truth value τ_{R_i} (a linguistic truth value of the implication). In contrast to the standard version of modus ponens, the entire inference process is completed at the higher conceptual level formed by the mechanisms of fuzzy logic (figure 8.6).

In general, the reasoning process invokes three phases:

1. Inverse linguistic truth qualification of available datum A, namely,

$$\tau_{A_i} = \text{compatibility of } A_i \text{ with respect to } A.$$

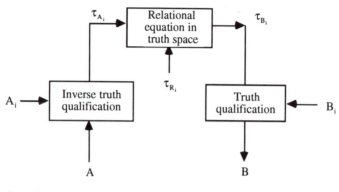

Figure 8.6
Reasoning with fuzzy logic

2. Fuzzy logic inference based on the fuzzy truth value of the rule and the antecedent exploiting the notion of the fuzzy implication defined by (8.11), as well as the solution of an inverse relational equation.

3. Inverse linguistic truth qualification producing conclusion B.

These basic processes are distributed at two conceptually distinct levels. The lower level, composed of two modules (inverse truth qualification and truth qualification), provides all necessary interfacing with the higher level, at which fuzzy logic inferencing takes place. The interaction mechanisms (inverse truth qualification and truth qualification) have already been discussed in detail, and we now devote more attention to the mechanism of fuzzy inference. Note that the truth values in the logical antecedent and conclusion space are governed by the expression

$$\tau_{R_i} = \tau_{A_i} \to \tau_{B_i}. \tag{8.12}$$

Here τ_{R_i} and τ_{A_i} are available; the fuzzy truth value of the conclusion τ_{B_i} is to be computed. Let us treat the above expression as a fuzzy-relational equation to be solved with respect to τ_{B_i}. To derive a solution, we rewrite the above expression in an equivalent form,

$$\tau_{R_i}(u) = \sup[\tau_{A_i}(z) \wedge \tau_{B_i} w)], \tag{8.13}$$

where the supremum is taken over all (z, w) in $[0,1]$ such that $u = z \to w$. Introduce a $\{0, 1\}$ (Boolean) relation,

$$M(z, w, u) = \begin{cases} 1, & \text{if } u = z \to w \\ 0, & \text{otherwise.} \end{cases} \tag{8.14}$$

This allows us to rewrite the above expression in an equivalent format,

$$\tau_{R_i}(u) = \sup[\tau_{A_i}(z) \wedge \tau_{B_i}(w) \wedge M(z, w, u)], \tag{8.15}$$

where now the supremum is taken over (z, w) in $[0,1]$. This eliminates the implication constraint, which has been absorbed in the form of the Boolean relation. Define the two-argument relation

$$G(w, u) = \sup_{z \in [0,1]} [\tau_{A_i}(z) \wedge M(z, w, u)]. \tag{8.16}$$

This implies the relation

$$\tau_{R_i}(u) = \sup_{w \in [0,1]} [G(w, u) \wedge \tau_{B_i}(w)]. \tag{8.17}$$

The last equation is simply a fuzzy-relational equation to be solved with respect to τ_{B_i}. Referring to the fundamental results in the theory of fuzzy-relational equations (chapter 4), the solution comes in the well-known format

$$\tau_{B_i} = G \, \varphi \, \tau_{R_i}, \tag{8.18}$$

meaning that

$$\tau_{B_i}(w) = \inf_{u \in [0,1]} [G(w, u) \, \varphi \, \tau_{R_i}(u)].$$

To illustrate how the inference scheme works, let us compute the truth values of τ_{B_i} for several truth values of the antecedent. Assume additionally that $\tau_{R_i}(v) = v$, so that the corresponding rule is regarded to be *true*. The relationship is obviously asymmetrical, meaning that τ_{A_i} and τ_{R_i} canot be interchanged. Consider a pointwise truth value of τ_{A_i},

$$\tau_{A_i}(z) = \delta(z - z_0).$$

The pointwise truth value located at z_0 shifts a piecewise linear membership function of τ_{B_i}. Refer to figure 8.7 with z_0 set to 0.5, 0.7, and 1.0,

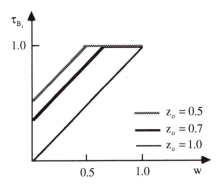

Figure 8.7
Fuzzy truth values of conclusion τ_{B_i} for various truth values of τ_{A_i}

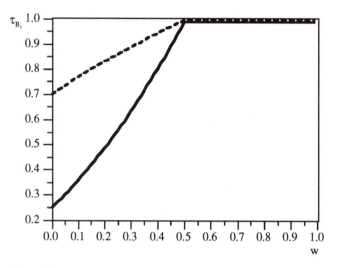

Figure 8.8
Fuzzy truth values for two truth values of the rules: $\tau_{R_i} = very\ true$ (solid line) and $\tau_{R_i} = more\ or\ less\ true$ (dotted line)

respectively. If $z_o = 0.0$, which corresponds to the binary *false*, then the fuzzy truth value of the conclusion is *unknown*.

If τ_{A_i} is equal to 1 over the entire unit interval (*unknown*) then the conclusion takes the form $\tau_{B_i} = true$; that is again in the vein of the implication (the truth of the antecedent is definitely weaker than the truth of the rule itself) and essentially states that the maximal truth value of the conclusion cannot exceed the truth value of the rule itself.

Now let us change the truth value of the rule, accepting successively *very true* and *more or less true* as the two standard linguistic qualifications. Figure 8.8 shows the conclusions, assuming that the truth value in the antecedent space is again pointwise and expressed as $\delta(z - 0.5)$.

EXERCISE 8.5   Consider several forms of implications to be used in inference schemes (Lukasiewicz, Gaines, etc; see chapter 10 for examples of implications). Derive the corresponding formulas for the fuzzy truth value of the conclusion.

To gain a better insight into numerical details, let us assume that $\tau_{A_i}(v) = v$, $\tau_{R_i}(u) = u$, and make the truth space discrete by specifying several equidistant truth values, say 0.0, 0.2, 0.4, 0.6, 0.8, and 1.0. The Godelian implication

$$z \rightarrow w = \begin{cases} 1, & \text{if } z \leq w \\ w, & \text{if } z > w, \end{cases}$$

is represented in a matrix form:

$$
\begin{array}{c}
 & & w & & & & \\
 & 0.0 & & & & & 1.0 \\
\begin{array}{c} 0.0 \\ \\ \\ z \\ \\ \\ 1.0 \end{array}
\begin{bmatrix}
1.0 & 1.0 & 1.0 & 1.0 & 1.0 & 1.0 \\
0.0 & 1.0 & 1.0 & 1.0 & 1.0 & 1.0 \\
0.0 & 0.2 & 1.0 & 1.0 & 1.0 & 1.0 \\
0.0 & 0.2 & 0.4 & 1.0 & 1.0 & 1.0 \\
0.0 & 0.2 & 0.4 & 0.6 & 1.0 & 1.0 \\
0.0 & 0.2 & 0.4 & 0.6 & 0.8 & 1.0
\end{bmatrix}.
\end{array}
$$

The three-dimensional matrix M is indexed by u, $u = 0.0, 0.2, 0.4$, etc.
The matrix values are directly obtained from (8.14):

$$
\begin{array}{cc}
\begin{array}{c}
 & w \\
 & 0.0 \qquad\quad 1.0 \\
\begin{array}{c} 0.0 \\ \\ \\ z \\ \\ \\ 1.0 \end{array}
\left|\begin{array}{ccc}
0 & 0 & \;\;0 \\
1 & 0 & \;\;0 \\
1 & 0 & \;\;0 \\
1 & 0 & \cdots\;\;0 \\
1 & 0 & \;\;0 \\
1 & 0 & \;\;0
\end{array}\right| \\
u = 0.0
\end{array}
&
\begin{array}{c}
 & w \\
 & 0.0 \qquad\quad 1.0 \\
\begin{array}{c} 0.0 \\ \\ \\ z \\ \\ \\ 1.0 \end{array}
\left|\begin{array}{cccc}
0 & 0 & 0 & \;\;0 \\
0 & 0 & 0 & \;\;0 \\
0 & 1 & 0 & \;\;0 \\
0 & 1 & 0 & \cdots\;\;0 \\
0 & 1 & 0 & \;\;0 \\
0 & 1 & 0 & \;\;0
\end{array}\right| \\
u = 0.2
\end{array}
\end{array}
$$

$$
\begin{array}{c}
 & w \\
 & 0.0 \qquad\qquad\quad 1.0 \\
\begin{array}{c} 0.0 \\ \\ \\ z \\ \\ \\ 1.0 \end{array}
\left|\begin{array}{ccccc}
0 & 0 & 0 & 0 & \;\;0 \\
0 & 0 & 0 & 0 & \;\;0 \\
0 & 0 & 0 & 0 & \;\;0 \\
0 & 0 & 1 & 0 & \cdots\;\;0 \\
0 & 0 & 1 & 0 & \;\;0 \\
0 & 0 & 1 & 0 & \;\;0
\end{array}\right|. \\
u = 0.4
\end{array}
$$

Finally, G becomes a diagonal matrix,

$$
\begin{bmatrix}
1 & 0 & 0 & 0 & 0 & 0 \\
0 & 1 & 0 & 0 & 0 & 0 \\
0 & 0 & 1 & 0 & 0 & 0 \\
0 & 0 & 0 & 1 & 0 & 0 \\
0 & 0 & 0 & 0 & 0 & 1
\end{bmatrix},
$$

and, as expected, τ_{B_i}, computed via (8.18), gives rise to a generic fuzzy truth value,

$$\tau_{B_i} = [0.0 \quad 0.2 \quad 0.4 \quad 0.6 \quad 0.8 \quad 1.0],$$

demonstrating clearly that this method of inference is in agreement with *modus ponens*. It may thus be regarded as its generalization.

The fuzzy *modus ponens* discussed so far has been based on a single rule. Practical circumstances call for a series of rules in which i takes on values from 1 to N. This makes an extra loop in the original scheme:

Repeat for all i from 1 to N
{
The fuzzy truth values are determined with respect to each A_i,

τ_{A_i} = inverse truth qualification of A_i with respect to A.

For given fuzzy truth values of the rules τ_{R_i}, determine fuzzy truth of the conclusion of the rule τ_{B_i}.
Obtain a fuzzy set of conclusion induced by B_i' by performing truth qualification

$$B_i'(y) = \tau_{B_i}(B_i(y)). \tag{8.19}$$

}

The fuzzy set of conclusion is produced by intersecting $B_i'(y)s$; this form of aggregation stems from the theory of fuzzy-relational equations,

$$B(y) = \bigwedge_{i=1}^{N} B_i'(y). \tag{8.20}$$

This type of rules summarization excludes rules that have not been activated. The notion of activation in this setting corresponds to linguistic truth values. When the linguistic truth value of A with respect to A_i is *false*, $\tau_{A_i}(v) = 1 - v$, then $\tau_{B_i} = 1$ and, subsequently, the inverse truth qualification yields

$$B_i'(y) = 1$$

for all y in $\mathbf{Y}$.

A second method for fuzzy inference in the truth space (Baldwin 1979a) consists of three steps:

1. Inverse linguistic truth qualification of available datum A, namely,

$$\tau_{A_i}(z) = \sup_{x:A_i(x)=z} A(x).$$

2. Fuzzy logic inference using an implication relation $R_i(z, w)$ of $[0, 1] \times [0, 1]$ and solving a direct-relational equation to compute $t_{B_i}(w)$, that is,

$$\tau_{B_i}(w) = \sup_{z \in [0,1]} [\tau_{A_i}(z) \wedge R_i(z, w)]. \tag{8.21}$$

3. Linguistic truth qualification producing conclusion B'_i, namely,

$$B'_i(y) = \tau_{B_i}(B_i(y)).$$

The fuzzy set of conclusion is now produced by the union of $B'_i s$, namely,

$$B(y) = \bigcup_{i=1}^{N} B'_i(y).$$

In (8.21), R_i is a fuzzy relation in the truth space capturing the meaning of $A_i \rightarrow B_i$. when the implication is truth-qualified through a fuzzy qualifier τ_i, the method remains essentially the same: Put $\tau_i(R_i(z, w))$ in the place of $R_i(z, w)$ in (8.21). Clearly, if $\tau_i = true$, then (8.21) is recovered. Note that this method has the same structure as that depicted in figure 8.6 and is conceptually analogous to the previous method except for the second step. Under appropriate choices of implication, it is also in agreement with modus ponens, since it can be easily viewed as repeating the numerical example provided above. This is left as an exercise.

8.6.2   Compositional Rule of Inference

Compositional inference, often referred to as direct approach for fuzzy reasoning, was originally proposed by Zadeh (1973). It has only one step; the conclusion $B'_i(y)$ is derived as follows:

$$B'_i = \sup_{x \in X} [A(x) \wedge R_i(x, y)],$$

where R is defined as an implication $R : [0, 1] \times [0, 1] \rightarrow [0, 1]$, a relation in the unit square. In the general case, the min can be replaced by a t-norm to perform sup-t composition:

$$B'_i = \sup_{x \in X} [A(x) \; t \; R_i(x, y)].$$

The compositional rule of inference is not in agreement with modus ponens for *any* choice of t-norms and implications. For instance, if we choose the min and the ordinary product as t-norms, then modus ponens holds for the Godel implication. This is not the case, however, if the Lukasiewicz implication is chosen.

The compositional rule of inference plays a fundamental role in rule-based computations, a topic to be pursued in detail in chapter 10. It is also a cornerstone of many applications of fuzzy systems, in engineering and in many other fields as well.

8.7   Some Remarks about Inference Methods

It is interesting to summarize some important conceptual characteristics of the fuzzy inference methods introduced. It has been shown, in Nojima and Mukaidono 1991, that if $A(x)$ is a function of $\mathbf{X}$ onto $[0,1]$—another terminology for this case: $A(x)$ is surjective—then the result produced by the direct truth space approach is the same as that originated by the direct method, namely, sup-min composition. We may draw a similar conclusion for the inverse truth space approach, because $A(x)$ must be onto for the inverse truth qualification to be meaningful. In contrast, in the compositional method, surjectiveness is not needed. Therefore, the compositional inference can do everything the truth space methods do, but the converse does not hold.

From the computational point of view, the one advantage of the truth space methods is that the actual inference step requires less computation than the direct approach, since the truth space has lower dimension. In general, inputs and antecedents of the implications are likely to be multi-dimensional, and the inverse truth qualification may become a complex computation (Tong and Efstathiou 1982) if they involve fuzzy sets. In many practical problems, however, inputs are pointwise and computations are much simpler. In addition we should be aware that computer technology has been growing fast, and efficient algorithms are being developed (Uehara and Fujise 1993) for exploring fuzzy sets' peculiarities in terms of their membership functions.

8.8   Chapter Summary

Fuzzy logic stems directly from the theory of fuzzy sets and hinges on the fabric of logic exploiting fuzzy truth values. In contrast to those of many-valued logics, the truth values of fuzzy logic are represented as fuzzy sets in the unit interval. All operations in fuzzy logic are constructed with the use of the extension principle. In this way fuzzy logic subsumes two-valued and many-valued logics. The chapter discussed modus ponens in depth, with an auxiliary discussion on rule truth qualification. It clarified some misunderstandings surrounding the term "fuzzy logic," which has

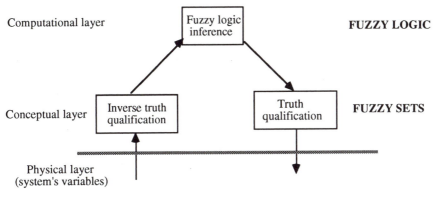

Figure 8.9
A hierarchy in fuzzy logic

obviously been intensively overused. For instance, the constructs claimed to be fuzzy logic controllers, fuzzy logic inference modules, and the like may have nothing in common with the semantics of fuzzy logic. We should be aware that fuzzy logic delivers an extra computational layer on top of the one formed originally by fuzzy sets, as portrayed in a hierarchy in figure 8.9. Fuzzy sets transform a basic layer of physical variables into conceptual chunks known as linguistic labels. These, in turn, are manipulated via the rules of fuzzy logic, and here emerges a profound role of the layer discussed in this chapter.

8.9   Problems

1. Assume the proposition "$(X$ is $A)$ is τ", τ being *true*. If we define *false* $= \tau(not\ A)$, how would you characterize the proposition "$(X$ is $A)$ is *false*"?

2. Consider the same propositions as in problem 1. Provide the truth qualification for the propositions "$(X$ is $A)$ is *very true*" and "$(X$ is $A)$ is *very false*." Hint: Recall the role of modifiers in linguistic variables.

3. Suppose that

    $\tau(A(x)) = 0.5/0.7 + 0.7/0.8 + 0.9/0.9 + 1.0/1.0$, and

    $\tau(B(x)) = 1.0/0.0 + 1.0/0.1 + 1.0/0.2 + 1.0/0.3 + 1.0/0.4 + 1.0/0.5 + 1.0/0.6$
    $\quad + 0.5/0.7 + 0.3/0.8 + 0.1/0.9.$

    Compute the truth value $\tau(A$ and $B)$ for several selected t-norms.

4. Given the fuzzy sets whose membership functions are depicted in figure 8.10, find $\tau_i(v)$.

5. Considering a single rule of the form "If A_i, then B," show step by step the fuzzy logic inference via inverse truth space method, assuming that A is a single numeric value situated at y, namely $A(x) = \delta(x - y)$ for all x in **R**. Compare the results with the direct truth space and the compositional methods.

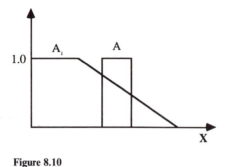

Figure 8.10

References

Baldwin, J. 1979a. A new approach to approximate reasoning using a fuzzy logic. *Fuzzy Sets and Systems* 2:309–25.

Baldwin, J. 1979b. Fuzzy logic and its application to fuzzy reasoning. In *Advances in Fuzzy Set Theory and Applications,* eds. M. Gupta, R. Ragade, and R. Yager, 93–114, Dordrecht, Netherlands, North-Holland.

Epstein, G. 1993. *Multiple-Valued Logic Design: An Introduction.* Bristol, U.K.: Institute of Physics Publishing.

Lukasiewicz, J. 1920. O logice trojwartosciowej. *Ruch Filozoficzny,* 5:169–70.

Muzio, J. C., and T. Wesselkamper. 1986. *Multiple-Valued Switching Theory.* Bristol, U.K., and Boston: A. Hilger.

Nojima, K., and M. Mukaidono. 1991. The direct approach includes the truth space approach in approximate reasoning systems. *Proc. of the fourth IFSA World Congress, Artificial Intelligence,* Brussels, Belgium, 161–64.

Novak, V. 1992. Fuzzy logic as a basis of approximate reasoning. In *Fuzzy Logic for the Management of Uncertainty,* ed. L. Zadeh and J. Kacprzyk, 247–64. New York: Wiley Interscience.

Pedrycz, W. 1995. *Fuzzy Sets Engineering.* Boca Raton, FL: CRC Press.

Rasiowa, H. 1992. Toward fuzzy logic. In *Fuzzy Logic for the Management of Uncertainty,* ed. L. Zadeh and J. Kacprzyk 5–25. New York: Wiley Interscience.

Rasiowa, H., and N. Cat Ho. 1992. LT-fuzzy logic. In *Fuzzy Logic for the Management of Uncertainty,* ed. L. Zadeh and J. Kacprzyk, 121–39. New York: Wiley Interscience.

Rescher, N. 1969. *Many-Valued Logics.* New York: McGraw-Hill.

Tong, R., and J. Efstathiou. 1982. A critical assessment of truth function modification and its use in approximate reasoning. *Fuzzy Sets and Systems* 7:103–8.

Tuskamoto, Y. 1979. An approach to fuzzy reasoning method. In *Advances in Fuzzy Set Theory and Applications,* ed. M. Gupta, R. Ragade, R. Yager, 137–49. Netherlands: North-Holland.

Uehara, K., and M. Fujise. 1993. Fuzzy inference based on families of α-level sets. *IEEE Trans. on Fuzzy Systems* 1(2):111–24.

Zadeh, L. A. 1973. Outline of a new approach to the analysis of complex systems and decision processes. *IEEE Trans. on Systems, Man, and Cybernetics* 3(1): 28–44.

Zadeh, L. A. 1975. The concept of a linguistic variable and its application to approximate reasoning, (part 2). *Information Sciences* 8:301–57.

Zadeh, L. A. 1979. A theory of approximate reasoning. In *Machine Intelligence,* ed. J. Hayes and J. Michie, vol. 9, 149–94. New York: Halstead Press.

Zadeh, L. A. 1988. Fuzzy logic. *IEEE Computer* 21(4):83–92.

9 Fuzzy Measures and Fuzzy Integrals

In this chapter we introduce the notion of fuzzy measures and fuzzy integrals and discuss their main properties. Fuzzy measures are examples of nonadditive measures. Fuzzy integrals emerge as formal constructs that support integration completed with respect to fuzzy measures. We also outline the use of fuzzy measures and integrals primarily in the realm of information fusion. In particular, sensor fusion is one area in which information acquired from various sensors, usually of very distinct characteristics, reliability, and accuracy, must be aggregated.

9.1 Fuzzy Measures

Before formally defining fuzzy measures themselves, we must introduce two ideas: measurable spaces and measures. By a measurable space, we mean a pair $(\mathbf{X}, \Omega)$ that consists of a universe $\mathbf{X}$ and a σ-algebra of measurable sets of $\mathbf{X}$.[1]

DEFINITION 9.1   A measure μ is a set function defined on a measurable space $(\mathbf{X}, \Omega)$ such that it assumes nonnegative values and satisfies the following postulates:

1. Boundary condition: $\mu(\varnothing) = 0$.

2. σ-additivity: For a family of pairwise disjoint sets, $A_i \in \Omega$; $A_i \cap A_j = \varnothing$; $i = 1, 2, \ldots, i \neq j$; and the measure of the union of A_is is equal to the sum of measures defined over the individual sets:

$$\mu\left(\bigcup_{i=1}^{\infty} A_i\right) = \sum_{i=1}^{\infty} \mu(A_i). \tag{9.1}$$

3. If $\{A_n\}, n = 1, 2, \ldots$ is an increasing sequence of measurable sets, then

$$\lim_{n \to \infty} \mu(A_n) = \mu\left(\lim_{n \to \infty} A_n\right). \tag{9.2}$$

Based on the above observations and taking into account the properties of σ-algebras, one can directly infer several basic properties:

1. If $A \subset B$, then $\mu(A) \leq \mu(B)$.
2. $\mu(\bar{A}) = 1 - \mu(A)$.
3. $\mu(X) = 1$.

The probability measure P is one of the most commonly applied examples of (additive) measures. More specifically, σ-additivity means that

$$P\left(\bigcup_{i=1}^{\infty} A_i\right) = \sum_{i=1}^{\infty} P(A_i) = \sum_{i=1}^{\infty} p_i,$$

with A_i being pairwise disjoint sets and $p_i = P(A_i)$.

A fuzzy measure, as introduced by Sugeno (1974, 1977) generalizes the above definition of the (additive) measure by replacing the additivity requirement with the condition of monotonicity, as we now defined more formally.

DEFINITION 9.2   A set function $g : \Omega \rightarrow [0, 1]$ is a fuzzy measure under the following conditions:

1. $g(\emptyset) = 0$; $g(\mathbf{X}) = 1$.

2. If $A \subset B$, then $g(A) \leq g(B)$.

3. For any increasing sequence of measurable sets $\{A_n\}$, $n = 1, 2, \ldots,$ we have

$$\lim_{n \rightarrow \infty} g(A_n) = g\left(\lim_{n \rightarrow \infty} A_n\right).$$

As a relaxation of the previous concept, in this definition, the monotonicity requirement has replaced the property of σ-additivity. The boundary conditions, however, remain the same. The third requirement is important only for infinite sequences of sets and can be ignored for any finite family of A_is.

EXAMPLE 9.1   Fuzzy measures take on an interesting interpretation in sensor fusion. Consider a system (e.g., an autonomous vehicle) equipped with a collection of sensors, say $x_1, x_2, \ldots, x_n$. The more sensors become active (provide information about the environment), the more complete the systems's recognition. With each sensor we associate a fuzzy measure,

$$g_1 = g(\{x_1\}), g_2 = g(\{x_2\}), \ldots, g_n = g(\{x_n\}),$$

such that the higher the g_i, the more relevant (important) the corresponding sensor. In this setting, the monotonicity condition assumes an evident interpretation,

$$g(\{x_1, x_2, \ldots, x_i\}) \leq g(\{x_1, x_2, \ldots, x_i, x_{i+1}\}),$$

indicating that adding more sensors to the system enhances its capability to recognize the environment. Similarly, the interpretation of the boundary condition becomes clear:

$$g(\text{all sensors inactive}) = 0; g(\text{all sensors active}) = 1.$$

Although sound, the above definition of a fuzzy measure is not operational, because it is not obvious how to compute a fuzzy measure of the union of two disjoint sets whose fuzzy measures, $g(A)$ and $g(B)$, are provided. To alleviate this deficiency and develop a fully operational version of this concept, Sugeno (1974) proposed a so-called λ-fuzzy measure. To avoid excessive notation, we will be denoting this measure by g—the context in which it will be used should help avoid misunderstanding. (In the literature, one also encounters the notion with an extension, such as g_λ.) For any two disjoint sets A and B, $A \cap B = \varnothing$, the value of the fuzzy measure it takes upon its union, $g(A \cup B)$, is computed as

$$g(A \cup B) = g(A) + g(B) + \lambda g(A)g(B), \qquad (9.3)$$

where $\lambda > -1$. The parameter of the fuzzy measure λ is used to describe an "interaction" between the components (sets) that are combined. If $\lambda = 0$, then the above expression reduces to the additive measure,

$$g(A \cup B) = g(A) + G(B).$$

If $\lambda > 0$, we obtain

$$g(A \cup B) = g(A) + g(B) + \lambda g(A)g(B) \geq g(A) + g(B).$$

This so-called super-additivity relationship quantifies a synergy effect, meaning that an evidence associated with the union of A and B is greater that the sum of the evidences arising from these two sources viewed separately (Gestalt principle). Simply, these two sources support each other. On the other hand, if $\lambda < 0$, leading to the sub-additivity effect, these two sources of evidence are in competition (or redundancy), and their effect translates into the form

$$g(A \cup B) \leq g(A) + g(B).$$

The value of the parameter of the λ-fuzzy measure is obtained from the normalization condition $g(\mathbf{X}) = 1$. To illustrate the detailed computations, let us proceed with $n = 2$ in example 9.1; here we get

$$1 = g_1 + g_2 + \lambda g_1 g_2.$$

The quantity $g_i = g(\{x_i\})$ is also referred to as the fuzzy density at x_i. The above relationship can be rewritten in the equivalent form

$$1 = \frac{1}{\lambda}[(1 + \lambda g_1)(1 + \lambda g_2) - 1].$$

Iterating for $n = 3$, we derive

$$g(\{x_1, x_2, x_3\}) = g(\{x_1, x_2\}) + g(\{x_3\}) + \lambda g(\{x_1, x_2\})g(\{x_3\})$$
$$= g(\{x_1\}) + g(\{x_2\}) + \lambda g(\{x_1\})g(\{x_2\}) + g(\{x_3\})$$
$$+ \lambda(g(\{x_1\}) + g(\{x_2\}) + \lambda g(\{x_1\})g(\{x_2\}))g(\{x_3\})$$
$$= g(\{x_1\}) + g(\{x_2\}) + g(\{x_3\}) + \lambda g(\{x_1\})g(\{x_2\})$$
$$+ \lambda g(\{x_1\})g(\{x_3\}) + \lambda g(\{x_2\})g(\{x_3\})$$
$$+ \lambda g(\{x_1\})g(\{x_2\})g(\{x_3\}).$$

Rearranging the terms as before, one has

$$g(\{x_1, x_2, x_3\}) = \frac{1}{\lambda}[(1 + \lambda g_1)(1 + \lambda g_2)(1 + \lambda g_3) - 1].$$

In general, it can be shown that

$$g(\{x_1, x_2, \ldots, x_n\}) = \frac{1}{\lambda}\left[\prod_{i=1}^{n}(1 + \lambda g_i) - 1\right], \quad \lambda \neq 0. \tag{9.4}$$

The value of λ is obtained through the boundary condition, $g(\mathbf{X}) = 1$. This yields a polynomial equation with respect to λ:

$$1 + \lambda = \prod_{i=1}^{n}(1 + \lambda g_i). \tag{9.5}$$

As Sugeno (1974) has shown, there exists a unique $\lambda \in (-1, \infty)$ different from zero and satisfying the above relationship. (See also Grabisch, Nguyen, and Walker 1995.) The fuzzy measure over the given set $k \subset \mathbf{X}$ is computed as

$$g(K) = \frac{1}{\lambda}\left[\prod_{x_i \in K}(1 + \lambda g_i) - 1\right].$$

EXAMPLE 9.2   We consider a collection of five sensors, each characterized by its own relevance factor that depends upon the quality of the measurements produced. The higher the value of the fuzzy density g_i, the more essential the sensor. Let the fuzzy densities reflecting the sensors' performance be determined as $g_1 = 0.60$, $g_2 = 0.35$, $g_3 = 0.05$, $g_4 = 0.21$, and $g_5 = 0.72$. The roots of the resulting fifth-order polynomial (9.5) are equal to $-20.13, -4.81 - 2.90i, -4.81 + 2.90i, -0.92$, and 0. The coefficient λ of the λ-fuzzy measure equals -0.92 (for obvious reasons, $\lambda = 0$ is excluded from the solution set). The computations of the fuzzy measure for some selected subsets of the sensors are summarized below:

Sensors	Fuzzy measure
$\emptyset$	0
$\{x_1\}$	0.60
$\{x_2\}$	0.35
$\{x_3\}$	0.05
$\{x_4\}$	0.21
$\{x_5\}$	0.720
$\{x_1, x_2\}$	0.757
$\{x_1, x_5\}$	0.936

Overall, the fuzzy measure of any subset of sensors $x_i s$, $\{x_i\}_{i \in \mathbf{I}}$ is computed using

$$g\left(\bigcup_{i \in \mathbf{I}}\{x_i\}\right) = \frac{1}{\lambda}\left[\prod_{i \in \mathbf{I}}(1 + \lambda g_i) - 1\right]. \tag{9.6}$$

This formula can be rewritten in a recurrent format. Define the collections of the sensors

$$A_i = \{x_1, x_2, \ldots, x_i\}, A_{i+1} = \{x_1, x_2, \ldots, x_i, x_{i+1}\}.$$

We then obtain

$$g(A_{i+1}) = g(A_i) + g(\{x_{i+1}\}) + \lambda g(A_i)g(\{x_{i+1}\}).$$

EXAMPLE 9.3 As in example 9.2, we consider five sensors whose fuzzy densities are equal to $g_1 = 0.20, g_2 = 0.07, g_3 = 0.01, g_4 = 0.09, g_5 = 0.69$. The roots of the polynomial equation are now equal to -100.16, $-15.74, -9.29i, -15.74 + 9.29i, -0.21$, and 0. Observe that $\lambda = -0.21$, which is greater than the value of the parameter obtained in example 9.2. In general, Grabisch, Nguyen, and Walker (1995) proved, the following relationship is valid: If the fuzzy densities satisfy the inclusion $g_i \leq g'_i$, $i = 1, 2, \ldots, n$, then the parameters of the corresponding λ-fuzzy measures satisfy the relation $\lambda > \lambda'$.

It is definitely of interest to examine associations between λ-fuzzy measures and probabilities (probability functions). Assume that the values of the fuzzy measure are available. We show that the function

$$P(A) = \frac{\log[1 + \lambda g(A)]}{\log(1 + \lambda)} \tag{9.7}$$

satisfies the axioms of probability (additive measure). It is easy to show that the boundary conditions are satisfied, $P(\emptyset) = 0$, $P(\mathbf{X}) = 1$. Clearly, since $g(\varnothing) = 0$,

$$P(\varnothing) = \frac{\log[1 + \lambda g(\varnothing)]}{\log(1 + \lambda)} = 0.$$

Similarly, from the fuzzy measure's second boundary condition, we obtain

$$P(\mathbf{X}) = \frac{\log[1 + \lambda g(\mathbf{X})]}{\log(1 + \lambda)} = 1.$$

Assume now that A and B are disjoint. Then

$$P(A \cup B) = \frac{\log[1 + \lambda g(A \cup B)]}{\log(1 + \lambda)}$$

$$= \frac{1}{\log(1 + \lambda)} \{\log[1 + \lambda g(A) + \lambda g(B) + \lambda^2 g(A)]\}$$

$$= \frac{1}{\log(1 + \lambda)} \{\log[1 + \lambda g(A)] + \log[1 + \lambda g(B)]\}$$

$$= \frac{1}{\log(1 + \lambda)} \{[\log(1 + \lambda)P(A)] + [\log(1 + \lambda)P(B)]\}$$

$$= P(A) + P(B).$$

9.2   Fuzzy Integrals

Having defined a fuzzy measure, we move now to the concept of a fuzzy integral, a nonlinear functional defined over all measurable sets of $\mathbf{X}$. More formally, we have

DEFINITION 9.3   Let $(\mathbf{X}, \Omega)$ be a measurable space. Let $h : \mathbf{X} \to [0, 1]$ be an Ω-measurable function. The fuzzy integral over $A \subset \mathbf{X}$ of function h taken with respect to a fuzzy measure g is expressed as

$$\int_A h(x) \circ g(\bullet) = \sup_{\alpha \in [0,1]} \{\min[\alpha, g(A \cap H_\alpha)]\}, \tag{9.8}$$

where H_α is the α-cut of the function to be integrated,

$$H_\alpha = \{x \mid h(x) \geq \alpha\}.$$

An interesting interpretation of the fuzzy integral arises in the context of the sensor fusion example. Consider, as before, that $g(\bullet)$ describes the relevance of the sources of information (sensors) and h denotes the results the sensors have reported. The fuzzy integral then combines (aggregates) nonlinearly the outcomes of all the sensors.

9.2.1 Basic Properties of Fuzzy Integration

The properties below highlight some interesting features of the fuzzy integral. The reader can easily contrast the features of this integral with the properties of standard integral calculus:

1. Integration of a constant function, $h(x) = c$: For all $x, c \in [0, 1]$, one gets

$$\int_x h \circ g(\bullet) = \int_x c \circ g(\bullet) = c. \tag{9.9}$$

2. Monotonicity of integration with respect to the integrand, the integration region, and the fuzzy measure:

- If $h_1(x) < h_2(x)$ for all $x \in [0, 1]$, then

$$\int_A h_1 \circ g(\bullet) \leq \int_A h_2 \circ g(\bullet) \tag{9.10}$$

- If $A_1 \subset A_2$, then

$$\int_{A_1} h \circ g(\bullet) \leq \int_{A_2} h \circ g(\bullet) \tag{9.11}$$

- If $g < g'$, then

$$\int_A h \circ g(\bullet) \leq \int_A h \circ g'(\bullet). \tag{9.12}$$

Computations involving fuzzy integrals are significantly simplified for finite spaces, $\text{Card}(\mathbf{X}) = n$. Without any loss of generality (since this requirement can be easily achieved by a straightforward rearrangement of the elements), let us assume that

$$h(x_1) \geq h(x_2) \geq \cdots \geq h(x_n).$$

Define an increasing sequence of sets

$$A_i = \{x_1, x_2, \ldots, x_i\}, \text{i.e., } A_1 \subset A_2 \subset \cdots \subset A_n.$$

The original definition of the fuzzy integral is then realized through the standard max-min composition of the two series $\{h(x_i)\}$ and $\{g(x_i)\}$ for each i, $i = 1, 2, \ldots, n$:

$$\int_{A_i} h \circ g(\bullet) = \max_{i=1,\ldots,n} \{\min[h(x_i), g(A_i)]\}. \tag{9.13}$$

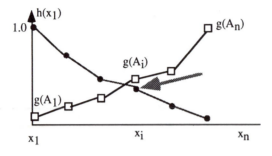

Figure 9.1
Calculations of a fuzzy integral

As figure 9.1 illustrates, computations involving the fuzzy integral require two steps:

1. taking the minimum of $h(x_i)$ and $g(A_i)$, and

2. determining the maximum of the sequence produced in (1).

A straightforward generalization of (9.13) involves the use of any t-norm (Weber 1984):

$$\int_{A_i} h \circ g(\bullet) = \max_{i=1,\ldots,n} [h(x_i) \text{ t } g(A_i)]. \tag{9.14}$$

EXAMPLE 9.4   As a continuation of example 9.1, we integrate the readings of the sensors arranged in vector form,

$$h = [0.1 \quad 0.4 \quad 0.3 \quad 0.7 \quad 0.05].$$

According to the above construct, the sensor readings need to be rearranged to produce a nonincreasing sequence. We simultaneously show the corresponding vector of the fuzzy densities:

$$h = [0.7 \quad 0.4 \quad 0.3 \quad 0.1 \quad 0.05];$$

$$g = [0.21 \quad 0.35 \quad 0.05 \quad 0.60 \quad 0.72].$$

The resulting fuzzy measures over the increasing sequence of the nested sets are $g(A_1) = 0.210$, $g(A_2) = 0.492$, $g(A_3) = 0.520$, $g(A_4) = 0.833$, $g(A_5) = 1.000$. Finally, the max-min composition of these two sequences provides the result of integration, equal to 0.4 (see figure 9.2).

As shown in Sugeno 1974, the fuzzy integral is a median:

$$\int_{A_i} h \circ g(\bullet) = \text{med}[h(x_1), h(x_2), \ldots, g(A_1), \ldots, g(A_n)]. \tag{9.15}$$

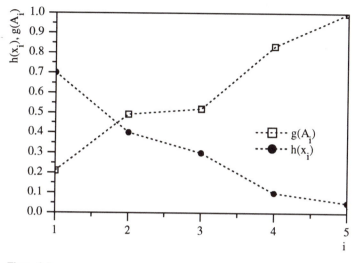

Figure 9.2
Determination of a fuzzy integral

Let us recall that, for a finite sequence $c_1, c_2, \ldots, c_n$,

$$\text{med}(c_1, \ldots, c_n) = \begin{cases} c_{((n+1)/2)} & \text{if } n \text{ is odd} \\ \frac{1}{2}(c_{(n/2)} + c_{(n/2+1)}) & \text{if } n \text{ is even.} \end{cases}$$

The Choquet (1953) integral is a closely related concept. This integral is defined as

$$\text{Ch}\int h \circ g = \sum_{i=1}^{n}[h(x_i) - h(x_{i+1})]g(A_i), \quad h(x_{i+1}) = 0. \tag{9.16}$$

It is interesting to observe the similarity of Choquet integral to our usual notion of integral.

EXAMPLE 9.5   The Choquet integral for the fuzzy measure and h defined as in example 9.4 are computed accordingly:

$$\text{Ch}\int h \circ h = (0.7 - 0.4)0.210 + (0.4 - 0.3)0.492 + (0.3 - 0.1)0.520$$

$$+ (0.1 - 0.05)0.833 + (0.05 - 0.0)1.0 = 0.3079$$

The primary applications of fuzzy measures and integrals emerge in the realm of information aggregation, including computer vision (Tahani and Keller 1990), prediction of wood strength (Ishii and Sugeno 1985), human reliability (Onisawa et al. 1986), and multiattribute decision making (Grabisch 1995).

9.2.2   Optimization Aspects of the Fuzzy Integral

As constructed in the previous sections, the fuzzy integral leaves no room for eventual parametric adjustments of the construct. Consider a sequence of discrepancies (differences) between the computed values of the integral and those obtained experimentally. Assume that we are provided with a series of such situations,

$$\left(h_1, \int h_1 \circ g, I_1 \right),$$

$$\left(h_2, \int h_2 \circ g, I_2 \right),$$

$$\vdots$$

$$\left(h_N, \int h_N \circ g, I_N \right),$$

where usually

$$\int h_k \circ g \neq I_k.$$

Because the fuzzy measure is equipped with a single parameter λ that has already been used to meet the normalization condition, there is obviously no parametric flexibility left that can eventually be used to adjust the construct to reflect the experimental data.

Among possible options that could alleviate the problem, one is to admit an additional parametric transformation of hs, say $f(h; \mathbf{a})$, with $\mathbf{a}$ being an adjustable vector of parameters a. The optimization calls for a solution of the mean squared errors problem of the form

$$V = \sum_{k=1}^{N} \left(\int_{\mathbf{X}} f[h_k; \mathbf{a}] \circ g - I_k \right)^2 \rightarrow \min \mathbf{a}. \tag{9.17}$$

The standard gradient-based minimization algorithm (see chapter 14) is

$$\mathbf{a}(\text{new}) = \mathbf{a}(\text{old}) + \alpha \left. \frac{\partial V}{\partial \mathbf{a}} \right|_{\mathbf{a}=\mathbf{a}(\text{old})}$$

The gradient is computed in a straightforward fashion:

$$\frac{\partial V}{\partial \mathbf{a}} = 2 \sum_{i=1}^{n} \left(\int_{\mathbf{X}} f[h_k; \mathbf{a}] \circ g - I_k \right) \frac{\partial}{\partial \mathbf{a}} f[h_k; \mathbf{a}] \circ g,$$

and

$$\frac{\partial}{\partial \mathbf{a}} f[h_k; \mathbf{a}] \circ g = \frac{\partial}{\partial \mathbf{a}} \left\{ \bigvee_{i=1}^{n} \{f[h_k; \mathbf{a}]_i \wedge g(A_i)\} \right\}.$$

The max-min composition of n argument series returns a single element, say

$$f[h_k; a]_{i_o} \wedge g(A_{i_o}).$$

Based on that we derive

$$\frac{\partial}{\partial \mathbf{a}} \left\{ \bigvee_{i=1}^{n} \{f[h_k; \mathbf{a}]_i \wedge g(A_i)\} \right\} = \frac{\partial}{\partial \mathbf{a}} \{f[h_k; a]_{i_o} \wedge g(A_{i_o})\}. \qquad (9.18)$$

Note that in general

$$\frac{\partial}{\partial x}(x \wedge \text{constant}) = \frac{\partial}{\partial x} \begin{cases} \text{constant}, & \text{if } x > \text{constant} \\ x, & \text{if } x \leq \text{constant} \end{cases}$$

$$= \begin{cases} 0, & \text{if } x > \text{constant} \\ 1, & \text{if } x \leq \text{constant}, \end{cases}$$

where x and the constant are in $[0,1]$. Therefore (9.18) produces a non-zero derivative when $f[h_k; \mathbf{a}]_{i_0}$ is less than the fuzzy measure defined on A_{i_0}. We then obtain

$$\frac{\partial V}{\partial \mathbf{a}} = 2 \sum_{i=1}^{n} \left(\int_{\mathbf{X}} f[h_k; \mathbf{a}] \circ g - I_k \right) \frac{\partial}{\partial \mathbf{a}} f[h_k; \mathbf{a}]_{i_o}.$$

Further calculations are carried out based on the specific form of the function f. For instance, we may consider a two-parameter family of sigmoid functions,

$$f[h; m, \sigma] = \frac{1}{1 + \exp(-(h - m)/\sigma)},$$

where $m \in (0, 1]$ and $\sigma \in \mathbf{R}$. More explicitly, $\mathbf{a} = [m\ \sigma]$. The gradient is computed with respect to m and σ, respectively.

9.3   Chapter Summary

Fuzzy measures and fuzzy integrals were introduced and their interpretations provided in terms of an important notion useful in many applications, namely, information fusion and information aggregation. The familiar concept of probability measures was shown to be an instance of fuzzy measures and, as such, to satisfy their properties. A procedure was provided for parametric adjustment of fuzzy integrals to data.

9.4   Problems

1.   The following table summarizes the relevance (importance) levels of five sensors.

Sensor no.	1	2	3	4	5
Fuzzy density	0.20	0.70	0.55	0.30	0.45

(a) Construct the corresponding λ-fuzzy measure.

(b) What is the relevance of information from any two sensors that are active (that is, providing information)?

(c) Specify which combination of three sensor provides the most relevant information.

2.   A collection of sensors is used to acquire information about the failures of a system. Each sensor produces information about the degree of failure, quantified in the [0,1] interval. Assuming that the sensors' relevancies are the same as in problem 9.1, compute the level of failure for the following data:

Sensor no.	1	2	3	4	5
h	0.90	0.42	0.37	0.11	0.45

What is the level of failure if sensors 1 and 3 are considered to be inactive (turned off)? What is the maximal level of failure observed there?

3.   Three-class classification problems use four classifiers representing different characteristics. The table provides each classifier's relevance with respect to the first and third class; note that all the classifiers exhibit the same quality of classification (0.5) with respect to the second class:

	Class 1	Class 2	Class 3
Classifier 1	0.40	0.50	0.20
Classifier 2	0.30	0.50	0.50
Classifier 3	0.20	0.50	0.70
Classifier 4	0.60	0.50	0.40

Determine class membership for the pattern for which the individual classifiers produced the following results:

(a) Classifier 1:   [0.9   0.0   0.1]?
(b) Classifier 2:   [0.3   0.5   0.2]?
(c) Classifier 3:   [0.0   0.0   1.0]?
(d) Classifier 4:   [0.1   0.7   0.2]?

As the aggregation mechanism, consider both the Sugeno and Choquet integrals.

4.   Consider several fragments of an image, as shown in figure 9.3. Subjectively assign relevance levels to each of these elements and construct a fuzzy measure. Identify a three-element combination of the fragments of the image for which the fuzzy measure achieves the highest value.

Figure 9.3
Fragments of an image

5. The estimates of relevance for different sources of information are given as

Source no.	1	2	3	4	5	6
g	0.05	0.40	0.10	0.60	0.40	0.55

The estimates of the fuzzy integral for $h_1 = [0.5\ 0.2\ 0.3\ 0.0\ 0.0]$, $h_2 = [0.5\ 0.7\ 0.4\ 0.2\ 0.1]$, and $h_3 = [0.2\ 0.0\ 0.4\ 0.3\ 0.7]$ are equal to 0.87, 0.52, and 0.13, respectively. Compute the fuzzy integrals. If the results differ significantly from the experimental data, propose a nonlinear mapping of the function to be integrated and adjust its parameters by minimizing the sum of squared errors $V = \sum_{k=1}^{3} \left[\int h_k \circ g - I_k \right]^2$.

Note

1. By σ-algebra of **X**, we mean a family of sets of **X** that is closed under countable union and complement.

References

Choquet, G. 1953. Theory of capacities. *Ann. Inst. Fourier* 5:131–295.

Grabisch, M. 1995. Fuzzy integral in multicriteria decision making. *Fuzzy Sets and Systems* 69:279–98.

Grabisch, M., H. Nguyen, and E. Walker. 1995. *Fundamentals of Uncertainty Calculi with Applications to Fuzzy Inference*, Dordrecht, the Netherlands: Kluwer Academic Publishers.

Ishii, K., and M. Sugeno. 1985. A model of human evaluation process using fuzzy measure. *Int. J. Man-Machine Studies* 22:19–38.

Onisawa, T., M. Sugeno, Y. Nishiwaki, H. Kawai, and Y. Harima. 1986. Fuzzy measure analysis of public attitude towards the use of nuclear energy. *Fuzzy Sets and Systems* 20:259–89.

Sugeno, M. 1974. Theory of fuzzy integrals and its applications. Ph.D. dissertation, Tokyo Institute of Technology.

Sugeno, M. 1977. Fuzzy measures and fuzzy integrals—A survey. In *Fuzzy Automata and Decision Processes*, eds. M. Gupta, G. Saridis, and B. Gaines, 89–102. Amsterdam: North Holland.

Tahani, H., and J. Keller. 1990. Information fusion in computer vision using the fuzzy integral. *IEEE Trans. on Systems, Man, and Cybernetics* 20:733–41.

Weber, S. 1984. ⊥-decomposable measures and integrals for Archimedean t-conorms ⊥. *J. Mathematical Analysis and Applications* 101:114–38.

II COMPUTATIONAL MODELS

10 Rule-Based Computations

Rules provide a formal way of representing directives and strategies and are often appropriate when domain knowledge results from empirical associations or experience. Rule-based systems are build upon a set of rules and use a collection of facts to make inferences. In this chapter rules are first introduced as a knowledge representation scheme. Their syntax and semantics are described and reasoning procedures presented for performing rule-based computations. Issues concerning completeness and consistency of a given set of rules are addressed from a system design perspective.

10.1 Rules in Knowledge Representation

In computer science and engineering, the term "knowledge" means the information a computational system needs before it can behave intelligently. Because knowledge is important for intelligent behavior, the representation of knowledge is a central issue. Knowledge representation embraces all faculties aimed at organizing pieces of evidence necessary to capture domain knowledge. A manipulable representation is one useful for computation. Information contained in manipulable representations is accessible to other system entities (e.g., inference procedures or interfaces) that use the representation as part of a computation. This information may take the form of fuzzy facts or propositions like

The temperature is high,

If the temperature is high, then the flow is low,

If the flow is low, then valve opening is high.

The first example, *The temperature is high*, is an unconditional fuzzy proposition in which *temperature* is viewed as an attribute of an object and *high* is viewed as its value. The *if* parts of if-then rules are generically called antecedents, whereas the *then* parts are their consequents. Similar notations such as condition and conclusion (action) are also used. The antecedent and consequent are both fuzzy propositions that in turn may be formed by disjunctions or conjunctions, respectively, of fuzzy propositions. Fuzzy rules are conditioned fuzzy propositions. Note that the proposition in the consequent of the first statement above is also an antecedent of the second. In such a case, the rules are said to be chained. Otherwise, the rules are called parallel rules.

10.1.1 Qualified Propositions

An important knowledge representation issue concerns uncertainty. Not all knowledge is known with certainty. Knowledge may be vague, conflicting, or incomplete. Yet it is often still desirable to reason and make decisions using such knowledge. Intelligent systems should be able to cope well with uncertain knowledge. For instance, sometimes there is a degree of uncertainty about a proposition. This can be expressed by certainty-qualified propositions like

The temperature is high with μ certainty,

If the temperature is high, then the flow is low with γ certainty,

If the flow is low, then valve opening is high with λ certainty,

where μ, γ, and λ are numbers in the unit interval denoting the degrees of certainty for the respective fuzzy propositions. Propositions of this sort are certainty-qualified propositions. They can also be truth quantified, as in the following examples:

The temperature is high is true,

If the temperature is high, then the flow is low is almost true,

If the temperature is low, then valve opening is high is quite true.

In addition, fuzzy propositions can be qualified by associating with the propositions modal or intensional operators leading to possibility qualification and probability qualification, respectively. For instance,

The temperature is high is possible,

If the temperature is low then valve opening is low is impossible,

The temperature is medium is likely,

If the temperature is low, then valve opening is low is unlikely,

in which modal operators *possible* and *impossible* are the qualifying fuzzy possibilities, and the qualifying fuzzy probabilities are *likely* and *unlikely*, respectively. Propositions may be more generally qualified, as in

Usually the temperature is high in tropical countries,

Usually if the speed is high, then travel time is short,

in which *usually* is interpreted as a fuzzy proportion.

10.1.2   Quantified Propositions

By contrast, propositions can be quantified by fuzzy quantifiers such as *most*, *frequently*, *many*, *several*, and *about five*, as in

Most big cars are heavy,

Frequently, if the temperature is medium, then the comfort level is good,

in which the quantifiers may be interpreted as fuzzy numbers that provide a characterization of the constraints defined for the respective variables.

In many circumstances, inferences can still be made even if not all propositions in a rule's antecedent are satisfied. For instance, the following situation may arise:

If at least half of the conditions are satisfied, then infer rule consequent.

In such circumstances, the rules are called antecedent-quantified rules because antecedent satisfaction is constrained by a fuzzy quantifier (*at least half* in the example above).

10.1.3   Unless Rules

Rules may also have exceptions as in, for example,

If the car speed is high, then travel time is short, unless the road is blocked.

This type of rule, often referred to as an unless rule, is important when dealing with situations where information available is incomplete, constraints must be taken into account, or exceptions are likely.

10.1.4   Gradual Rules

In many instances, gradual relationships between objects, properties or concepts are also encountered. For example, it is often the case that

The more sophisticated the car is, the more expensive it is.

This statement can be regarded as a gradual rule expressing the progressive change of a car price as its degree of sophistication varies. Gradual rules are frequent in commonsense reasoning.

10.1.5   Potential Inconsistency and Conflicting Rules

Consider the following rule base.

If temperature is high, then flow is high,

If flow is high, then temperature is low.

Although no conflict exists, there is the potential for one if we get high temperature data. Such a set of rules is said to be potentially inconsistent and may contain mutually exclusive knowledge. Implemental and/or contradictory information in rule protocols may lead to unexpected and unsatisfactory results. A specific instance of potential inconsistency occurs when two or more rules have the same antecedent but diverse consequents, as in

If temperature is high, then flow is high,

If temperature is high, then flow is low.

Rules of this sort are referred to as conflicting rules and express contradictory knowledge.

10.1.6   Categorical and Dispositional Propositions

The various types of fuzzy rules introduced above can be broadly classified into two main groups: categorical propositions and dispositional propositions. Propositions containing no fuzzy quantifiers and no fuzzy probabilities are called categorical propositions; others are referred to as dispositional propositions, that is, propositions that are preponderantly, but not necessarily always, true (Zadeh 1989).

Generally, knowledge may be viewed as a collection of propositions in a language. For purposes of manipulable representation, each proposition is interpreted as a collection of fuzzy constraints or, equivalently, fuzzy relations. Representing propositions in manipulable representations involves the following steps:

Step 1: Identification of the variables whose values are constrained by the proposition. (Usually these variables are implicit rather than explicit in the propositions.)

Step 2: Identification of the constraints induced by the propositions.

Step 3: Characterization of each constraint by means of fuzzy relations.

Clearly, any choice concerning how to represent knowledge will depend on the type of problem to be solved and the inference procedure to be adopted. Other knowledge representation schemes are available, including first-order predicate logic, modal logics, fuzzy logic, frames, and associative networks.

10.2   Syntax of Fuzzy Rules

The basic information unit representing assertions and fuzzy rules is a proposition of the type

The (attribute) of (object) is (value).

For example, *The temperature* of *oven* is *high* and *The pressure* of *oven* is *low* are propositions that have this general form. The ideas of attribute and object can be combined in the concept of a variable. Thus in the examples above, the oven temperature and pressure can be considered variables. Therefore, the basic information units, or alternatively atomic propositions, can be written in a canonical form,

$p : X$ is A,

where X is a linguistic variable, that is, the attribute (object) pair, and A is its value. For instance, the meaning of the proposition *The temperature of oven is high* might be expressed as

The temperature of oven is high : *temperature*(*oven*) is *high*,

where *temperature*(*oven*) is the variable and *high* is its value. If we denote *temperature*(*oven*) by T and *high* by H, for short, we get

$p : T$ is H.

In general the variable values are linguistic labels attached to fuzzy relations in appropriate universes that play the role of fuzzy constraints on those relations. Therefore, variables in propositions can be viewed as fuzzy variables in which the meaning of the linguistic values is established by fuzzy relations.

Compound propositions can be constructed through conjunctions and/or disjunctions of propositions to form new propositions of the form

$p : X_1$ is A_1 and X_2 is $A_2, \ldots,$ and X_n is A_n,

$q : X_1$ is A_1 or X_2 is $A_2, \ldots,$ or X_n is A_n,

where $A_1, A_2, \ldots, A_n$ are fuzzy sets in the universes $\mathbf{X}_1, \mathbf{X}_2, \ldots, \mathbf{X}_n$, and $X_1, X_2, \ldots, X_n$ are variables. They induce fuzzy relations P and Q over $\mathbf{X}_1 \times \mathbf{X}_2 \times \cdots \times \mathbf{X}_n$ such that

$$P(x_1, \ldots, x_n) = \mathop{T}_{i=1}^{n} [A_i(x_i)],$$

$$Q(x_1, \ldots, x_n) = \mathop{S}_{i=1}^{n} [A_i(x_i)],$$

where T and S denote triangular norms and conorms, viewed as conjunctive and disjunctive operators, respectively. Thus propositions p and q can be expressed as

$p : (X_1, X_2, \ldots, X_n)$ is P,

$q : (X_1, X_2, \ldots, X_n)$ is Q.

Accordingly, qualified and quantified propositions can be expressed by adding the qualifying or the quantifying labels. For instance, certainty-qualified propositions may be written as

$p : X$ is A with certainty μ.

In particular, qualified propositions may be transformed into propositions of the form X is B with implied certainty 1. If A and B are fuzzy sets of $\mathbf{X}$, then the proposition X is B is such that (Yager 1984)

$B(x) = [\mu \text{ t } A(x)] + (1 - \mu)$.

EXERCISE 10.1   Assume X is A with certainty 1. Show that in this case $B = A$. Otherwise, show that if X is A with certainty 0, then $B = \mathbf{X}$.

It can be shown (see problem 1 of this chapter) that the act of qualifying a proposition with a certainty reduces the specificity of its unqualified equivalent.

Typically, a fuzzy rule has the general format of a conditioned proposition,

$p :$ If antecedent, then consequent,

where antecedent and consequent are fuzzy propositions. For instance

$p :$ If X_1 is A_1 and X_2 is $A_2, \ldots,$ and X_n is $A_n,$ then Y_1 is B_1 and Y_2 is $B_2, \ldots,$ and Y_m is $B_m,$

$p :$ If X_1 is A_1 or X_2 is $A_2, \ldots,$ or X_n is $A_n,$ then Y_1 is B_1 or Y_2 is $B_2, \ldots,$ or Y_m is $B_m,$

where $A_1, A_2, \ldots, A_n$ and $B_1, B_2, \ldots, B_m$ are fuzzy sets in the universes $\mathbf{X}_1, \mathbf{X}_2, \ldots, \mathbf{X}_n$ and $\mathbf{Y}_1, \mathbf{Y}_2, \ldots, \mathbf{Y}_m,$ respectively, and $X_1, X_2, \ldots, X_n$ and $Y_1, Y_2, \ldots, Y_m$ are variables. Rules p and q are seen to induce relations P and Q on $\mathbf{X}_1 \times \mathbf{X}_2 \times \cdots \times \mathbf{X}_n \times \mathbf{Y}_1 \times \mathbf{Y}_2 \times \cdots \times \mathbf{Y}_m$ such that

$P(x_1, x_2, \ldots, x_n, y_1, y_2, \ldots, y_m) = f(P_a(x_1, x_2, \ldots, x_n), P_c(y_1, y_2, \ldots, y_m)),$

$Q(x_1, x_2, \ldots, x_n, y_1, y_2, \ldots, y_m) = f(Q_a(x_1, x_2, \ldots, x_n), Q_c(y_1, y_2, \ldots, y_m)),$

where f can be an implication operator or a triangular norm, and P_a, P_c, Q_a, and Q_c are the relations induced by the antecedent and consequent of rules p and q, respectively. Therefore, rules can be expressed as the propositions

$p : (X_1, X_2, \ldots, X_n, Y_1, Y_2, \ldots, Y_m)$ is P,

$q : (X_1, X_2, \ldots, X_n, Y_1, Y_2, \ldots, Y_m)$ is Q.

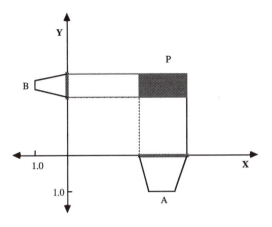

Figure 10.1
Relation induced by an if-then rule

In the simplest case, $n = m = 1$, and it will be frequently sufficient to focus on this case. Therefore the rule "If X is A, then Y is B" can be understood as the proposition $p : (X, Y)$ is P, where $P(x, y)$ is a fuzzy relation on $\mathbf{X} \times \mathbf{Y}$, as figure 10.1 illustrates.

It should be noted that certainty qualification can easily be applied to rules. Consider the rule

p : If X is A and Y is B, then Z is C with certainty μ.

If $Q(x, y, z)$ is a fuzzy relation of $\mathbf{X} \times \mathbf{Y} \times \mathbf{Z}$, the certainty-qualified rule transforms into its unqualified equivalent with implied certainty 1:

$q : (X, Y, Z)$ is Q,

where

$$Q(x, y, z) = [P(x, y, z) \ \mathrm{t} \ \mu] + (1 - \mu).$$

Like the other types of rules, antecedent-quantified rules also induce fuzzy relations, the essential difference lying in the determination of the relation $P_a(x_1, x_2, \ldots, x_n)$ due to the antecedent. Before we present a method for determining P_a, we recall that a linguistic quantifier can be expressed as a fuzzy set. In particular, two kinds of quantifiers are of interest: absolute quantifiers, such as *about seven* and *at least four*, and relative quantifiers, such as *almost all* and *at least half*. Absolute quantifiers can be represented by a fuzzy set of the nonnegative reals, whereas relative quantifiers can be expressed as a fuzzy set of the unit interval. Let Q be either an absolute or relative quantifier, defined in an appropriate universe $\mathbf{X}$. Thus Q is said to be monotonically nondecreasing if for any

$x_1, x_2 \in \mathbf{X}$ such that $x_1 > x_2$, then $\mathbf{Q}(x_1) \geq \mathbf{Q}(x_2)$. The method described below for determining P_a is restricted to monotonically nondecreasing quantifiers (Yager 1984); refer also to chapter 7.

For any point $(x_1, x_2, \ldots, x_n) \in \mathbf{X}_1 \times \mathbf{X}_2 \times \cdots \times \mathbf{X}_n$ where $\mathbf{X}_i$ is the universe of A_i, $P_a(x_1, x_2, \ldots, x_n)$ is obtained as follows. Let $D(x_1, x_2, \ldots, x_n) = \{A_1(x_1), A_2(x_2), \ldots, A_n(x_n)\}$ and $D_i(x_1, x_2, \ldots, x_n)$ the ith largest element in the set $D(x_1, x_2, \ldots, x_n)$. Thus for any absolute quantifier Q

$$P_a(x_1, x_2, \ldots, x_n) = \max_{i=1,\ldots,n} [Q(i) \wedge D_i(x_1, x_2, \ldots, x_n)].$$

If Q is a relative quantifier, we replace $\mathbf{Q}(i)$ by $\mathbf{Q}(i/n)$. Having obtained $P_a(x_1, x_2, \ldots, x_n)$ for every $(x_1, x_2, \ldots, x_n) \in \mathbf{X}_1 \times \mathbf{X}_2 \times \cdots \times \mathbf{X}_n$, we define

$$P(x_1, x_2, \ldots, x_n, y_1, y_2, \ldots, y_m) = f(P_a(x_1, x_2, \ldots, x_n), P_c(y_1, y_2, \ldots, y_m)),$$

which becomes the induced relation from the antecedent-quantified rule

If $\mathbf{Q}(X_1$ is A_1, X_2 is $A_2, \ldots, X_n$ is $A_n)$, then Y_1 is B_1 and Y_2 is $B_2, \ldots$, and Y_m is B_m.

EXERCISE 10.2   Extend the approach for handling antecedent-quantified rules to the case where both antecedent and consequent are quantified. Can the method be generalized using triangular norms and conorms instead of min and max?

When considering chaining of fuzzy rules, certain consistency conditions may be required; that is, the rules

If X is A, then Y is B

If Y is B, then Z is C

If X is A, then Z is C

must be compatible. If this is the case, the last rule can be deduced from the other two, and the syllogism is said to hold. In other words, if $P(x, y)$, $Q(y, z)$ and $R(x, z)$ are the fuzzy relations induced by the rules, if the following equality is satisfied, then the syllogism holds:

$$R = P \circ Q,$$

where $\circ$ is the composition operator. Generally, two chained rules will not be equivalent to a single rule.

A collection of N fuzzy rules of the form

If X is A_i, then Y is B_i, $i = 1, \ldots, N,$

composes a collection of parallel rules. Parallel rules are most commonly

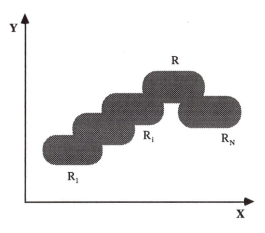

Figure 10.2
Relation induced by a collection of parallel rules

interpreted by viewing each rule as inducing a fuzzy relation R_i, and the set of rules as a fuzzy relation that is an aggregation of the individual relations. Therefore, the overall relation is given by

$$R = \mathop{A}_{i=1}^{N} R_i,$$

where A is an aggregation operator typically characterized by an s-norm (figure 10.2), t-norms and averaging operators being common alternatives.

Now assume that there is an exception to the rule "If X is A, then Y is B," that is, given that "X is A" holds, one cannot expect that Y is B when it is known that another proposition, "Z is C," also holds. As discussed earlier, this is expressed by an unless rule of the form

If X is A, then Y is B, unless Z is C.

One way to treat this case is to consider the three rules separately:

If X is A and Z is $\bar{C}$, then Y is B,

If X is A and Z is C, then Y is $\bar{B}$,

If X is A and Z is Z, then Y is B.

Therefore, depending on which of the antecedent propositions holds, the corresponding rule can be considered. When the number of exceptions to be described is large, however, it may be desirable to find a single relation $R(x, y, z)$ that represents the above three rules together (Driankov and

Hellendorn 1992). Whenever such a relation can be found, it enhances computational performance. The existence of $R(x, y, z)$ depends on the meaning attached to the rules, that is, rule semantics, and on the type of aggregation that must be performed to get a single relation from the individual ones. Section 10.3 discusses this issue and a solution to this case in more detail.

Let us turn to the issue of representation of gradual rules by means of fuzzy sets (Dubois and Prade 1992), that is, syntax of gradual rules. Assume two fuzzy sets A and B in the universes $\mathbf{X}$ and $\mathbf{Y}$, respectively. Again, the gradual rule

The more X is A, the more Y is B

establishes a relationship $R(x, y)$ in $\mathbf{X} \times \mathbf{Y}$ between variables X and Y. However $R(x, y)$ must be defined in such a way that the constraint

$$B(y) \geq A(x), \quad \forall x \in \mathbf{X} \text{ and } y \in \mathbf{Y},$$

is taken into account. This is a natural requirement, as the following example demonstrates. Suppose A and B are fuzzy sets of $\mathbf{X} = \mathbf{Y}$ and $A \subseteq B$. Clearly $B(x) \geq A(x)$, $\forall x \in \mathbf{X}$. Thus fuzzy set inclusion is equivalent to the gradual rule, that is, $\forall x \in \mathbf{X}$,

The more x is A, the more x is B.

Several syntactic forms may be devised for the representation of fuzzy rules. Using the BNF notation, we employ the following syntax to summarize the classes of rules just discussed.

⟨if_then_rule⟩ ::= if ⟨proposition⟩ then ⟨proposition⟩

⟨unless_rule⟩ ::= ⟨if_then_rule⟩ unless ⟨proposition⟩

⟨gradual_rule⟩ ::= the ⟨word⟩⟨proposition⟩ the ⟨word⟩⟨proposition⟩

⟨proposition⟩ ::= ⟨disjunction⟩ {and⟨disjunction⟩}

⟨disjunction⟩ ::= ⟨variable⟩ {or⟨variable⟩}

⟨variable⟩ ::= ⟨attribute⟩ is ⟨value⟩

⟨word⟩ ::= more | less

10.3   Semantics of Fuzzy Rules and Inference

Essentially, statements of the form "If X is A, then Y is B" describe a relation between the fuzzy variables X and Y. This suggests that a fuzzy rule can be defined as a fuzzy relation R, with the membership grade $R(x, y)$ representing the degree to which $(x, y) \in \mathbf{X} \times \mathbf{Y}$ is compatible with the relation between the variables X and Y involved in the given rule. If A and B are fuzzy sets of $\mathbf{X}$ and $\mathbf{Y}$, respectively, then the relation R on $\mathbf{X} \times \mathbf{Y}$ can be determined by the relational assignment equation

$$R(x, y) = f(A(x), B(y)), \quad \forall (x, y) \in \mathbf{X} \times \mathbf{Y},$$

where f is a function of the form $f : [0, 1]^2 \to [0, 1]$. In general, the fuzzy relations induced are derived from three main classes of f functions: fuzzy conjunction, fuzzy disjunction, and fuzzy implication, with conjunction and implication being the most common. Fuzzy conjunction (f_t) and fuzzy disjunction (f_s) may be viewed as dual generalizations of the fuzzy Cartesian product via triangular norms and co-norms, whereas fuzzy implication (f_i) may be associated with generalization of implications in multiple-valued logic.

DEFINITION 10.1   A fuzzy conjunction is a function $f_t : [0, 1]^2 \to [0, 1]$ defined by

$$f_t(A(x), B(y)) = A(x) \text{ t } B(y), \quad \forall (x, y) \in \mathbf{X} \times \mathbf{Y}.$$

Typical examples of fuzzy conjunction include

$$f_c(A(x), B(y)) = A(x) \wedge B(y), \quad \forall (x, y) \in \mathbf{X} \times \mathbf{Y}, \qquad \text{(Mamdani)}$$

$$f_p(A(x), B(y)) = A(x) . B(y), \quad \forall (x, y) \in \mathbf{X} \times \mathbf{Y}, \qquad \text{(Larsen)}$$

where the min operator $\wedge$ and the algebraic product . are taken as the respective triangular norms.

DEFINITION 10.2   A fuzzy disjunction is a function $f_s : [0, 1]^2 \to [0, 1]$ defined as

$$f_s(A(x), B(y)) = A(x) \text{ s } B(y), \quad \forall (x, y) \in \mathbf{X} \times \mathbf{Y}.$$

DEFINITION 10.3   A fuzzy implication is a function $f_i : [0, 1]^2 \to [0, 1]$ with the following properties:

1. Monotonicity in the second argument: $B(y_1) \le B(y_2)$ implies $f_i(A(x), B(y_1)) \le f_i(A(x), B(y_2))$.
2. Dominancy of falsity: $f_i(0, B(y)) = 1$.
3. Neutrality of truth: $f_i(1, B(y)) = B(y)$.

Two additional properties are usually added:

4. Monotonicity in the first argument: $A(x_1) \le A(x_2)$ implies $f_i(A(x_1), B(y)) \ge f_i(A(x_2), B(y))$.
5. Exchange: $f_i(A(x_1), f_i(A(x_2), B(y))) = f_i(A(x_2), f_i(A(x_1), B(y)))$.

Further properties that can be also considered include

6. Identity: $f_i(A(x), A(x)) = 1$.
7. Boundary conditions: $f_i(A(x), B(y)) = 1$ if and only if $A(x) \leq B(y)$.
8. Contraposition: $f_i(A(x), B(y)) = f_i(\bar{B}(y), \bar{A}(x))$.
9. Continuity: $f_i(A(x), B(y))$ is a continuous function.

Actually, most known fuzzy implications do not fulfill all these properties, with the exception of the Lukasiewicz implication (f_l), where $\forall (x, y) \in \mathbf{X} \times \mathbf{Y}$,

$$f_l(A(x), B(y)) = \min[1, 1 - A(x) + B(y)],$$

and its parameterized forms,

$$f_\lambda(A(x), B(y)) = \min\left[\frac{1, 1 - A(x) + (\lambda + 1)B(y)}{1 + \lambda A(x)}\right], \quad \lambda > -1,$$

$$f_w(A(x), B(y)) = \min[1, (1 - A(x)^w + B(y)^w)^{1/w}], \quad w > 0.$$

Further examples of fuzzy implications include

1. $f_a(A(x), B(y)) = \begin{cases} 1, & \text{if } A(x) \leq B(y) \\ 0, & \text{if } A(x) > B(y). \end{cases}$ (Gaines)

2. $f_g(A(x), B(y)) = \begin{cases} 1, & \text{if } A(x) \leq B(y) \\ B(y), & \text{if } A(x) > B(y). \end{cases}$ (Gödel)

3. $f_\Delta(A(x), B(y)) = \begin{cases} 1, & \text{if } A(x) \leq B(y) \\ \dfrac{B(y)}{A(x)}, & \text{if } A(x) > B(y). \end{cases}$ (Goguen)

4. $f_b(A(x), B(y)) = \max[1 - A(x), B(y)]$. (Kleene)
5. $f_r(A(x), B(y)) = 1 - A(x) + A(x)B(y)$. (Reichenbach)
6. $f_k(A(x), B(y)) = 1 - A(x) + A(x)^2 B(y)$. (Klir-Yuan)
7. $f_m(A(x), B(y)) = \max[1 - A(x), \min(A(x), B(y))]$. (Zadeh)

Implication can be broadly classified into two categories, S-implications and R-implications, characterized by the following definitions (Klir and Yuan 1995).

DEFINITION 10.4   An S-implication is defined as

$$f_{is}(A(x), B(y)) = \bar{A}(x) \text{ s } B(y), \quad \forall (x, y) \in \mathbf{X} \times \mathbf{Y}.$$

This category of implication arises from the formalism of classical logic $p \rightarrow q = \bar{p} \vee q$ and possesses properties one through five of fuzzy implications. The Lukasiewicz, Kleene, and Reichenbach implications are typical examples.

DEFINITION 10.5   An R-implication is obtained by residuation of a continuous triangular norm t,

$$f_{ir}(A(x), B(y)) = \sup_{c \in [0,1]} [A(x) \, t \, c \leq B(x)], \quad \forall (x, y) \in \mathbf{X} \times \mathbf{Y}.$$

This class of implication functions arises from the formalism of intuitionist logic and possesses properties one through five of fuzzy implications. The Lukasiewicz, Gödel, and Goguen implications are well-known examples.

EXERCISE 10.3   The following identities are semantically equivalent within classical propositional logic: $p \rightarrow q = \bar{p} \vee q = \bar{p} \vee (p \wedge q) = (\bar{p} \wedge \bar{q}) \vee q$. Based on these identities, suggest additional ways to assign a fuzzy relation with a rule. What conditions should be fulfilled for this purpose? Are they equivalent from the point of view of fuzzy set theory?

At this point let us emphasize that fuzzy rules represent functional dependencies and relations. In particular, expressing the meaning of the rules through fuzzy conjunctions brings us to the idea of granular representation of functional and relational dependencies by fuzzy graphs. To introduce this idea and related concepts, recall that we express the meaning of the rule "If X is A, then Y is B" as a fuzzy relation induced by a fuzzy constraint on the joint variable (X, Y). More specifically, if the fuzzy constraint is understood as a fuzzy conjunction, that is, a generalization of the Cartesian product concept, then the if-then rule becomes "(X, Y) is $A \times B$," where $A \times B$ is the Cartesian product of A and B whose membership function $(A \times B)(x, y)$ is given by

$$(A \times B)(x, y) = A(x) \, t \, B(y), \quad \forall (x, y) \in \mathbf{X} \times \mathbf{Y}.$$

As suggested by the min t-norm, $A \times B$ may be viewed as a fuzzy point (granule or patch) in the space $\mathbf{X} \times \mathbf{Y}$, as figure 10.3 shows (Zadeh 1975, 1994a; Kosko 1994).

In the case of a collection of parallel rules "If X is A_i, then Y is B_i," $i = 1, \ldots, N$, the meaning may be defined as

$$(X, Y) \text{ is } (A_1 \times B_1 + A_2 \times B_2 + \cdots + A_N \times B_N)$$

or, equivalently,

$$(X, Y) \text{ is } \left(\sum_{i=1}^{N} A_i \times B_i \right).$$

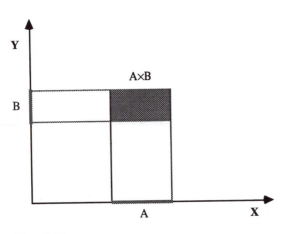

Figure 10.3
Fuzzy point or granule in $\mathbf{X} \times \mathbf{Y}$

The expression $(\sum_{i=1}^{N} A_i \times B_i)$ is interpreted as an aggregation, via disjunction in this case, of fuzzy points composing an overall relation F^* induced by the N rules, as shown in figure 10.4. It represents a coarse characterization of the dependency between the fuzzy variables X and Y in $\mathbf{X} \times \mathbf{Y}$, and for this reason it is called a fuzzy graph. The concept of a fuzzy graph is in close analogy with its conventional counterpart, that is, the graph of a nonfuzzy relation or function.

More generally, the meaning of each rule in a collection is defined by a relation R_i through any of the appropriate functions, the collection in turn being represented by "(X, Y) is R," with $R = A_{i=1}^{N} R_i$. Therefore, an essential issue in designing rule-based fuzzy systems concerns the task of assigning appropriate semantics to the rules, which involves consideration of practical issues and fundamental requirements. Let $y = f(x)$ be a function $f : \mathbf{X} \to \mathbf{Y}$. Recall that the graph of f is the set $F = \{(x, y) \mid y = f(x), x \in \mathbf{X}, y \in \mathbf{Y}\}$, as depicted in figure 10.5(a). As a generalization of this concept, a fuzzy graph F^* of a functional dependency $f : \mathbf{X} \to \mathbf{Y}$ between the fuzzy variables X and Y in $\mathbf{X}$ and $\mathbf{Y}$, respectively, is defined (Zadeh 1994a, 1994b) as an approximate, granular representation of f in the form $F^* = (\sum_{i=1}^{N} A_i \times B_i)$, as depicted in figure 10.5(b). In general terms, a fuzzy graph is a fuzzy set F^* whose membeship function is

$$F^*(x, y) = \underset{i=1}{\overset{N}{S}} [A_i(x) \text{ t } B_i(x)], \quad \forall (x, y) \in \mathbf{X} \times \mathbf{Y}.$$

Now assume we are given a function $y = f(x)$ and a single value for x, say $x = a$. Then from $y = f(x)$ and $x = a$ we can compute $y = f(a) = b$.

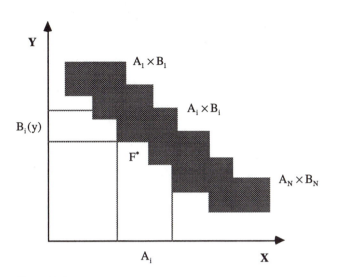

Figure 10.4
Fuzzy graph

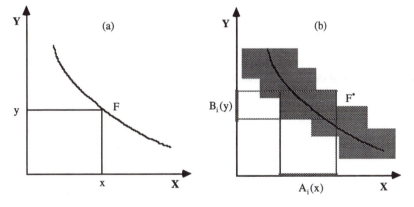

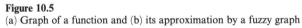

Figure 10.5
(a) Graph of a function and (b) its approximation by a fuzzy graph

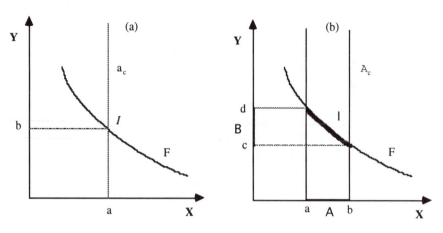

Figure 10.6
Computing function values with operations on sets

This familiar procedure, illustrated in figure 10.6(a), can be symbolically written as

$$\frac{\begin{array}{l} x = a \\ y = f(x) \end{array}}{y = b}$$

and involves the following steps:

1. A vertical line a_c is drawn on the $\mathbf{X} \times \mathbf{Y}$ plane, starting at the point $x = a$. In other words, the cylindrical extension a_c of x with base a is determined.

2. The intersection I of the cylindrical extension a_c with the graph F of f, is found.

3. The intersection point I in $\mathbf{X} \times \mathbf{Y}$ is projected onto $\mathbf{Y}$ by drawing a horizontal line from I. If x is a set $\mathbf{A} \subseteq \mathbf{X}$, for example, an interval $\mathbf{A} = [a, b]$, we find by repeating the procedure above for all points in $[a, b]$ that the corresponding ys are points of the interval $\mathbf{B} = f(\mathbf{A}) = [c, d]$, as shown in figure 10.6(b), where $\mathbf{A}_c \subseteq \mathbf{X} \times \mathbf{Y}$ is the cylindrical extension of $\mathbf{A}$.

When we are given a statement of the form "x is $\mathbf{A}$," from (x, y) is F (here $\mathbf{A} \subseteq \mathbf{X}$ and F is a relation, $F \subseteq \mathbf{X} \times \mathbf{Y}$), we still can infer that y is $\mathbf{B}, \mathbf{B} \subseteq \mathbf{Y}$, following the very same procedure, as shown in figure 10.7(a). Note that the computing scheme in this case involves sets and retains the symbolic form

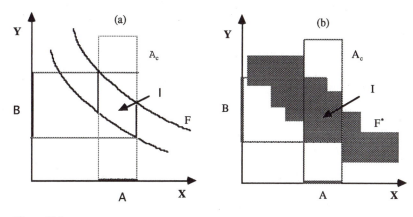

Figure 10.7
Composition operation with relations

x is **A**

(x, y) is F

y is **B**.

In the more general case when a collection of fuzzy rules is interpreted as a functional dependency F^* between the fuzzy variables X and Y, the problem of computing the value of Y given a value of X can be also expressed as the inference scheme

X is A

(X, Y) is F^*

Y is B.

Therefore, if the operations involved in computing B are understood within the fuzzy set theory, inference can clearly be pursued analogously to above. Thus the problem of assigning a linguistic value A to X and computing the corresponding value of Y can be summarized as follows:

1. Find the cylindrical extension A_c with base A.

2. Find I, the intersection of A_c with F^*.

3. Project I onto **Y** to get B.

These steps are illustrated in figure 10.7(b). Actually, this procedure to compute B is the essence of the compositional rule of inference introduced by Zadeh (1975, 1988).

The compositional rule of inference also applies when interpreting each fuzzy rule of a given collection as a fuzzy relation $R_i, i = 1, \ldots, N$, induced by, for example, any of the fuzzy implication functions. The meaning of

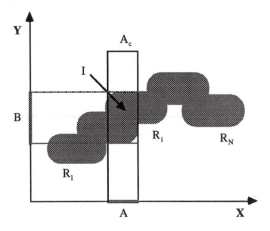

Figure 10.8
Compositional rule of inference

the collection is "(X, Y) is $(\sum_{i=1}^{N} A_i \times B_i)$," or "$(X, Y)$ is R" for short, and the compositional rule of inference reads

X is A

(X, Y) is R

Y is B.

As discussed above, B is found by projecting $I = A_c \cap R$ on to **Y**. In other words

$B = \mathrm{Proj}_Y(A_c \cap R),$

as shown in figure 10.8. Recalling that the intersection operator can be any triangular norm t and that projection is accomplished by the sup operation, the membership function of the inferred fuzzy set B is

$$B(y) = \sup_x [A_c(x) \text{ t } R(x, y)] = \sup_x [A(x) \text{ t } R(x, y)].$$

Thus it is evident that $B = A \circ B$, and the inference pattern becomes

X is A

(X, Y) is R

Y is $A \circ R$.

At this point it becomes obvious that a central point in designing fuzzy rule–based systems concerns the task of assigning appropriate semantics to the rules, involving consideration of fundamental issues as discussed above. We elaborate further, considering a semantics for unless and gradual rules.

10.3.1   Semantics of Unless Rules

An unless rule is a rule in which an exception condition is added. It has the form

If X is A, then Y is B, unless Z is C,

whose interpretation is a ternary relation $R(x, y, z)$. Let us denote by $R_{\bar{c}}, R_c,$ and R_z the relations induced by the following three rules (Driankov and Hellendorn 1992; Dogherty, Driankov, and Hellendorn 1993):

If X is A and Z is $\bar{C}$, then Y is B,

If X is A and Z is C, then Y is $\bar{B}$,

If X is A and Z is $\mathbf{Z}$, then Y is B.

In this case, if it reasonable, if $R = R_{\bar{c}} \cap R_c \cap R_z$, to make the following inferences:

$$(A(x) \wedge C(z)) \circ R(x, y, z) = \bar{B}(y),$$

$$(A(x) \wedge \bar{C}(z)) \circ R(x, y, z) = B(y),$$

$$(A(x) \wedge Z(z)) \circ R(x, y, z) = B(y).$$

In general, however, these inferences do not hold if the fuzzy relations $R_{\bar{c}},$ $R_c,$ and R_z are defined by *any* implication. Nevertheless, if A, B, and C are normal, and if $R_{\bar{c}}$ and R_c are based on f_g, namely,

$$R_{\bar{c}}(x, y, z) = \begin{cases} 1, & \text{if } \min[A(x), \bar{C}(x)] \leq B(y) \\ B(y), & \text{otherwise,} \end{cases}$$

$$R_c(x, y, z) = \begin{cases} 1, & \text{if } \min[A(x), C(x)] \leq \bar{C}(y) \\ \bar{B}(y), & \text{otherwise,} \end{cases}$$

and $R = R_{\bar{c}} \cap R_c$, then the two first inferences do hold but the third does not, because even if we set $R = R_{\bar{c}} \cap R_c \cap R_z$, then $[A(x) \wedge Z(z)] \circ R(x, y, z) \neq B(y)$. But if we let

$$\tilde{A}(x) = \begin{cases} 0, & \text{if } \min[A(x_i) \wedge Z(z_k)), r_{ijk}] \geq B(y_j) \\ A(x) \wedge Z(z), & \text{otherwise,} \end{cases}$$

where r_{ijk} is the (i, j, k)th element of R, then we get $\tilde{A}(x, z) \circ R(x, y, z) = B(y)$, with $R = R_{\bar{c}} \cap R_c$, which can be regarded as an alternative for the third inference. Under the assumptions just discussed, then, a semantics for unless rules can be formulated.

10.3.2   Semantics of Gradual Rules

To complete our discussion, consider now gradual rules of the form

The more X is A, the more Y is B,

or rules of similar form that appear when "less" is substituted for one or both occurrences of "more." Given the fuzzy sets A and B on $\mathbf{X}$ and $\mathbf{Y}$, a gradual rule translates into the constraint (Dubois and Prade 1991, 1992)

$$B(y) \geq A(x), \quad \forall x \in \mathbf{X} \text{ and } \forall y \in \mathbf{Y},$$

which defines a relation between the variables involved. Clearly, the constraint expresses the idea behind the rule: Whenever the degree of membership of x in A increases, the degree of membership of y in B also increases. Actually, the constraint associates to each value of x a subset $B_{R(x)} = \{y \in \mathbf{Y} \mid B(y) \geq A(x)\}$ of values y viewed as possible when x is given, as figure 10.9 illustrates. Therefore we can express the meaning of gradual rules by means of a relation defined by

$$R(x, y) = \begin{cases} 1, & \text{if } B(y) \geq A(x) \\ 0, & \text{otherwise.} \end{cases}$$

Clearly R is, in this case, a relation equal to f_a, but in general other choices of implication functions can be made; for example f_g or f_Δ.

EXERCISE 10.4   Assume the following rules:

(a) The more X is A and the more Y is B, the more Z is C.

(b) The more X is A or the more Y is B, the more Z is C.

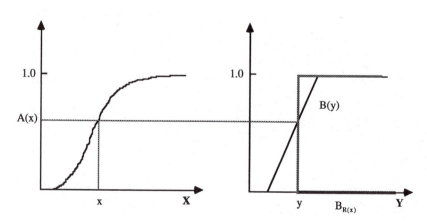

Figure 10.9
Constraint associated with a gradual rule

In these rules, the conjunction and disjunction are, for instance, the min and max operators, respectively. What constraints do they imply? Develop a semantics by choosing appropriate relations to represent the rules.

Aggregation of a collection of gradual rules should be treated carefully. For instance, the disjunction of two gradual rules,

The more X is A, the more Z is C,

The more Y is B, the more Z is C,

entails the rule "The more X is A and the more Y is B, the more Z is C," but the converse is not ture. (See problem 4 at the end of the chapter).

If the gradual rule "The more X is A, the less Y is B" is understood as "the more X is A, the more Y is $\bar{B}$," with $\bar{B}(y) = 1 - B(y)$, then it can be represented by the constraint

$$1 - B(y) \geq A(x), \quad \forall x \in \mathbf{X} \text{ and } \forall y \in \mathbf{Y}.$$

EXERCISE 10.5   Which constraint would you suggest to represent, assuming the complement as one-complement, the rule "The less X is A, the more Y is B"?

After an interpretation for gradual rules has been defined, inference proceeds according the compositional rule of inference. More specifically, given the propositions

X is A',

The more X is A, the more Y is B,

we infer that Y is B', where $B' = A' \circ R$. If R is defined as above, that is, via f_a, then the compositional rule of inference becomes (see problem 5 at the end of the chapter)

$$B'(y) = \sup_{u \in B_{R(x)}} A'(u).$$

A detailed computational procedure is provided in section 10.4.

10.4   Computing with Fuzzy Rules

Fuzzy rule–based computing is prompted by rule semantics and the compositional rule of inference. Computations can be greatly simplified, however, if we make certain assumptions concerning rule aggregation operators, the intersection operation in the compositional rule, and rule semantics. Naturally, this has great impact in practical applications and for this reason, the most important assumptions and the respective inference patterns are presented next.

Assume a collection of N fuzzy rules of the form

If X is A_i and Y is B_i, then Z is C_i,

denoted symbolically by "A_i and $B_i \rightarrow C_i$," where A_i, B_i, and C_i, $i = 1, \ldots, N$, are fuzzy sets in the universes $\mathbf{X}$, $\mathbf{Y}$, and $\mathbf{Z}$, respectively. Suppose that rule aggregation is performed via the max operator, and that either the min or algebraic product is chosen as a t-norm in the compositional rule of inference. Thus if a proposition "X is A and Y is B" is given, the value of C inferred from the complete collection of the N rules is equivalent to the aggregated value derived from the individual rules. More formally, if the proposition is $P(x, y) = A(x) \, \mathrm{t} \, B(y)$, that is, $P = A \times B$, then

$$C = \bigcup_{i=1}^{N} C_i',$$

$$C_i' = (A \times B) \circ (A_i \text{ and } B_i \rightarrow C_i) = (A \times B) \circ R_i.$$

In addition, if each R_i is defined as either $R_i = f_c$ or $R_i = f_p$, then under sup-min composition, with $(A_i \text{ and } B_i) \, (x, y) = A_i(x) \wedge B_i(y)$,

$$C_i' = (A \circ (A_i \rightarrow C_i)) \cap (B \circ (B_i \rightarrow C_i)), \quad \text{if } P(x, y) = A(x) \wedge B(y).$$

In particular, when the inputs A and B are pointwise, computations can be made even simpler. For instance, if $m_i = P(a, b) = A_i(a) \wedge B_i(b)$, $a \in \mathbf{X}$ and $b \in \mathbf{Y}$, then the following holds:

$$C(z) = \max[m_i \wedge C_i(z), i = 1, \ldots, N], \quad \forall z \in \mathbf{Z},$$

when rules are interpreted according to f_c, and

$$C(z) = \max[m_i \cdot C_i(z), i = 1, \ldots, N], \quad \forall z \in \mathbf{Z},$$

when rules are interpreted according to f_p. The numbers m_i and n_i measure the contribution of the ith rule to the overall inference, and for this reason they are often called activation degree or firing strength. For the particular cases above, simple geometric interpretations can be provided as shown, for rules i and j, in figures 10.10 and 10.11.

In the general case, the composition rule dictates the steps to be performed in computing. For instance, assuming andwise atomic propositions in the rule's antecedent and consequents (analogous steps can be readily devised for the orwise case, remembering that *or* operators are s-norms) the reasoning scheme is, for $k = 1, \ldots, N$,

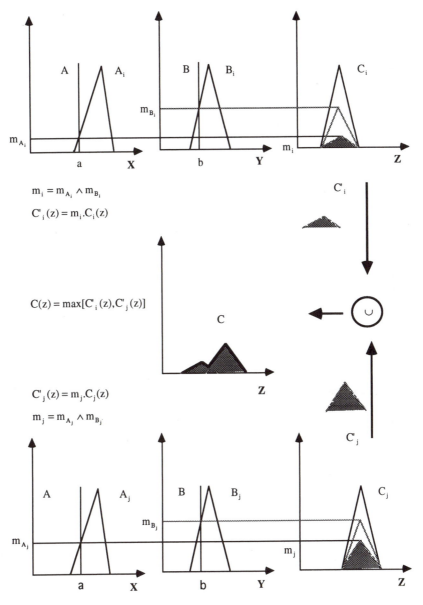

$$m_i = m_{A_i} \wedge m_{B_i}$$

$$C'_i(z) = m_i.C_i(z)$$

$$C(z) = \max[C'_i(z), C'_j(z)]$$

$$C'_j(z) = m_j.C_j(z)$$

$$m_j = m_{A_j} \wedge m_{B_j}$$

Figure 10.10
Fuzzy reasoning with min in P, sup-mim composition, and f_p

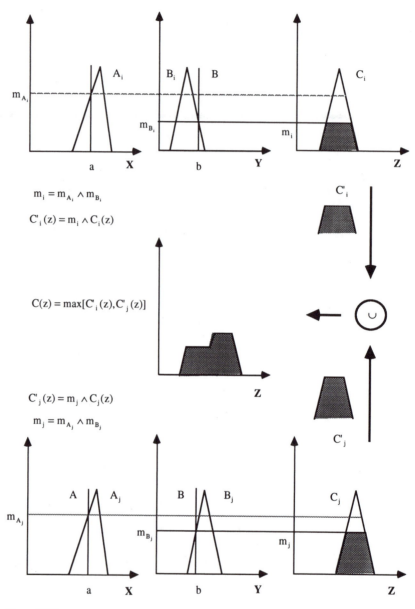

$$m_i = m_{A_i} \wedge m_{B_i}$$

$$C'_i(z) = m_i \wedge C_i(z)$$

$$C(z) = \max[C'_i(z), C'_j(z)]$$

$$C'_j(z) = m_j \wedge C_j(z)$$

$$m_j = m_{A_j} \wedge m_{B_j}$$

Figure 10.11
Fuzzy reasoning with min in P, sup-min composition, and f_c

- Input proposition: X_1 is A_1 and X_2 is $A_2 \ldots$ and X_n is A_n.
- Rules: If X_1 is A_1^k and X_2 is $A_2^k \ldots$ and X_n is A_n^k, then Y_1 is B_1^k and Y_2 is $B_2^k \ldots$ and Y_m is B_m^k.
- Inferred proposition: Y_1 is B_1 and Y_2 is $B_2 \ldots$ and Y_m is B_m.

The steps are as follows:

Step 1: Choose an appropriate t-norm to define the *and* conjunction, an appropriate f function to define each rule k meaning, an appropriate t-norm for the compositional rule of inference, and an aggregation operator for rule combination.

Step 2: For $k = 1, \ldots N$, compute

$$P_{ak}(x_1, x_2, \ldots, x_n) = \mathop{T}_{i=1}^{n} A_i^k(x_i),$$

$$P_{ck}(y_1, y_1, \ldots, y_m) = \mathop{T}_{i=1}^{m} B_i^k(y_i),$$

$$R_k(x_1, x_2, \ldots, x_n, y_1, y_2, \ldots, y_m)$$
$$= f(P_{ak}(x_1, x_2, \ldots, x_n), P_{ck}(y_1, y_2, \ldots, y_m)).$$

Set $P_a(x_1, x_2, \ldots, x_n) = T_{i=1}^{n} A_i(x_i)$.

Step 3: Compute $R(x_1, x_2, \ldots, x_n, y_1, y_2, \ldots, y_m) = \mathop{A}_{k=1}^{N}(R_k(x_1, x_2, \ldots, x_n, y_1, y_2, \ldots, y_m))$.

Step 4: Find the inferred proposition P_c by

$$P_c(y_1, y_2, \ldots, y_m)$$
$$= \sup_{x_i, i=1 \ldots n} [P_a(x_1, x_2, \ldots, x_n) \, t \, R(x_1, x_2, \ldots, x_n, y_1, y_2, \ldots, y_m)].$$

In the procedure just outlined, more specifically, in steps 3 and 4, the compositional rule of inference is applied after the overall relation associated with the rule base is found; that is,

$$P_c = P_a \circ \mathop{A}_{k=1}^{N} (R_k).$$

Alternatively, the compositional rule can be applied locally to each relation R_k and the resulting fuzzy sets aggregated to provide the inferred proposition; that is,

$$P_c = \mathop{A}_{k=1}^{N} (P_a \circ R_k).$$

Therefore, steps 3 and 4 can be modified accordingly,

Step 3a: For $k = 1 \ldots N$, compute

$$P'_{ck}(y_1, y_2, \ldots, y_m)$$

$$= \sup_{x_i, 1 \ldots n} \; [P_a(x_1, x_2, \ldots, x_n) \; t \; R_k(x_1, x_2, \ldots, x_n, y_1, y_2, \ldots, y_m)].$$

Step 4a: Find the inferred proposition P_c by

$$P_c(y_1, y_2, \ldots, y_m) = \overset{N}{\underset{k=1}{\mathrm{A}}} \; (P'_{ck}(y_1, y_2, \ldots, y_m)).$$

In general $P_a \circ A_{k=1}^N (R_k) \neq A_{k=1}^N (P_a \circ R_k)$, but under sup-min or any sup-t composition with continuous t-norms or any union (max) or intersection (min) aggregation, the following relationships hold:

$$P_c \circ \left(\bigcap_{k=1}^N R_k \right) \subseteq \bigcap_{k=1}^N (P_c \circ R_k) \subseteq P_c \circ \left(\bigcup_{k=1}^N R_k \right) = \bigcup_{k=1}^N (P_c \circ R_k).$$

Note that the particular, simplified inference cases discussed at the begining of this section are just-instances of this equality.

EXERCISE 10.6   Assume rule antecedent propositions as disjunctively related, that is, they are of the form "X is A_i or Y is B_i," $i = 1 \ldots N$. Provide a graphical interpretation for the inference scheme just described, assuming pointwise inputs, sup-min composition, and rules as f_c and f_p.

At this point, it is instructive to detail a special case of compositional inference, namely, the sup-min composition, when considering

- Input proposition: X is A and Y is B,
- Rules: If X is A_k and Y is B_k, then Z is C_k, $k = 1 \ldots N$,

with the meaning of each rule defined by the assignment $R_i = f_c$ and the input proposition viewed as $P(x, y) = A(x) \wedge B(y)$. For each rule i, we get

$$C'_k(z) = \sup_{x,y} \{ [A(x) \wedge B(y)] \wedge [A_k(x) \wedge B_k(y)] \wedge C_k(z) \}$$

$$= \sup_{x,y} \{ [A(x) \wedge B(y) \wedge A_k(x) \wedge B_k(y) \wedge C_k(z) \}$$

$$= \sup_{x,y} \{ [A(x) \wedge A_k(x)] \wedge [B(y) \wedge B_k(y)] \wedge C_k(z) \}$$

$$= \{ \sup_x [A(x) \wedge A_k(x)] \wedge \sup_y [B(y) \wedge B_k(y)] \} \wedge C_k(z).$$

Thus, four main steps emerge, namely:

Step 1: $m_{Ak} = \sup_x [A(x) \wedge A_k(x)] = \mathrm{Poss}(A, A_k), k = 1 \ldots N,$

$\qquad\quad m_{Bk} = \sup_y [B(y) \wedge B_k(y)] = \mathrm{Poss}(B, B_k), k = 1 \ldots N.$

In this case, matching operations are performed because possibility measures are computed for the fuzzy sets appearing in the proposition and rule antecedents, respectively.

Step 2: $m_k = \min[m_{Ak}, m_{Bk}], k = 1 \ldots N.$

Step 3: $C_k'(z) = m_k \wedge C_k(z), k = 1 \ldots N.$

Step 4: $C(z) = \bigcup_{k=1}^{N} C_k'(z) = \max[C_k'(z), k = 1 \ldots N], \forall z \in \mathbf{Z}.$

This type of inference pattern suggests an alternative to compositional inference (called here scaled inference, for short) once the pattern is extended to other possible combinations of the conjunction, rule semantics, and aggregation operators. However, a word of caution is necessary: There is, in general, no direct relationship between the scaled-inference approach and compositional inference. Additional examples include (Mizumoto 1994):

1. Product-algebraic sum method:

$$C_i'(z) = A_i(x_o). B_i(y_o). C_i(z).$$

$$C(z) = \overset{N}{\underset{k=1}{S_4}} [C_k'(z)].$$

2. Bounded product–bounded sum method:

$$C_i'(z) = A_i(x_o)t_2 B_i(y_o)t_2 C_i(z).$$

$$C(z) = \overset{N}{\underset{k=1}{S_2}} [C_k'(z)],$$

with, for simplicity, $x_o \in \mathbf{X}$ and $y_o \in \mathbf{Y}$, and t_2 and s_2 taken with $p = 0$. Included in this class of inference method is the additive form (Kosko 1994, Mizumoto 1994), as, for example:

- Product-sum method:

$$C_i'(z) = A_i(x_o). B_i(y_o). C_i(z).$$

$$C(z) = \sum_{k=1}^{N} [C_k'(z)].$$

- Min-sum method:

$$C_i'(z) = A_i(x_o) \wedge B_i(y_o) \wedge C_i(z).$$

$$C(z) = \sum_{k=1}^{N} [C_k'(z)].$$

Clearly, generalizations can be readily devised for non pointwise inputs and other operators.

The general scaled inference scheme can be outlined as follows:

Step 1: Antecedent matching: For each rule, compute the degree of matching between each atomic proposition of the rule antecedent and the corresponding atomic proposition in the given input proposition.

Step 2: Antecedent aggregation: For each rule, compute the rule activation level by conjunctively or disjunctively operating on the corresponding degrees of matching depending on whether the atomic propositions of the rule antecedent are conjunctively or disjunctively related, respectively.

Step 3: Rule result derivation: For each rule, compute the corresponding inferred value based on its antecedent aggregation and the rule semantics chosen.

Step 4: Rule aggregation: Compute the inferred value from the complete set of rules by aggregating the result of the inferred values derived from individual rules.

Further modifications are possible when, for example, the membership functions involved are monotonic (Tsukamoto 1979) or when the rule consequents are functions of the inputs (Takagi and Sugeno 1983), especially when the inputs are points. Figures 10.12 and 10.13 exhibit the inference process involved. Note that in this case rules take the following format:

If X is A_i and Y is B_i, then $z = f_i(x, y)$,

where $f_i : \mathbf{X} \times \mathbf{Y} \rightarrow \mathbf{Z}$. The overall result is found by the weighted combination of the individual rule contributions. In other words, aggregation is performed via the weighted-average aggregation operator.

EXERCISE 10.7   Provide a graphical interpretation of the scaled-inference method considering rules i and j with two andwise atomic propositions in the antecedent and one in the consequent, assuming triangular fuzzy sets in both rules and inputs, the min as the conjunction, and rules interpreted as f_p. Consider first the max s-norm, and second the ordinary sum as rules-aggregation operators. Compare your results and elaborate on how they would contrast with those derived through compositional inference.

Let us now discuss inference with gradual rules interpreted, as an example, by f_a, Gaines implication. Here we will restrict ourselves to compositional inference only. Similar treatments can be developed for alternative rule interpretations.

Consider the case of gradual rules of the form "the more X is A, the more Y is B," and a proposition "X is A'." When $A(x)$ and $B(y)$ are

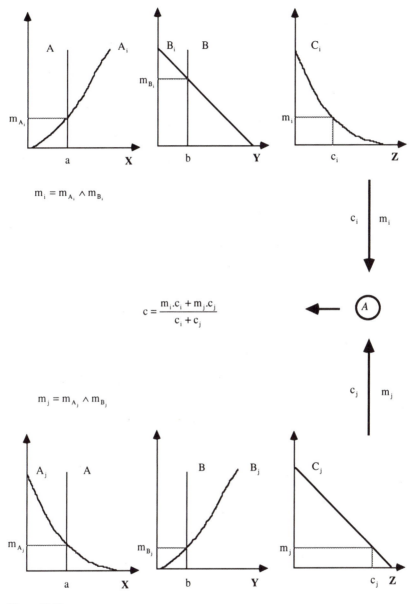

Figure 10.12
Fuzzy reasoning with monotonic membership functions

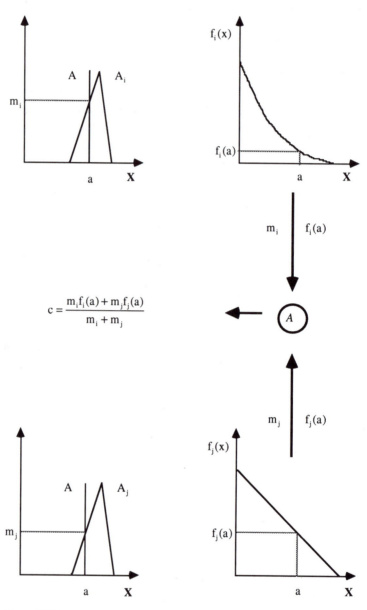

Figure 10.13
Fuzzy inference with consequents as a function of input variables

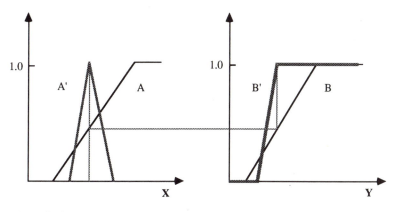

Figure 10.14
Inference with gradual rules

monotonically increasing functions on closed intervals, and $A'(x)$ is unimodal and upper semicontinuous, the compositional inference translates into the following procedure (Dubois and Prade 1992), as illustrated in figure 10.14:

Step 1: For any value of y, compute $v = B(y)$.

Step 2: If $v = 1$, then $B'(y) = 1$; if $v < 1$, then compute $a = A^{-1}(v)$.

Step 3: If $a \geq \inf\{x, A'(x) = 1\}$, then $B'(y) = 1$; otherwise $B'(y) = A'(a)$.

As is evident from figure 10.14, in general, the inferred result B' does not always contain B. However, if $A' \subseteq A$ but $A' \neq A$, then $B' \subseteq B$ but $B' \neq B$.

10.5   Some Properties of Fuzzy Rule–Based Systems

Frequently, the output of a fuzzy rule–based system must be nonfuzzy, an essential requirement in many engineering problems, for example, fuzzy modeling and control applications. A defuzzification stage is needed, and the decoding mechanisms introduced in section 3.6 are of interest.

The first important property *to be considered concerns function approximation: Fuzzy systems can be qualified to uniformly approximate continuous functions to any degree of accuracy on closed and bounded (compact) sets. Function approximation capabilities become intuitively apparent if we look at a fuzzy rule–based system as a fuzzy graph, as in figure 10.4. Some relevant results are worth reviewing.

Wang and Mendel (1992) showed that the following are universal approximators: fuzzy systems in which rules are viewed as the product fuzzy conjunction f_p, antecedent andwise propositions defined by the product, sup-min composition, ordinary sum rules aggregation, pointwise inputs, CoG defuzzification, and Gaussian membership functions. Later, Kosko (1994) proved that fuzzy additive rule systems have the same property under similar conditions, except for Gaussian membership functions. His result also admits rule interpretation via the min conjunction f_c, but symmetry of the consequent fuzzy sets may be required. Castro (1995) showed that the following are also universal approximators: fuzzy rule–based systems with fuzzy conjunctions modeled by an arbitrary t-norm, rules interpreted as R-implications f_{ir} or fuzzy conjunctions f_c with CoG defuzzification and with triangular or trapezoidal membership functions. Castro and Delgado (1996) also identified the following as universal approximators: fuzzy systems characterized by a fixed fuzzy inference, fuzzy relations satisfying a smooth property, and a fixed defuzzification method verifying the requirement of producing a point in the support of the original fuzzy set. These results are important because they provide a sound justification for using fuzzy systems in many real world applications. But since the proofs are not constructive, the results are just existence theorems. In contrast with alternative approximators (e.g., neural networks), however, fuzzy systems are uniquely suited to incorporating linguistic information naturally and systematically.

Another important property concerns the interpolation capabilities of fuzzy systems (Zadeh 1988). Again, the intuitive meaning of interpolation can be appreciated by considering F^* in figure 10.7(b) as a fuzzy graph. The interpolation property is also shared with gradual rules, as Dubois and Prade (1992) pointed out.

10.6   Rule Consistency and Completeness

It is quite natural to accept the idea that inconsistent information in any rule–based system may lead to unexpected and unsatisfactory results. When inconsistency is detected among a collection of propositions, those found to be inconsistent must be handled properly to produce a consistent system. In detecting the existence of inconsistencies within a fuzzy rule–based system, a central issue is how to determine the degree of matching between two ruzzy propositions.

A convenient methodology for inconsistency checking, based on a type of backward inference scheme, is the reflecting-on-the-input approach (Yager and Larsen 1991) and its extensions (Scarpelli and Gomide 1994). The main idea is as follows. Suppose X is an input variable to a given collection of fuzzy propositions. If no data is supplied to X and the value of X is required, the process of reflecting on the input is applied to run the system backwards. When there are potential inconsistencies, the inferred value of X, rather than being the whole domain of X, is some subset of it.

DEFINITION 10.6   A fuzzy set is considered inconsistent if it is subnormal, that is, hgt $(A) < 1$.

DEFINITION 10.7   The degree of inconsistency of a fuzzy set A in a universe $\mathbf{X}$ is defined as

$c = \mathrm{Nec}(X \text{ is } \varnothing, X \text{ is } A).$

Since $\mathrm{Nec}(X \text{ is } \varnothing, X \text{ is } A) = 1 - \mathrm{Poss}(\mathbf{X}, A)$, it follows that

$c = 1 - \sup_x A(x).$

The degree of inconsistency may therefore be viewed as a measure of how far the value of the variable is from being a normal set. Thus, if A is normal, there is no inconsistency and $c = 0$. In the following analysis of the process of reflection on the input, the fuzzy sets appearing in antecedent and consequent atomic propositions are called input and output sets, respectively.

Consider first the simplest case: propositions with a single condition in both the antecedent and the consequent, such as

If X is A_1, then Y is B_1,

If X is A_2, then Y is B_2,

X is A,

where the As and Bs are fuzzy sets in the universes $\mathbf{X}$ and $\mathbf{Y}$, respectively. Assume rules conjunctively aggregated through the min operator and interpreted according to the implication function f_b, and inference performed by the sup-min compositional rule. Thus

$R_1(x, y) = \max(1 - A_1(x), B_1(y)) = \bar{A}_1(x) \vee B_1(y),$

$R_2(x, y) = \max(1 - A_2(x), B_2(y)) = \bar{A}_2(x) \vee B_2(y),$

whose aggregation is $R = R_1 \cap R_2$ or, equivalently, $R(x, y) = R_1(x, y) \wedge R_2(x, y)$. The inferred value of Y is B, $B = A \circ R$,

$B(y) = \sup_x [R(x, y) \wedge A(x)],$

which can easily be rewritten as (see problem 6 at the end of the chapter)

$$B(y) = [\text{Poss}(\bar{A}_1 \wedge \bar{A}_2, A)] \vee [\text{Poss}(\bar{A}_1, A) \wedge B_2(y)]$$
$$\vee [\text{Poss}(\bar{A}_2, A) \wedge B_1(y)] \vee [B_1(y) \wedge B_2(y)].$$

To obtain a degree of inconsistency of α for the variable taking the value B, it must be the case that $B(y) \leq 1 - \alpha$, $\forall y \in \mathbf{Y}$. Hence

$$\text{Poss}(\bar{A}_1 \wedge \bar{A}_2, A) \leq 1 - \alpha,$$

$$[\text{Poss}(\bar{A}_1, A) \wedge B_2(y)] \vee [\text{Poss}(\bar{A}_2, A) \wedge B_1(y)] \leq 1 - \alpha,$$

$$B_1(y) \wedge B_2(y) \leq 1 - \alpha, \quad \forall y \in \mathbf{Y},$$

and, from the last inequality, $\text{Poss}(B_1, B_2) \leq 1 - \alpha$. In addition, assuming B_1 and B_2 are normal and consistent, it follows from the second inequality above that

$$\text{Poss}(\bar{A}_1, A) \leq 1 - \alpha,$$

$$\text{Poss}(\bar{A}_2, A) \leq 1 - \alpha.$$

These inequalities guarantee that $\text{Poss}(\bar{A}_1 \wedge \bar{A}_2, A) \leq 1 - \alpha$. Assuming A is normal, there is at least one $x^* \in \mathbf{X}$ such that $A(x^*) = 1$, and from $\text{Poss}(\bar{A}_1, A) \leq 1 - \alpha$, it readily follows that $\bar{A}_1(x^*) \leq 1 - \alpha$. Similarly $\bar{A}_2(x^*) \leq 1 - \alpha$. These two conditions mean that $A_1(x^*) \geq \alpha$ and $A_2(x^*) \geq \alpha$. Thus $A_1(x^*) \wedge A_2(x^*) \geq \alpha$, which implies that $\text{Poss}(A_1, A_2) \geq 0$. Therefore two conditions are necessary for any inconsistency to exist: for some $\alpha \in (0, 1]$,

1. $\text{Poss}(B_1, B_2) \leq 1 - \alpha < 1$.
2. $\text{Poss}(A_1, A_2) \geq \alpha > 0$.

Consider now only the two rules whose conjunction results in "(X, Y) is R", with

$$R(x, y) = [\bar{A}_1(x) \vee B_1(y)] \wedge [\bar{A}_2(x) \vee B_2(y)].$$

If X is reflected through the projection of R, the set E is found as

$$E(x) = \sup_y R(x, y) = \bar{A}_1(x) \wedge \bar{A}_2(x) \wedge \text{Poss}(B_1, B_2).$$

If no potential inconsistencies exist, that is, if $\text{Poss}(B_1, B_2) = 1$ or $\text{Poss}(A_1, A_2) = 0$, then $E(x) = 1$, $\forall x \in \mathbf{X}$, meaning that X can be any element in its universe. On the other hand, if the necessary conditions for potential inconsistencies are satisfied, that is, $\text{Poss}(B_1, B_2) = \alpha_1 < 1$ and $\text{Poss}(A_1, A_2) = \alpha_2 > 0$, then $E(x) = \bar{A}_1(x) \wedge \bar{A}_2(x) \wedge \alpha_1$, and there exists at least one $x \in \mathbf{X}$ such that $\bar{A}_1(x) \wedge \bar{A}_2(x) = 1 - \alpha_2 < 1$. Therefore, for some $x \in \mathbf{X}$, $E(x) < 1$. This is an indicator that a potential inconsistency exists.

A similar result can be obtained when the antecedent contains multiple propositions, as in

If X_1 is A_1 and ... and X_n is A_n, then Y is B_1,

If X_1 is C_1 and ... and X_n is C_n, then Y is B_2.

If $A = A_1 \times \cdots \times A_n$ and $C = C_1 \times \cdots \times C_n$, then the necessary conditions for potential inconsistencies are

1. $\text{Poss}(B_1, B_2) \leq 1 - \alpha$,
2. $\text{Poss}(A, C) \geq \alpha$.

Note that the situation above is a particular case of, with $k < n$,

If X_1 is A_1 and ... and X_n is A_n, then Y is B_1,

If X_1 is C_1 and ... and X_k is C_k, then Y is B_2.

If the absence of variables $X_{k+1}, \ldots, X_n$ means that it does not matter what values are given to them, one could fill the remaining atomic propositions in the antecedent with the propositions "X_{k+1} is $C_{k+1}, \ldots, X_n$ is C_n" to recover the necessary conditions, considering $C = C_1 \times \cdots \times C_n \times \mathbf{X}_{k+1} \times \cdots \times \mathbf{X}_n$.

The second necessary condition involves joint variables, which may require compound calculi in consistency checking procedures. In this situation, a reflection on any individual antecedent variable X may not produce a less restricted space than X. As long as the aim is to achieve conditions that require simple calculations, the second necessary condition should be relaxed and use should be made of the fact that if $\text{Poss}(A, C) \geq \alpha$, then $\text{Poss}(A_1, C_1) \geq \alpha$, $\text{Poss}(A_2, C_2) \geq \alpha, \ldots$, $\text{Poss}(A_n, C_n) \geq \alpha$. The derived conditions are weaker but certainly are also necessary conditions for existing potential inconsistencies and do not require computations over compound sets. When rules have different variables in the antecedent, such as

If X_1 is A, then Y is B_1,

If W is C, then Y is B_2,

the condition $\text{Poss}(B_1, B_2) \leq 1 - \alpha$ is sufficient for potential inconsistency.

So far, we have analyzed conditions for inconsistency only at the local level since they involve variables in a collection of parallel rules. Whenever rules are chained, however, inconsistency should be considered at the global level, that is, after rule chaining. For instance, assume the following propositions:

If X is A_1, then Y is B_1,

If Y is G_2, then W is S_2,

If X is A_3, then W is S_3,

X is A.

Proceeding as before, we find that the following conditions are necessary:

1. $\text{Poss}(S_2, S_3) \le 1 - \alpha$.
2. $\text{Poss}(A_1, A_3) \ge \alpha$.
3. $\text{Nec}(G_2, B_1) \ge \alpha$.

A more complicated situation arises when the antecedent includes several conditions. The analysis to be performed is the same, however, and analogous conditions to those above are derived. (See problem 7 at the end of the chapter for an example.)

To address computational procedures for consistency analysis at both the local and global level, the idea of a condition test matrix, introduced in Scarpelli and Gomide (1994), is essential. For this purpose, consider a collection of N rules in which *each* rule (rule indexes in the input and output sets are omitted to simplify notation) is of the form

If X_1 is A_1 and ... and X_k is A_k, then Y is B.

Let M be the maximum number of variables appearing in the collection of N rules. Consider the following example:

1. If V_1 is A_1 and V_2 is C_1, then V_5 is B_1.
2. If V_1 is A_2 and V_2 is C_2, then V_6 is J_1.
3. If V_6 is J_2 and V_7 is D_1 and V_8 is E_1, then V_{10} is F_1.
4. If V_2 is C_3, then V_9 is G_1.
5. If V_2 is C_4, then V_9 is G_2.
6. If V_2 is C_5, then V_9 is G_3.
7. If V_3 is H_1 and V_4 is I_1, then V_9 is G_4.

There are seven rules and ten variables, so $N = 7$ and $M = 10$.

DEFINITION 10.8   The condition test matrix T is a $M \times N$ matrix such that each entry t_{ij}, $i = 1, \ldots, M$ and $j = 1, \ldots, N$ is given by

$$t_{ij} = \begin{cases} -A_i, & \text{if } A_i \text{ is an input set for rule } j \\ B, & \text{if } B \text{ is an output set for rule } j \\ 0, & \text{otherwise.} \end{cases}$$

The condition test matrix (CT-matrix) for the example above is

	R_1	R_2	R_3	R_4	R_5	R_6	R_7
V_1	$-A_1$	$-A_2$	0	0	0	0	0
V_2	$-C_2$	$-C_2$	0	$-C_3$	$-C_4$	$-C_5$	0
V_3	0	0	0	0	0	0	$-H_1$
V_4	0	0	0	0	0	0	$-I_1$
V_5	B_1	0	0	0	0	0	0
V_6	0	J_1	$-J_2$	0	0	0	0
V_7	0	0	$-D_1$	0	0	0	0
V_8	0	0	$-E_1$	0	0	0	0
V_9	0	0	0	G_1	G_2	G_3	G_4
V_{10}	0	0	F_1	0	0	0	0

DEFINITION 10.9   Let $Q = (q_i)$, $i = 1, \ldots, M$, and $T = [t_{ij}]$, $i = 1, \ldots, M$ and $j = 1, \ldots, N$, be a vector and a matrix, respectively, whose elements are fuzzy sets over the appropriate universes. The matrix composition $Q^*T = (m_j)$, $j = 1, \ldots, N$, is such that

$$m_j = \sum_{i=1}^{M} S(q_i, t_{ij})$$

and

$$S(a, b) = \begin{cases} \text{Poss}(a, b), & \text{if } a \text{ and } b \text{ are nonnull and have the same sign} \\ 0 & \text{otherwise,} \end{cases}$$

where $\sum$ means a formal sum of values.

In the context of consistency checking, matrix composition is used to compare new rules with existing rules. In each row of $m = (m_j)$, the number of nonzero components in the formal sum indicates how many propositions in the rule being added interfere with the existing ones.

Next, procedures for consistency checking at local and global levels are introduced. Initially a consistent rule set is assumed, to which a new rule R of the form

If V_1 is A_1 and ... and V_k is A_k, then U is B

is to be added.

Procedure: Consistency checking at the local level.

Input: $T_{M \times N}$, rule R, degree of inconsistency $\alpha \in (0, 1]$.

Output: Message.

Step 1: Construct a CT-matrix T' from T, corresponding to rules with U in the consequent. Assume U is the lth variable, which corresponds to the lth row of T.

For $j = 1, \ldots, N$, do

If $t_{ij} \neq 0$, then include column t_j in T'.

Step 2: Construct a test vector $V = (v_i)$ for rule R.

For $i = 1, \ldots, M$, do

If rule R has the proposition "V_i is G," then if "V_i is G" is in the antecedent

$$\text{then } v_i = -G$$

$$\text{else } v_i = G$$

$$\text{else } v_i = 0.$$

Step 3: Apply matrix-composition $V^{T*}T' = F$; V^T is the transpose of V.

Step 4: Analyze the values of F against $\alpha \in (0, 1]$.

For each element $f_j = s_1 + s_2 + \cdots + s_M$, do

Find $f_j' = r_1, r_2, \cdots, r_k$, $k \geq 1$, with all and only nonzero elements of f_j.

If $r_k \leq 1 - \alpha$ and $r_1 > \alpha$, and $\ldots r_{k-1} > \alpha$,

Then return the message *Necessary conditions for inconsistency hold*

Else return the message *Necessary conditions for inconsistency do not hold.*

For instance, suppose the rule "If V_2 is C_6, then V_9 is G_5" is to be added to the seven rules in the previous example. Thus, the test vector is

$$V^T = (0 \ -C_6 \ 0 \ 0 \ 0 \ 0 \ 0 \ G_5 \ 0).$$

The matrix composition yields $V^{T*}T' = F$: more specifically,

$$V^{T*} \begin{bmatrix} 0 & 0 & 0 & 0 \\ -C_3 & -C_4 & -C_5 & 0 \\ 0 & 0 & 0 & -H_1 \\ 0 & 0 & 0 & -I_1 \\ 0 & 0 & 0 & 0 \\ 0 & 0 & 0 & 0 \\ 0 & 0 & 0 & 0 \\ 0 & 0 & 0 & 0 \\ G_1 & G_2 & G_3 & G_4 \\ 0 & 0 & 0 & 0 \end{bmatrix} = F,$$

$$F = \begin{bmatrix} 0 + \text{Poss}(C_6, C_3) + 0 + 0 + 0 + 0 + 0 + 0 + \text{Poss}(G_5, G_1) + 0 \\ 0 + \text{Poss}(C_6, C_4) + 0 + 0 + 0 + 0 + 0 + 0 + \text{Poss}(G_5, G_2) + 0 \\ 0 + \text{Poss}(C_6, C_5) + 0 + 0 + 0 + 0 + 0 + 0 + \text{Poss}(G_5, G_3) + 0 \\ 0 + 0 + 0 + 0 + 0 + 0 + 0 + 0 + \text{Poss}(G_5, G_4) + 0 \end{bmatrix}.$$

Each row of the vector F is a formal sum of possibility values in which the last nonzero element corresponds to output sets and all other elements correspond to input sets. The last row indicates that there are no common variables among the antecedents of rules seven and eight. Now, consider input and output sets defined in the same universe $\mathbf{X} = \{1, 2, 3\}$, and let

$C_3 = \{1/1, 1/2, 0.5/3\}, \quad C_4 = \{0.7/1, 0.3/2, 1/3\},$

$C_5 = \{1/1, 0.2/2, 0/3\}, \quad C_6 = \{0.6/1, 0.4/2, 1/3\},$

$G_1 = \{1/1, 0.5/2, 0.6/3\}, \quad G_2 = \{0/1, 0.2/2, 1/3\},$

$G_3 = \{0.8/1, 0.7/2, 0.2/3\}, \quad G_4 = \{1/1, 0.4/2, 0.3/3\},$

$G_5 = \{1/1, 0.7/2, 0.2/3\}.$

The resulting vector F is

$$F = \begin{bmatrix} 0 + 0.6 + 0 + 0 + 0 + 0 + 0 + 0 + 1 + 0 \\ 0 + 1 + 0 + 0 + 0 + 0 + 0 + 0 + 0.2 + 0 \\ 0 + 0.6 + 0 + 0 + 0 + 0 + 0 + 0 + 0.8 + 0 \\ 0 + 0 + 0 + 0 + 0 + 0 + 0 + 0 + 1 + 0 \end{bmatrix}.$$

The second row indicates potential inconsistency between rules five and eight considering any value $\alpha \leq 0.8$, because $\text{Poss}(C_6, C_4) = 1 > \alpha$, and $\text{Poss}(G_5, G_2) = 0.2 \leq 1 - \alpha$.

The computational procedure for consistency checking at the global level is similar to that just described except that the necessary condition due to rule chaining, which involves expressions of the form $\text{Nec}(J_1, J_2)$, should be verified. Therefore, the procedure requires searching the CT-matrix for necessary information and evaluating expressions in terms of the complement of sets, since $\text{Nec}(J_1, J_2) = 1 - \text{Poss}(\bar{J}_1, J_2)$. This either strengthens or eliminates the potential inconsistencies. To test for global inconsistency, it becomes necessary first to identify the relevant portion of the rules collection affected by the rule added—in other words, the minimal cover for the new rule. The minimal cover of a rule is defined as the part of the rules collection that minimally covers all dependencies in the new rule. The basic idea is due to Agarwal and Taniru (1992). The procedure is as follows.

Procedure: Consistency checking at the global level.

Input: $T_{M \times N}$, rule R, degree of inconsistency $\alpha \in (0, 1]$.

Output: Message.

Step 1: Construct a CT-matrix T' form T, corresponding to the minimal cover of R. Assume U is the lth variable, which corresponds to the lth row of T.

For $j = 1, \ldots, N$, do

if $t_{ij} \neq 0$, then include column t_j in T'.

Step 2: Construct a test vector $V = (v_i)$ for rule R.

For $i = 1, \ldots, M$, do

If rule R has the proposition "V_i is G," then if "V_i is G" is in the antecedent

then $v_i = -G$

else $v_i = G$

else $v_i = 0$.

Step 3: Apply matrix-composition $V^{T*} T' = F$; V^T is the transpose of V.

Step 4: Analyze the values of F against $\alpha \in (0, 1]$.

For each element $f_j = s_1 + s_2 + \cdots + s_M$, do

Find $f_j' = r_1, r_2, \ldots, r_k, k \geq 1$, with all and only nonzero elements of f_j.

If $r_k \leq 1 - \alpha$ and $r_1 > \alpha$, and $\cdots r_{k-1} > \alpha$

then Go To step 5

else return the message *Necessary conditions do not hold*, and stop.

Step 5: Check the inclusion grade between values of chained variables.

Let $p_i, i = 1, \ldots, l$ be the rows of T' with nonnull elements S_i and Q_i of opposite signs.

If $\text{Nec}(S_1, Q_1) \geq \alpha$ and, $\ldots$, and $\text{Nec}(S_l, Q_l) \geq \alpha$

then return the message *Necessary conditions for potential inconsistencies hold.*

else return the message *Necessary conditions do not hold.*

For an example of global consistency, see problem 8 at the end of the chapter.

Let us consider the issue of completeness of a fuzzy rule–based system. By completeness, we mean that for any input proposition, the rule–based system can generate an answer or, equivalently, an inferred proposition. More formally (Pedrycz 1993), a collection of N rules

If X is A_i, then Y is B_i,

where A_i and B_i are fuzzy sets of $\mathbf{X}$ and $\mathbf{Y}$, respectively, is complete if $\forall x \in \mathbf{X}$ there exists at least one $i, 1 \leq i \leq N$, such that

$$A(x) > \varepsilon, \quad \varepsilon \in (0, 1].$$

This condition requires that, if rules are aggregated via a union operator,

$$\left(\bigcup_{i=1}^{N} A_i(x) \right) > \varepsilon.$$

Is not difficult to comply with this condition. The requirements seem to be natural, since fuzzy sets attached to linguistic labels usually overlap. The parameter ε describes this overlap. The inequality is violated only if some labels are missing, which could be the case if a relevant piece of knowledge has been omitted.

10.7   Chapter Summary

In this chapter the notion of fuzzy propositions in general and fuzzy rules in particular was introduced as an important vehicle for knowledge representation and processing when uncertainty, viewed as fuzziness, is the central issue. Several classes of rules, for example, rules with certainty factors, exceptions, and gradual statements, as well as quantified and qualified rules, were presented. Their syntax and meaning were discussed with the aid of the notion of fuzzy relations, which provide meaningful geometrical interpretations of the ideas involved, such as fuzzy graphs, intersections and projections. When coupled, these ideas clarify the very important reasoning scheme known as the compositional rule of inference. Computational inference procedures for typical situations of the most relevant practical applications were detailed. In addition, the central issues of consistency checking and completeness of a collection of rules were analyzed and computational procedures provided.

10.8   Problems

1. If A is a normal fuzzy set of the finite universe $\mathbf{X}$, then the specificity of A, $S(A)$, is defined as (Yager 1984) $S(A) = \int_0^1 (1/\text{Card } F) \, d\alpha$. Show that if the proposition "X is A with certainty μ" is transformed into "X is B," then $S(A) \geq S(B)$. If "X is A_1 with certainty μ_1" transforms into "X is B_1," and "X is A_2 with certainty μ_2" transforms into "X is B_2," what can you say about $S(B_1)$ and $S(B_2)$?

2. Consider the quantified rule "If most (X_1 is A_1, X_2 is A_2, X_3 is A_3), then Y is B," where A_1, A_2, A_3, and B are fuzzy sets of $\mathbf{X}_1 = \{a, b\}$, $\mathbf{X}_2 = \{c, d\}$, $\mathbf{X}_3 = \{e, f\}$,

and $Y = \{g, h\}$, with $A_1 = \{1/a, 0/b\}$, $A_2 = \{1/c, 0/d\}$, $A_3 = \{0/e, 1/f\}$ and $B = \{1/g, 0/h\}$. Let *most* be defined by $Q(0) = 0$, $Q(1/3) = 0$, $Q(2/3) = 0.5$, and $Q(1) = 1$. Find the relation induced by the rule.

3. Assume the rule "If X is A, then Y is B, unless Z is C." Show that, if $R = R_{\ddot{c}} \cap R_c$, then

   (a) $(A(x) \wedge C(z)) \circ R(x, y, z) = \bar{B}(y)$.

   (b) $(A(x) \wedge \bar{C}(z)) \circ R(x, y, z) = B(y)$.

   (c) $(A(x) \wedge \mathbf{Z}(z)) \circ R(x, y, z) \neq B(y)$.

4. Show that the logical disjunction of two gradual rules ("The more X is A, the more Y is B") or ("The more Z is C, the more Y is B") entails, but is not equivalent to, "The more X is A and Z is C, the more Y is B," because it is true that

   $$(\forall x \in \mathbf{X}, A(x) \leq B(y)) \text{ or } (\forall z \in \mathbf{Z}, C(z) \leq B(y))$$

   $$\Rightarrow \forall x \in \mathbf{X}, \forall z \in \mathbf{Z}, \min[A(x), C(z)] < B(y),$$

   but the converse implication is false.

5. Given the proposition "X is A'" and the rule "The more X is A, the more Y is B," which is interpreted by

   $$R(x, y) = \begin{cases} 1, & B(y) \geq A(x) \\ 0, & \text{otherwise,} \end{cases}$$

   show that the compositional rule of inference reduces to $B'(y) = \sup_{u \in B_{R(x)}} A'(u)$.

6. Assume the following inference scheme:

   X is A
   If X is A_1, then Y is B_1
   If X is A_2, then Y is B_2

   Y is B, ,

   where rules are conjunctively aggregated through the min operator and interpreted by, for $i = 1, 2$,

   $$R_i(x, y) = \max[1 - A_i(x), B_i(y)] = \bar{A}_i(x) \vee B_i(y).$$

   Show that the inferred value of Y is such that

   $$B(y) = [\text{Poss}(\bar{A}_1 \wedge \bar{A}_2, A)] \vee [\text{Poss}(\bar{A}_1, A) \wedge B_2(y)]$$

   $$\vee [\text{Poss}(\bar{A}_2, A) \wedge B_1(y)] \vee [B_1(y) \wedge B_2(y)].$$

7. Suppose that the following collection of rules is provided:

   If V_1 is A_1 and V_2 is A_2, then U_1 is B_1.

   If U_1 is B_2 and U_2 is B_3 and U_3 is B_4, then W is S_2.

   If V_2 is A_3 and U_3 is B_5, then W is C_2.

   Show that the necessary conditions for potential global inconsistencies are:

   1. $\text{Poss}(C_2, C_3) \leq 1 - \alpha$,
   2. $\text{Poss}(A_2, A_3) \geq \alpha$,
   3. $\text{Poss}(B_4, B_5) \geq \alpha$, and
   4. $\text{Nec}(B_2, B_1) \geq \alpha$.

8.   Consider the example presented in section 10.6, and assume that a new rule of the form

If V_2 is C_6 and V_9 is D_2, then V_{10} is F_2,

is to be added to the rules presented in that example. Let $C_6 = \{0.5/1, 0.4/2, 1/3\}$, $C_2 = \{0.4/1, 1/2, 0.5/3\}$, $D_1 = \{0.8/1, 1/2, 0/3\}$, $D_2 = \{0.7/1, 1/2, 0.8/3\}$, $F_1 = \{0.3/1, 1/2, 0/3\}$, $F_2 = \{0.2/1, 0/2, 1/3\}$, and $\alpha = 0.5$. Show that in this case, a potential inconsistency exists when the rule above is inserted.

References

Agarwal, L., and M. Taniru. 1992. A Petri-net based approach for verifying the integrity of production systems. *Int. J. Man-Machine Studies* 36:447–68.

Castro, J. 1995. Fuzzy logic controllers are universal approximators. *IEEE Trans. Systems, Man, and Cybernetics* 25(4):629–35.

Castro, J., and M. Delgado. 1996. Fuzzy systems with defuzzification are universal approximators. *IEEE Trans. on Systems, Man, and Cybernetics* 26:149–52.

Dogherty, P., D. Driankov, and H. Hellendorn. 1993. Fuzzy if-then-unless rules and their implementation. *Int. J. of Uncertainty, Fuzziness and Knowledge Based Systems* 1(2):167–82.

Driankov, D., and H. Hellendorn. 1992. Fuzzy Logic with Unless-Rules. Report IDA-RKL-92-TR 50, Laboratory for Representation of Knowledge in Logic, Dept. of Computer and Information Science, Linkoping University, Sweden.

Dubois, D., and H. Prade. 1991. Fuzzy sets in approximate reasoning (parts 1 and 2). *Fuzzy Sets and Systems* 40:143–244.

Dubois, D., and H. Prade. 1992. Gradual inference rules in approximate reasoning. *Information Sciences* 61:103–22.

Klir, G., and B. Yuan. 1995. *Fuzzy Sets and Fuzzy Logic: Theory and Applications*. Englewood Cliffs, NS: Prentice Hall.

Kosko, B. 1994. Fuzzy systems are universal approximators. *IEEE Trans. on Computers* 43(11):329–32.

Mizumoto, M. 1994. Fuzzy controls under product-sum-gravity methods and new fuzzy control methods. In *Fuzzy Control Systems*, eds. A. Kandel and G. Langholz. Boca Raton, Florida: CRC Press.

Pedrycz, W. 1993. *Fuzzy Control and Fuzzy Systems*. New York: RSP Press.

Scarpelli, H., and F. Gomide. 1994. Discovering potential inconsistencies in fuzzy knowledge bases using high level nets. *Fuzzy Sets and systems* 64:175–93.

Takagi, T., and M. Sugeno. 1983. Derivation of fuzzy control rules from human operator's control actions. *Proc. of the IFAC Symp. on Fuzzy Information, Knowledge Representation and Decision Analysis*, Marseilles, 55–60.

Tsukamoto, Y. 1979. An approach to fuzzy reasoning methods. In *Advances in Fuzzy Set Theory*, eds. R. Ragade and R. Yager, 137–49. Amsterdam: North-Holland.

Wang, L., and J. Mendel. 1992. Fuzzy basis functions, universal approximation, and orthogonal least-squares learning. *IEEE Trans. on Neural Networks* 3(5):807–14.

Yager, R., H. Larsen. 1991. On discovering potential inconsistencies in validating uncertain knowledge bases by reflecting on the input. *IEEE Trans. on Systems, Man, and Cybernetics* 21(4):790–801.

Yager, R. 1984. Approximate reasoning as a basis for rule-based expert systems. *IEEE Trans. on Systems, Man, and Cybernetics* 14(4):636–43.

Zadeh, L. A. 1975. The concept of a linguistic variable and its application to approximate reasoning (parts 1 and 2). *Information Sciences* 8:199–249, 301–57.

Zadeh, L. A. 1988. Fuzzy logic. *IEEE Computer* 21(4):83–93.

Zadeh, L. A. 1989. Knowledge representation in fuzzy logic. *IEEE Trans. on Knowledge and Data Engineering* 11(1):89–100.

Zadeh, L. A. 1994a. Fuzzy logic, neural networks, and soft computing. *Comm. of the ACM* 37(3):77–84.

Zadeh, L. A. 1994b. Soft computing and fuzzy logic. *IEEE Software* 11(6):48–56.

Fuzzy Neurocomputation

This chapter aims to expose the reader to fuzzy neural networks and fuzzy neurocomputation, which are regarded as a new paradigm of intelligent computation. This paradigm hinges on the concepts of fuzzy sets, especially their logic operations, which within this setting are converted into flexible computational tools. We begin with a brief introduction to neural networks, a concise prerequisite to this class of parallel processing. These networks are then contrasted with logic-oriented building blocks (fuzzy neurons) of fuzzy neural networks. Afterwards, we discuss some architectures of fuzzy neural networks and show how they can be directly established at a structural level by relying on available qualitative domain knowledge. The chapter includes a series of detailed learning schemes.

11.1 Neural Networks: Basic Notions, Architectures, and Learning

It is instructive to begin with a brief treatment of neural networks and contrast them with the systems emerging in the setting of fuzzy neurons. Neural networks are massive parallel and distributed computational structures composed of numerous simple processing units called (artificial) neurons. Each neuron is regarded as an n-input single-output processing unit combining inputs $x_1, x_2, \ldots, x_n$ and producing a single output y. The processing is static and described by the relationship

$$y = f(w_0 + w_1 x_1 + \cdots + w_n x_n),$$

where $w_0, w_1, \ldots, w_n$ are the neuron's connections (weights). The first weight, w_0, is usually referred to as a bias. The function f is specified as a nonlinear and monotonically increasing function from $\mathbf{R}$ to $[0,1]$ or $[-1,1]$. Table 11.1 summarizes some commonly used instances of this mapping.

Any neural network is constructed using a collection of these neurons by organizing them in a certain layered topology. Several layers are usually hidden between the input and output layer, as figure 11.1 shows.

Neural networks are universal approximators; theorem 11.1 articulates this property (Hecht-Nielsen 1991; Grossberg 1988). The theorem has its roots in Kolmogorov's mathematical findings.

THEOREM 11.1   A continuous function

$$f : [0, 1]^n \to \mathbf{R}$$

can be approximated to any desired accuracy be a three-layer feedforward neural network.

The theorem is clearly existential in nature; it states that a network can

Table 11.1
Examples of nonlinear functions used in neurons

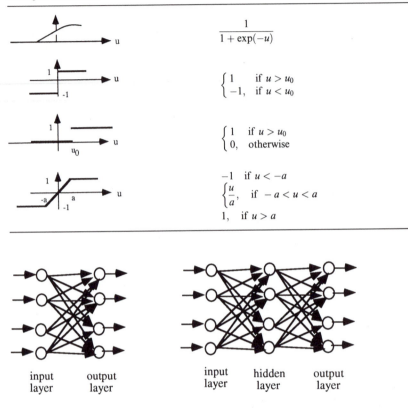

$$\frac{1}{1 + \exp(-u)}$$

$$\begin{cases} 1 & \text{if } u > u_0 \\ -1, & \text{if } u < u_0 \end{cases}$$

$$\begin{cases} 1 & \text{if } u > u_0 \\ 0, & \text{otherwise} \end{cases}$$

$$\begin{aligned} & -1 \quad \text{if } u < -a \\ & \begin{cases} \dfrac{u}{a}, & \text{if } -a < u < a \\ 1, & \text{if } u > a \end{cases} \end{aligned}$$

input   output   input   hidden   output
layer   layer   layer   layer   layer

Figure 11.1
Neural networks with and without hidden layers

be constructed realizing any nonlinear mapping, but it offers no hint on *how* this can efficiently be accomplished.

Depending on how the network processes the flow of information, two classes of architectures are distinguished: feedforward neural networks and neural networks with feedback. In feedforward neural networks, signals are propagated only in one direction, starting from the input layer, going through the successive hidden layers, and reaching the output layer. In neural networks with feedback, signals are directed back as the inputs of the network or some of their processing elements in the hidden layer, as figure 11.2 depicts. Supervised learning in these networks is commonly concentrated on a parametric level. Given a fixed architecture for the network, the crux of this type of learning is to modify the neurons' connections (weights) to minimize a given performance index.

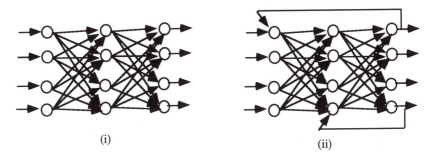

(i)   (ii)

Figure 11.2
Examples of neural networks (i) feedforward and (ii) with feedback

11.2   Logic-Based Neurons

In this section we introduce and study basic properties of fuzzy neurons developed with the aid of logic operations (Pedrycz 1991, 1993a, 1993b; Pedrycz and Rocha 1993). (See chapter 2.) Let us consider a collection of inputs x_i, $i = 1, 2, \ldots, n$, and arrange them in a vector form ($\mathbf{x}$). They can subsequently be viewed as the elements of a unit hypercube, $\mathbf{x} \in [0, 1]^n$. The connections of the neurons distributed again in the unit hypercube are denoted by $\mathbf{w}, \mathbf{v}, \ldots$, and so forth. The first class of neurons (called aggregative logic neurons) realizes an aggregation of the input signals, whereas the second is oriented toward referential processing. Figure 11.3 offers a more detailed diagram outlining this taxonomy.

11.2.1   Aggregative OR and AND Logic Neurons

The OR neuron realizes a mapping $[0, 1]^n \to [0, 1]$ and is described as

$$y = \text{OR}(\mathbf{x}; \mathbf{w}), \tag{11.1}$$

with its coordinatewise description given by

$$y = \text{OR}[x_1 \text{ AND } w_1, x_2 \text{ AND } w_2, \ldots, x_n \text{ AND } w_n],$$

where $\mathbf{w} = [w_1, w_2, \ldots, w_n] \in [0, 1]^n$ summarizes a collection of the neuron's connections (weights).

The standard implementation of fuzzy set connectives involves triangular norms, which means that the OR and AND operators are realized by some s- and t-norms, respectively. This produces the following expression:

$$y = \mathop{S}_{i=1}^{n} [x_i \text{ t } w_i].$$

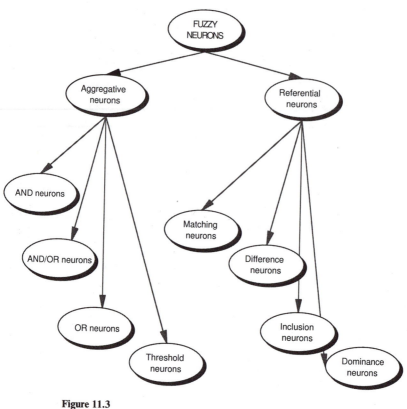

Figure 11.3
Fuzzy neurons: a taxonomy

In the AND neuron, the OR and AND operators are reversed, producing the following expression,

$$y = \text{AND}(\mathbf{x}; \mathbf{w}),$$ (11.2)

which, using the notation of triangular norms, is expressed as

$$y = \overset{n}{\underset{i=1}{T}} [x_i \ s \ w_i]$$

The AND and OR neurons realize "pure" logic operations on the input values. The role of the connections is to differentiate between particular levels of impact that the individual inputs may have on the result of aggregation. In particular, let us consider all the connections of the neuron equal to 0 or 1. For the OR neuron with $\mathbf{w} = 1$, its formula reduces to the form

$$y = x_1 \ \text{OR} \ x_2 \ \text{OR} \dots \text{OR} \ x_n,$$

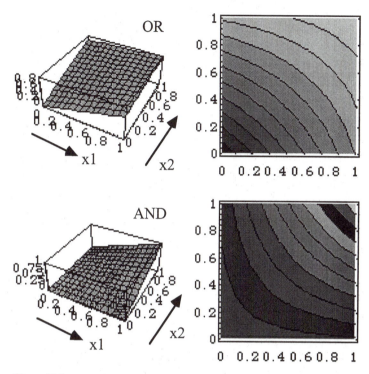

Figure 11.4
Characteristics of OR and AND neurons

whereas the connections $\mathbf{w} = 0$ yield $y = 0$. For the AND neuron, one gets

- For $\mathbf{w} = 0$: $y = x_1$ AND x_2 AND $\ldots$ AND x_n.
- For $\mathbf{w} = 1$: $y = 1$.

In general, despite the form of the triangular norms used in these constructs, the characteristics of the neurons coincide at the vertices of the unit hypercube and produce the logical values consistent with the two-valued OR or AND connective. Figure 11.4 shows some characteristics of logic neurons for several selected triangular norms.

Because of triangular norms' boundary conditions, we conclude that higher values for the connections in the OR neuron emphasize that the corresponding inputs exert a stronger influence on the neuron's output. The opposite weighting (ranking) effect takes place in the case of the AND neuron: Here the values of w_i close to 1 make the influence of x_i almost negligible.

EXERCISE 11.1   Determine (plot) characteristics of logic neurons for some other triangular norms (such as, e.g., Lukasiewicz connectives).

EXERCISE 11.2   In decision making, we are instantaneously faced with problems of an inherently logical character. For instance, assume that a decision depends on three factors (criteria): energy efficiency, reliability, and accuracy. Assume that all are quantified in [0,1], where 1 connotes the highest level of the corresponding criterion. What neuron would be appropriate to handle this problem? Identify its inputs and outputs. Justify your choice. If the criterion of efficiency is far more important than that of reliability, what are the values of the neuron's corresponding connections?

11.2.2 OR/AND Neurons

As a straightforward extension of the two aggregative neurons discussed so far, we introduce a neuron with intermediate logical characteristics. The OR/AND neuron (Hirota and Pedrycz 1994) is constructed by bundling several AND and OR neurons into a single two-layer structure, as shown in figure 11.5.

The main reason for combining several neurons and considering them as a single computational entity (which, in fact, constitutes a small fuzzy neural network) lies in this combination's ability to synthesize intermediate logical characteristics. These characteristics can be located anywhere between the "pure" OR and AND characteristics generated by the neurons previously discussed. The influence of the OR (AND) part of this neuron can be properly balanced by selecting suitable values of the connections v_1 and v_2 during the neuron's learning. At one limit, when $v_1 = 1$ and $v_2 = 0$, the OR/AND neuron operates as a pure AND neuron. At the other extreme, when $v_1 = 0$ and $v_2 = 1$, the structure

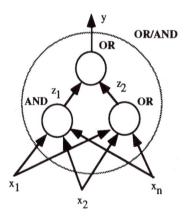

Figure 11.5
Architecture of an OR/AND neuron

functions as a pure OR neuron. The notation

$$y = \text{OR}/\text{AND}(\mathbf{x}; \mathbf{w}, \mathbf{v})$$

clearly emphasizes the nature of the intermediate characteristics the neuron produces. The relevant detailed formulas describing this architecture read as

$$y = \text{OR}([z_1 z_2]; \mathbf{v}),$$

$$z_1 = \text{AND}(\mathbf{x}; \mathbf{w}_1),$$

$$z_2 = \text{OR}(\mathbf{x}; \mathbf{w}_2),$$

with $\mathbf{v} = [v_1 v_2]$, $\mathbf{w}_i = [w_{i_1} w_{i_2}]$, $i = 1, 2$, being the corresponding neurons' connections. We can encapsulate the above expressions in a single formula,

$$y = \text{OR}/\text{AND}(\mathbf{x}; \textbf{connections}), \tag{11.3}$$

where the **connections** summarize all the network's connections.

11.2.3   Conceptual and Computational Augmentations of Fuzzy Neurons

Two further enhancements of the fuzzy neurons are aimed at making these logical processing units more flexible from a conceptual as well as a computational point of view.

11.2.3.1   Representing Inhibitory Information

Because the coding range commonly encountered in fuzzy sets constitutes the unit interval, the inhibitory effect to be conveyed by some variables can be achieved by including their complements instead of the direct variables themselves, say $\bar{x}_i = 1 - x_i$. Hence, the higher the value of x_i, the lower the activation level associated with it. Thus the original input space $[0, 1]^n$ is augmented, and the neurons are now described as follows

- OR neuron:

$$y = \text{OR}([x_1 x_2 \ldots x_n \bar{x}_1 \bar{x}_2 \ldots \bar{x}_n]; [w_1 w_2 \ldots w_n w_{n+1} w_{n+2} \ldots w_{2n}])$$

- AND neuron:

$$y = \text{AND}([x_1 x_2 \ldots x_n \bar{x}_1 \bar{x}_2 \ldots \bar{x}_n]; [w_1 w_2 \ldots w_n w_{n+1} w_{n+2} \ldots w_{2n}])$$

The reader familiar with two-valued digital systems and their design can easily recognize that any OR neuron acts as a generalized maxterm (Schneeweis 1989) summarizing the x_i and their complements,.whereas the AND neurons can be viewed as the generalization of miniterms (product terms) encountered in digital circuits. Symbolically, the complemented variable (input) is denoted by a small dot, as depicted in figure 11.6(a).

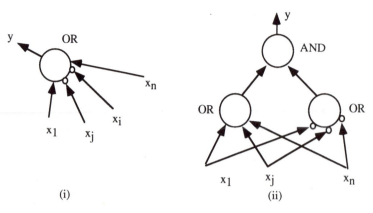

Figure 11.6
Representing inhibitory and excitatory information in logic neurons

EXERCISE 11.3   Returning to the problem posed in exercise 11.2, consider an additional criterion such as price. Again this variable is normalized to the unit interval, with 1 denoting the maximal cost of the alternative. How could the criterion be added to the neuron? Draw the relevant neuron.

Another way of representing inhibitory information in the neuron (or the network) is to split all the inputs (both inhibitory and excitatory) instead of combining them and then expand the architecture by adding a few extra neurons. Figure 11.6(b) illustrates the concept. The inhibitory inputs are first aggregated in the second neuron. The third neuron, situated in the outer layer, combines the neurons' outputs ANDwise. Consider that all the inhibitory inputs are each equal to 1. The output z_2 is then equal to 0. Assuming that the connections of the AND neuron are equal to $\emptyset$, this reduces the value of the output (y) to $\emptyset$ as well.

11.2.3.2   Computational Enhancements of Fuzzy Neurons

Despite fuzzy neurons' well-defined semantics, the main concern one may eventually raise about these constructs happens to be numerical. Once the connections (weights) are set (after learning), each neuron realizes an *in* (rather than an *on*) mapping between the unit hypercubes, which means that the values of the output y for all possible inputs cover a subset of the unit interval but not necessarily the entire interval $[0, 1]$. More specifically, for the OR neuron, the values of y are included in $[0, S_{i=1}^{n} w_i]$ whereas the accessible range of the output values for the AND neuron is limited to $[T_{i=1}^{n} w_i, 1]$. As figure 11.7 shows, augmenting the neuron with

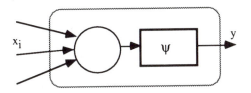

Figure 11.7
Fuzzy neuron with a nonlinear processing element

a nonlinear element placed in series with the previous logical component can alleviate this observed shortcoming. Neurons obtained in this manner are formalized accordingly:

$$y = \psi \left(\overset{n}{\underset{i=1}{S}} (x_i \ t \ w_i) \right) = \psi(u), \tag{11.4}$$

$$y = \psi \left(\overset{n}{\underset{i=1}{T}} (x_i \ s \ w_i) \right) = \psi(u), \tag{11.5}$$

where $\psi : [0, 1] \rightarrow [0, 1]$ is a nonlinear monotonic mapping. In general, we can even introduce mappings whose monotonicity is restricted to some regions of the unit interval.

A useful two-parametric family of sigmoidal nonlinearities is specified in the form

$$y = \psi(u) = \frac{1}{1 + \exp[-(u - m)\sigma]},$$

$u, m \in [0, 1]$, $\sigma \in \mathbf{R}$. By adjusting the parameters of the function (that is, σ and m), various forms of the nonlinear characteristics of the element can be easily obtained. In particular, the values of σ determine the characteristics of the obtained neuron as either increasing or decreasing, whereas the second parameter (m) shifts all the characteristics along the unit interval.

The incorporation of this nonlinearity changes the neuron's numerical characteristics, but its essential logical behavior is sustained. Figures 11.8 and 11.9 summarize some of the static input-output relationships encountered in such neurons (with the triangular norms set up as the product and probabilistic sum).

The nonlinearities of logic neurons can be treated as efficient models of linguistic modifiers (Zadeh 1983). Since their very inception, the issue of linguistic modifiers (hedges) in fuzzy computations has been studied in various contexts (Yager 1983; Kacprzyk 1985). The calculus of these objects has been also developed in different ways (cf. Kacprzyk 1986). In

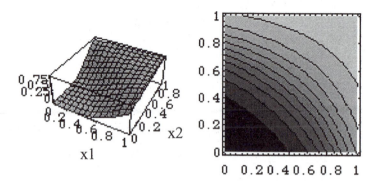

Figure 11.8
Nonlinear characteristics of the OR neuron, with $w_1 = 0.7$, $w_2 = 0.8$; $m = 0.6$; $\sigma = 10$

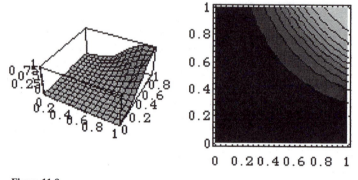

Figure 11.9
Nonlinear characteristics of the AND neuron with $w_1 = 0.1$, $w_2 = 0.05$; $m = 0.6$; $\sigma = 10$

fuzzy neurocomputations, the nonlinear element used in the neurons discussed can be viewed as the linguistic modifier defined over [0,1]. Depending upon the specific value of the parameter σ, we are looking at the modifiers of the type *at least* (realized as a monotonically increasing function, $\sigma > 0$) or *at most* (a monotonically decreasing function, $\sigma < 0$). The modifier is calibrated as necessary, based on the data provided, through the learning of the neuron as clarified before. This, in particular, pertains to the parameters of this nonlinearity (m and σ). One can model some other linguistic modifiers in the same way; figure 11.10 includes some additional examples of the quantifiers.

In a nutshell, the use of the modifier produces a "weightless" neuron in which all the connections are equal, $w_i = w$, $i = 1, 2, \ldots, n$. Hence the neuron's flexibility resides to a significant extent within the modifier's parameters.

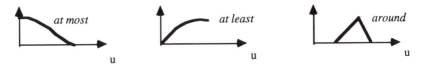

Figure 11.10
Examples of linguistic modifiers

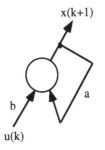

x(k+1)

a

b

u(k)

Figure 11.11
A logic neuron with feedback

11.3   Logic Neurons and Fuzzy Neural Networks with Feedback

The logic neurons studied so far realize a static, memoryless nonlinear mapping in which the output depends solely upon the neuron's inputs. In this form, the neurons are not capable of handling dynamical (memory-based) relationships between the inputs and outputs, a capacity which might however, be essential in a proper problem description.

As an example, consider the following diagnostic problem. A decision about a system's failure should be issued while one of the systems' sensors is providing information about an abnormal (elevated) temperature. Observe that the duration of the phenomenon has a primordial impact on expressing confidence about the failure. As the elevation of the temperature is prolonged, confidence about the failure rises. On the other hand, some short temporary temperature elevations (spikes) reported by the sensor might be almost ignored (filtered out) and need not have any impact on the decision about failure. To properly capture this dynamical effect, one must equip the basic logic neuron with a certain feedback link. Figure 11.11 illustrates schematically a straightforward extension of this nature.

A neuron with feedback is described as follows:

$$x(k+1) = [b \text{ OR } u(k)] \text{ AND } [a \text{ OR } x(k)] \tag{11.6}$$

The feedback connection (a) uniquely defines the dynamics of the neuron obtained in this fashion as the connection determines the speed of evidence accumulation, $x(k)$. The initial condition, $x(0)$, expresses a priori confidence associated with x. After a sufficiently long period of time $x(k + 1)$ could take on higher vales in comparison to the level of the original evidence present at input. Figure 11.12 summarizes the neuron's dynamical behavior, with the positive and negative feedback displayed in successive discrete time instants.

High-order dynamical dependencies to be accommodated by the network, if necessary, must be accommodated via a feedback loop consolidating several pieces of a temporal information; for example,

$$x(k + 2) = [b \text{ OR } u(k)] \text{ AND } [a_1 \text{ OR } x(k)] \text{ AND } [a_2 \text{ OR } x(k + 1)].$$

$$(11.7)$$

One can also refer to (11.6) and (11.7) as examples of *fuzzy difference equations*.

The limit analysis of the neuron (or network) with feedback allows us to explore some general dynamical properties of the system. As an example, let us study the neuron described by the relationship

$$x(k + 1) = \bar{x}(k) \text{ OR } a,$$

where a characterizes the feedback loop, $a \in [0, 1]$. Additionally, let the s-norm be given as the probabilistic sum. This yields

$$x(k + 1) = a + (1 - x(k)) - a(1 - x(k)).$$

Iterating from $x(0)$, that is, computing $x(1), x(2), \ldots$, we can unveil the system's steady-state behavior (assuming that it exists). The above example is simple enough to analyze it in detail. Let the steady state be $x(\infty)$. We then get

$$x(\infty) = a + \bar{x}(\infty) - a\bar{x}(\infty).$$

Rearranging the terms, one obtains

$$x(\infty)(2 - a) = 1$$

and finally

$$x(\infty) = \frac{1}{2 - a}.$$

As the simulations in figure 11.13 reveal, for the connection in (0,1), the neuron does not converge to any specific value but oscillates continuously, assuming the states between 0 and 1. In fact, the value $x(\infty)$

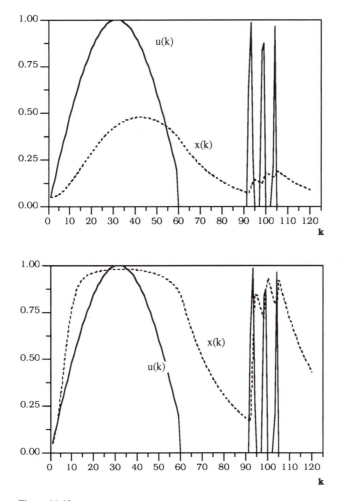

Figure 11.12
Dynamics of the neurons with positive or negative feedback in discrete time: (top) positive
feedback $x(k+1) = \mathrm{OR}([ab][x(k)u])$; (bottom) negative feedback $x(k+1) = \mathrm{OR}([ab][\bar{x}(k)u])$;
t-norm: product; s-norm: probabilistic sum; $a = 0.6$; initial condition, $x(0) = 0.3$

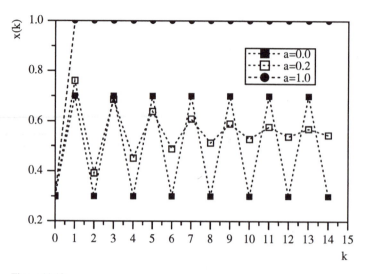

Figure 11.13
States of a fuzzy neuron $x(k+1) = a$ OR $\bar{x}(k)$ for several values of the feedback connection; initial condition $x(0) = 0.3$

becomes an average of the states at which the neuron can reside. The same observation holds for the remaining values of the feedback parameter, except that lower values of a yield oscillations of a more profound amplitude.

Let us now study the neuron with feedback governed by the expression

$$x(k+1) = x(k) \text{ OR } a.$$

Confining ourselves to the probabilistic sum, we derive

$$x(k+1) = a + x(k) - ax(k).$$

Steady-state analysis leads to the following expression:

$$x(\infty) = 1.$$

In this case the neuron converges to $x(\infty)$, as confirmed by the simulation experiments shown in figure 11.14. Here the speed of convergence depends on the feedback connection.

EXERCISE 11.4   A system is described as follows: "The concentration of substance depends on its previous level and is inhibitively affected by substrate x_1. The other substrate whose concentration is denoted by x_2 has a positive influence on the concentration of this substance."

Convert this description into a logic model using fuzzy neurons. Experiment with the network. In particular, analyze it for several triangular

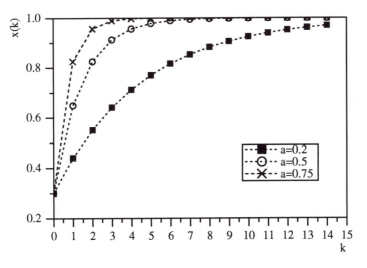

Figure 11.14
States of a fuzzy neuron $x(k+1) = a$ OR $x(k)$ for several values of the feedback connection; initial condition $x(0) = 0.3$

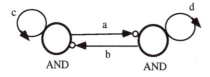

AND   AND

Figure 11.15
Logic Oscillator

norms and selected values of the connections. What is the concentration's steady state? Can you determine this value analytically, without resorting to simulations? Verify your findings experimentally.

EXERCISE 11.5   The network shown in figure 11.15 can be regarded as a logic oscillator. Write a formal description of it in the form of a fuzzy neural network. Assuming some values of the connections (a, b, c, d), analyze the behavior of the network for time (k) approaching infinity. Run some experiments to confirm your findings.

11.4   Referential Logic-Based Neurons

In comparison to the AND, OR and OR/AND neurons, which perform operations of an aggregative character, referential logic-based neurons are useful in performing reference computations. The main idea behind these neurons is that the input signals are not directly aggregated, since this has been done in the aggregative neuron, but rather than that they are analyzed first (e.g., compared) with respect to a given reference point. The results of this analysis (involving such operations as matching, inclusion, difference, and dominance) are afterwards summarized in the aggregative part of the neuron as described before. In general, one can describe the reference neuron as a disjunctive form of aggregation,

$$y = \text{OR}(\text{REF}(\mathbf{x}; \textbf{reference\_point}), \mathbf{w}),$$

or a conjunctive form of aggregation,

$$y = \text{AND}(\text{REF}(\mathbf{x}; \textbf{reference\_point}), \mathbf{w}),$$

where REF(.) stands for the reference operation carried out with respect to the point of reference provided.

The neuron's functional behavior is described accord to the reference operation performed (All formulas below pertain to the disjunctive form of aggregation):

1. MATCH neuron:

$$y = \text{MATCH}(\mathbf{x}; \mathbf{r}, \mathbf{w}) \tag{11.8}$$

or equivalently,

$$y = \overset{n}{\underset{i=1}{S}} \left[w_i \text{ t } (x_i \approx r_i) \right],$$

where $\mathbf{r} \in [0, 1]^n$ stands for a reference point defined in the unit hypercube, as discussed in chapter 2. To emphasize the referential character of the processing this neuron carries out, one can rewrite (11.8) as

$$y = \text{OR}(\mathbf{x} \approx \mathbf{r}; \mathbf{w}).$$

See also figure 11.16. The use of the OR neuron indicates the final aggregation has an "optimistic" (disjunctive) character. The pessimistic form of this aggregation is produced by using the AND operation.

2. DIFFER neuron: The difference neuron combines degrees to which $\mathbf{x}$ is different from the given reference point $\mathbf{g} = [g_1, g_2, \ldots, g_n]$. The output is interpreted as a global level of difference observed between the input $\mathbf{x}$ and this reference point (Pedrycz 1990),

$$y = \text{DIFFER}(\mathbf{x}; \mathbf{w}, \mathbf{g}), \tag{11.9}$$

that is

$$y = \overset{n}{\underset{i=1}{S}} \left[w_i \text{ t } (x_i \approx| g_i) \right],$$

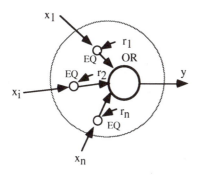

Figure 11.16
Referential neuron as a superposition of referential and aggregative computation

where the difference operator is defined as a complement of the equality index,

$$a \approx\!\!\mid b = 1 - (a \approx b).$$

As before, the referential character of processing is emphasized by noting that

$$\text{DIFFER}(\mathbf{x}; \mathbf{w}, \mathbf{g}) = \text{OR}(\mathbf{x} \approx\!\!\mid \mathbf{g}; \mathbf{w})$$

3. INCL neuron: The inclusion neuron summarizes the degrees to which **x** is included in the reference point **f**,

$$y = \text{INCL}(\mathbf{x}; \mathbf{w}, \mathbf{f}),$$

$$y = \overset{n}{\underset{i=1}{S}} [w_i \, t \, (x_i \rightarrow f_i)],$$

Inclusion is expressed in the sense of the pseudocomplement operation (implication). The properties of the φ-operator (implication),

If $a < b$, then $a \, \varphi \, b = 1$,

If $a > b' > b$, then $a \, \varphi \, b' \geq a \, \varphi \, b$,

$a, b, b' \in [0, 1]$, assure that the neuron's output becomes a monotonic function of the degree of satisfaction of the inclusion property.

4. DOM neuron: The dominance neuron expresses a relationship complementary to that carried out by the inclusion neuron

$$y = \text{DOM}(\mathbf{x}; \mathbf{w}, \mathbf{h}),$$

where **h** is a reference point. In other words, the dominance relationship generates degrees to which **x** dominates **h** (or, equivalently, **h** is dominated by **x**). The coordinatewise notation of the neuron reads as

$$y = \mathop{S}_{i=1}^{n} \left[w_i \, t \, (h_i \rightarrow x_i) \right] .$$

Figure 11.17 portrays some characteristics of the referential neurons.

To model complex situations, the referential neurons can be encapsulated into a form of neural network. An example is a tolerance neuron, which consists of DOM and INCL neurons placed in the hidden layer and a single AND neuron in the output layer, as figure 11.18 shows. The tolerance neuron generates a tolerance region, as figure 11.19 shows.

EXERCISE 11.6   Propose a referential neuron that implements the weighted Tschebyschev distance

$$\|x - y\| = \max(w_1|x_1 - y_1|, w_2|x_2 - y_2|, \ldots, w_n|x_n - y_n|),$$

where $\mathbf{x} = [x_1 x_2 \ldots x_n]$, $\mathbf{y} = [y_1 y_2 \ldots y_n]$, $\mathbf{w} = [w_1 w_2 \ldots w_n]$.

11.5   Fuzzy Threshold Neurons

Fuzzy threshold neurons, constituting a straightforward generalization of threshold computing units (threshold gates) (Muroga 1971), are formed by a serial composition of the aggregative neuron followed by the inclusion operator equipped with a one-dimensional reference element (λ). As such, the fuzzy threshold neuron is a straightforward generalization of a two-valued threshold element,

$$y = \text{INCL}(\lambda; \text{OR}(\mathbf{x}; \mathbf{w})) = \lambda \rightarrow \text{OR}(\mathbf{x}; \mathbf{w}),$$

where $\lambda \in [0, 1]$ denotes a threshold level. The values of the $\text{OR}(\mathbf{x}; \mathbf{w})$ unit that exceed this threshold are elevated to 1 (see figure 11.20). In particular, when the threshold λ approaches $\varnothing$, the neuron behaves very much like an on-off device.

11.6   Classes of Fuzzy Neural Networks

We have proposed several clearly distinct types of fuzzy neurons, which could potentially give rise to a tremendous abundance of networks (fuzzy neural networks). The following sections show the variety of such schemes. We introduce and study two specific architectures that, because of their functional characteristics, are encountered in many applications forming an essential part of overall processing structures: logic processors (networks performing tasks of logic-oriented approximation) and referential processors (networks for mapping referential properties between input and output spaces).

11.6.1   Approximation of Logical Relationships—Development of the Logic Processor

An important class of fuzzy neural networks concerns approximation of mappings between the unit hypercubes (namely, from $[0, 1]^n$ to $[0, 1]^m$ or $[0,1]$ for $m = 1$) realized in a logic-based format. To comprehend the fundamental idea behind this architecture fully, let us recall some very simple yet powerful concepts from the realm of two-valued systems. The well-known Shannon's theorem (Schneeweiss 1989) states that any Boolean function $\{0, 1\}^n \rightarrow \{0, 1\}$ can be uniquely represented as a logical sum (union) of minterms (a so-called SOM representation) or, equivalently, a product of some maxterms (known as a POM representation). By a minterm, we mean an AND combination of all input variables of the function; they could appear in either a direct or a complemented (negated) form. Similarly, a maxterm consists of the variables in their OR combination. A complete list of minterms and maxterms for Boolean functions of two variables includes

- Minterms: $\bar{x}_1$ AND $\bar{x}_2$, x_1 AND $\bar{x}_2$, $\bar{x}_1$ AND x_2, x_1 AND x_2.
- Maxterms: $\bar{x}_1$ OR $\bar{x}_2$, x_1 OR $\bar{x}_2$, $\bar{x}_1$ OR x_2, x_1 OR x_2.

From a functional point of view, the minterms can be identified with the AND neurons, and the OR neurons can be used to produce the corresponding maxterms. The connections of these neurons are restricted to the two-valued set $\{0, 1\}$, making these neurons two-valued selectors. Considering how Boolean functions are represented, two complementary (dual) architectures are envisioned. In the first, the network includes a single hidden layer, constructed with the aid of the AND neurons, followed by the output layer, consisting of the OR neurons (the SOM version of the network). In the second, dual type of the network, the hidden layer has some OR neurons, and the output layer is formed by the AND neurons (the POM version of the network). The generalization of these networks for the continuous case of the input-output variables is referred to as a logic processor. Analogously to those of the networks sketched so far in the Boolean situation, we are interested in the two topologies of the logic processor (LP), namely, its POM and SOM versions (figure 11.21).

Depending on the values of m we will be referring to a scalar or vector version of the logic processor. Its scalar version, $m = 1$, could be viewed as a generic LP architecture.

Logic processors in their continuous and two-valued versions have at least one crucial difference:

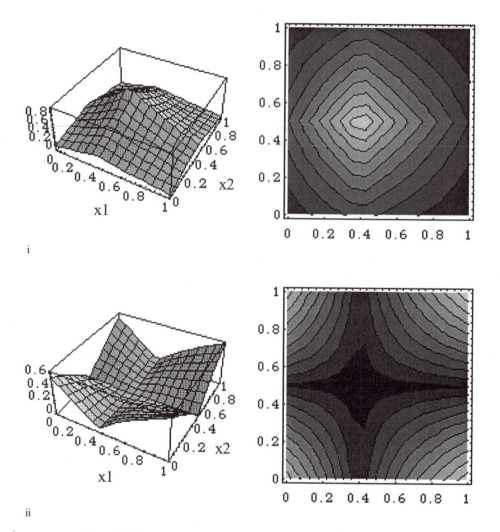

i

ii

Figure 11.17
Characteristics of referential neurons: (i) equality neuron; (ii) difference neuron; (iii) inclusion neuron; (iv) dominance neuron; s-norm: probabilistic sum, t-norm: product, implication induced by product $w = [0.05 \ 0.05]$ $r = [0.4 \ 0.5]$

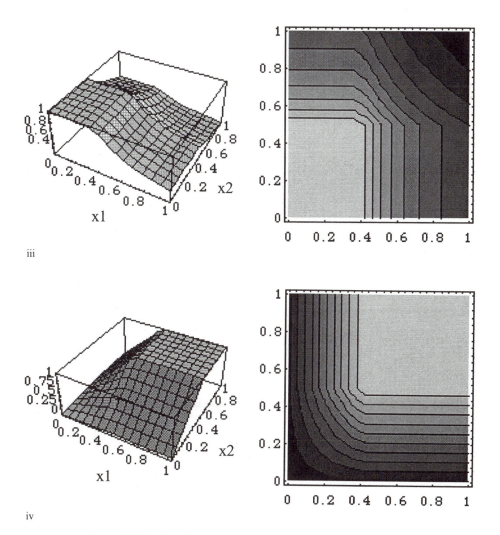

iii

iv

Figure 11.17 (continued)

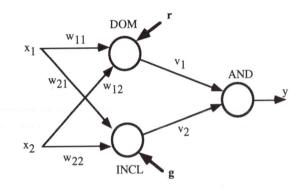

Figure 11.18
Tolerance neuron

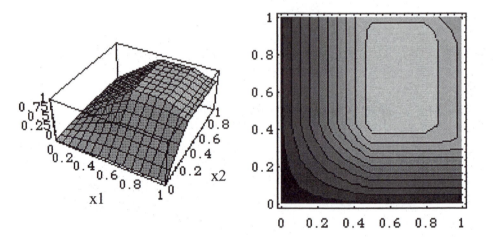

Figure 11.19
2D and 3D characteristics of a tolerance neuron: AND neuron: min operator; INCL and DOM neuron: $a \rightarrow b = \min(1, b/a), a, b \in [0, 1]$; $w_{ij} = 0.05, v_i = 0.0$; Reference points: INCL neuron: $\mathbf{r} = [0.8 \;\; 0.9]$, DOM neuron: $\mathbf{g} = [0.5 \;\; 0.4]$

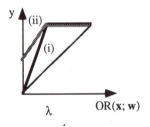

Figure 11.20
Characteristics of a single-input threshold neuron: (i) $\lambda \rightarrow OR(\mathbf{x}; \mathbf{w}) = \min(1, OR(\mathbf{x}; \mathbf{w})/\lambda)$;
(ii) $\lambda \rightarrow OR(\mathbf{x}; \mathbf{w}) = \min(1, 1 - \lambda + OR(\mathbf{x}; \mathbf{w}))$

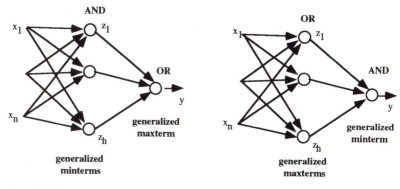

Figure 11.21
SOM and POM versions of a logic processor

1. The logic processor used for Boolean data *represents* or *approximates* data. Assuming that all the input combinations are different, we are talking about a representation of the corresponding Boolean function. The POM and SOM versions of the logic processors for the same Boolean function are equivalent.

2. The logic processor used for fuzzy (continuous) data approximates a certain unknown fuzzy function. The equivalence of the POM and SOM types of the obtained LPs is not at all guaranteed.

When necessary, we use a concise notation $LP(\mathbf{x}; \mathbf{w}, \mathbf{v})$ to describe the network with the connections $\mathbf{w}$ and $\mathbf{v}$ standing between the successive layers. The detailed formulas are

• SOM version:

$z_i = \text{AND}(\mathbf{x}; \mathbf{v}_i),$

$y = \text{OR}(\mathbf{z}; \mathbf{w}),$

$i = 1, 2, \ldots, h.$

• POM version:

$z_i = \text{OR}(\mathbf{x}; \mathbf{v}_i),$

$y = \text{AND}(\mathbf{z}; \mathbf{w}),$

$i = 1, 2, \ldots, h.$

In both versions, $\mathbf{v}_1, \mathbf{v}_2, \ldots, \mathbf{v}_h$ are the connections between the input and hidden layer, whereas $\mathbf{w}$ summarizes the connections between the hidden and output layer.

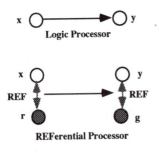

Figure 11.22
Logic processing and referential processing

11.7   Referential Processor

Whereas the logic processor's role is to implement the logic-based approximation of mappings between the unit hypercubes, the essence of the processing conducted by a referential processor is concerned with the mapping of some referential properties between the input and output spaces. Figure 11.22 highlights these differences in more detail.

Among the various types of referential computations, one definitely worth discussing deals with analogical reasoning. This form of reasoning involves inferring similarities between certain prototypes and current inputs. In fact, this scheme of reasoning has been found useful in a number of different fields, including pattern recognition, machine learning, reasoning by analogy, and the like. Let us consider a reference pair $(\mathbf{r}, \mathbf{g})$ as a given pair of associations, $\mathbf{r} \in [0, 1]^n$, $\mathbf{g} \in [0, 1]^m$. Qualitatively, the scheme reads as

$\mathbf{x}$ and $\mathbf{r}$ are *similar*
$\mathbf{r}, \mathbf{g}$ are associated

$\mathbf{y}$ and $\mathbf{g}$ are *similar*

and entails two steps:

1. determination (quantification) of similarity between $\mathbf{y}$ and $\mathbf{g}$

2. determination of $\mathbf{y}$ based on $\mathbf{g}$ and the level of similarity computed in step 1

Naturally, one could anticipate that the more similar $\mathbf{x}$ and $\mathbf{r}$ are, the higher the similarity level that can be expected for the objects situated in the output space ($\mathbf{y}$ and $\mathbf{g}$). Figure 11.23 depicts the architecture of the referential (in particular, the analogical) processor in this case. This processor is in fact based upon the logic processor now used to transform

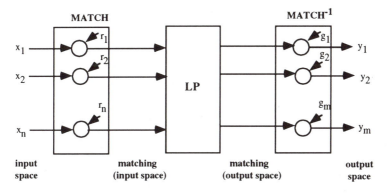

Figure 11.23
General architecture of the analogical processor

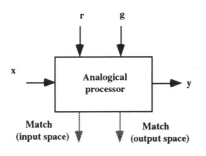

Figure 11.24
Analogical processor: a functional view

the referential property of matching. Symbolically, one can express this function as

$$(\text{matching})_{\text{output space}} = LP((\text{matching})_{\text{input space}}, \textbf{connections}).$$

In comparison to the plain logic processor, this architecture is augmented by two additional layers. The input layer (MATCH) carries out matching (realized through some matching neurons); the output layer (MATCH^{-1}) converts the level of matching into the objects in the output space. From a functional point of view, one can look at the matching (analogical) processor as a static input-output structure (figure 11.24) with additional layers for preprocessing and postprocessing.

Both the analogical processor and the logic processor can be encapsulated in the form of a single universal structure (figure 11.25) in which an additional SELECT signal switches between the structures; more precisely, the switch turns the layers of the referential neurons off or on, allowing the signals to bypass these layers.

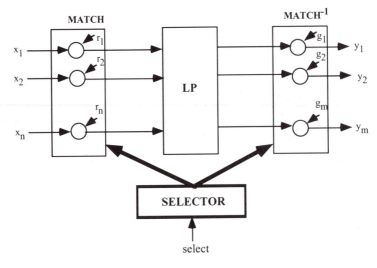

Figure 11.25
Encapsulation of logical and analogical processors

11.8   Fuzzy Cellular Automata

Fuzzy cellular automata and computations can be regarded as a characteristic approach for modeling complex systems (Weisbuch 1991; Wolfram 1994). We intend here to implement this idea with fuzzy computing. Just as cellular automata can be regarded as collections of two-valued logic processing units cooperating with their close neighbors, fuzzy cellar automata are built as collections of fuzzy neurons linked with their closest neighbors. For modeling purposes, we acknowledge that the connections are both inhibitory and excitatory and allow for some feedback links formed within the individual neurons. A simple example of a fuzzy cellular automaton is formed as a 20×20 grid of two types of referential neurons performing matching and difference computations. The neurons are distributed regularly across the grid; if the sum of the indices $(i + j)$ corresponding to the position of the neuron is even, we assign to that node a matching neuron; an odd sum of the indices places a difference neuron. Each neuron interacts with its four closest neighbors according to the relationship

$$x_{(ij)}(k + 1) = \text{REF}(\mathbf{x}_{(ij)})(k), (\mathbf{w}),$$

where

$$\mathbf{x}_{(ij)}(k) = [x_{(ij+1)}(k) x_{(ij-1)}(k) x_{(i+1,j)}(k) x_{(i-1,j)}(k)],$$

Figure 11.26
Fuzzy cellular automaton—initial activation levels

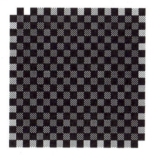

Figure 11.27
Fuzzy cellular automaton—activation levels at $k = 9$

and **w** denotes a vector of connections (weights). The neuron's position is identified by two indices (i and j), $i, j = 1, 2, \ldots, 20$. The t- and s-norms are selected as the product operator and probabilistic sum, respectively. The matching neuron uses the OR-type combination of the inputs with all the connections equal to 0.7. For the difference neuron, realized as the AND-type of aggregation of the partial results (the synaptic level), the connections are set to 0.4.

The computations begin with an initial randomly selected activation level for the neurons ($k = 0$) as shown in figure 11.26. The darker the shading of a particular node of the grid, the higher the value of the output of its associated neuron. After a few iterations, the system produces a regular and stable chessboard-like pattern of cell activations (figure 11.27).

11.9   Learning

The discussion of learning in this section is concerned with the supervised mode of learning. In general, all learning tasks can be partitioned into two groups, depending on whether they pertain to parametric or structural meaning. Most existing learning schemes are preoccupied with parametric learning, whose aim is to optimize the parameters of the fuzzy neural network. On the other hand, structural learning, which is definitely more demanding, is devoted to optimizing the network's structure. Structural learning can be accomplished in many different ways, for example, by changing the number of layers or adding, replacing, or deleting individual neurons. The techniques falling into this category require nonparametric methods, such as those coming form genetic programming and genetic algorithms.

A principal idea of parametric learning, no matter how it is implemented, can be portrayed as follows. For a given collection of input-output pairs of data $(\mathbf{x}_1, \mathbf{t}_1), \ldots, (\mathbf{x}_N, \mathbf{t}_N)$, modify the network's parameters (connections as well as reference points) to minimize the furnished performance index Q. The general scheme of learning can be qualitatively described as

$$\Delta\_\mathbf{connections} = -\xi \frac{\partial Q}{\partial\ \mathbf{connections}},$$

where ξ denotes a learning rate. Quite often the above formula is augmented by a so-called momentum term whose role is to "filter" high frequencies in the changes of the connections and assure smooth modifications of the connections. The update of the connections is governed by

$$\Delta\_\mathbf{connections}(\text{iter} + 1) = -\xi \frac{\partial Q}{\partial\ \mathbf{connections}} + \beta\Delta\_\mathbf{connections}(\text{iter}),$$

where the actual increments of the connections (at time iter $+ 1$) depend also on the previous increments at the previous iteration; β denotes a momentum rate. Subsequently, the parameters of the network are adjusted following these increments:

$$\mathbf{new\_connections} = \mathbf{connections} + \Delta\_\mathbf{connections}.$$

The learning scheme's relevant details can be fully specified once the network's topology as well as some other details regarding the form of triangular norms have been made available. Before getting into general structures of fuzzy neural networks, let us consider learning methods for a single fuzzy neuron. The network's parametric learning can be augmented by admitting neurons realized with the use of parameterized t- and s-norms. This option, signaled in Pedrycz and Rocha 1993, has since been studied in detail by Stoica (1996), who proposed several basic criteria for selecting an appropriate triangular norm:

1. Differentiability: This feature supports gradient-based learning.

2. Parametric form of the triangular norm: This supports substantial learning abilities.

3. Inclusion property: The norm should include the max and min operations as their limit cases for some specific values of the parameter used in the definition of the t- or s-norm.

The so-called fundamental t- and s-norms,

$$x \; t_s \; y = \log_s \left[\frac{1 + (s^x - 1)(s^y - 1)}{s - 1} \right], \quad s \geq 0$$

$$x \; s_s \; y = 1 - \log_s \left[\frac{1 + (s^{1-x} - 1)(s^{1-y} - 1)}{s - 1} \right], \quad s \geq 0$$

satisfy the above requirements. We have

$$x \; t_s \; y = \begin{cases} \min(x, y), s = 0 \\ xy, s = 1 \\ \max(0, x + y - 1), s = \infty \\ \log_s \left[\dfrac{1 + (s^x - 1)(s^y - 1)}{s - 1} \right], \text{otherwise,} \end{cases}$$

$$x \; s_s \; y = \begin{cases} \max(x, y), s = 0 \\ x + y - xy, s = 1 \\ \min(1, x + y), s = \infty \\ 1 - \log_s \left[\dfrac{1 + (s^{1-x} - 1)(s^{1-y} - 1)}{s - 1} \right], \text{otherwise.} \end{cases}$$

Moreover, these families of functions are continuous with respect to their parameters (Butnariu and Klement 1993) in the sense that

$$\lim_{v \to v_0} x \; t_v \; y = x \; t_{v_0} \; y$$

and

$$\lim_{v \to v_0} x \; s_v \; y = x \; s_{v_0} \; y.$$

11.9.1 Learning in a Single Neuron

The standard learning procedure for a single fuzzy neuron pertains to the parametric modifications of the connections

$$\frac{\partial d}{\partial w_i} = \frac{\partial d}{\partial z_1} \frac{\partial z_1}{\partial w_i}, \quad i = 1, 2,$$

$$\frac{\partial d}{\partial w_i} = \frac{\partial d}{\partial z_2} \frac{\partial z_2}{\partial w_i}, \quad i = 3, 4,$$

$$\frac{\partial d}{\partial v_i} = \frac{\partial \text{OR}([z_1 z_2]; [z_1 z_2])}{\partial v_i}, \quad i = 1, 2,$$

$$\frac{\partial d}{\partial z_i} = \frac{\partial \text{OR}([z_1 z_2]; [z_1 z_2])}{\partial z_i}, \quad i = 1, 2,$$

$$\frac{\partial z_1}{\partial w_i} = \frac{\partial \text{AND}([x_1 \bar{x}_2]; [w_1 w_2])}{\partial w_i}, \quad i = 1, 2,$$

$$\frac{\partial z_2}{\partial w_i} = \frac{\partial \text{SIM}([x_2 x_3]; [w_3 w_4], [r_3 r_4])}{\partial w_i}, \quad i = 3, 4.$$

Let us discuss the similarity (equality) neuron in more depth. Considering its ORwise form of aggregation, one gets

$$z_2 = \text{OR}([x_2 x_3] \approx [r_3 r_4]; [w_3 w r_4]).$$

The logic operations are instantiated accordingly: OR-maximum, AND-product. The similarity operation is induced by the Lukasiewicz implication, giving rise to the expression

$$a \approx b = \begin{cases} 1 - a + b, & \text{if } a \geq b \\ 1 - b + a, & \text{if } a < b. \end{cases}$$

Now

$$\frac{\partial z_2}{\partial w_i} = \frac{\partial}{\partial w_i}[(x_i \approx r_i)w_i \vee (x_j \approx r_j)w_j],$$

$i, j = 3, 4$, which finally produces the expression,

$$\frac{\partial}{\partial w_i}[(x_i \approx r_i)w_i \vee (x_j \approx r_j)w_j] = (x_i \approx r_i)[(x_j \approx r_j)w_j \; \varphi \; (x_i \approx r_i)w_i]$$

EXERCISE 11.7   Experiment with learning by a logic processor; analyze how different values of the learning rate affect the learning process of learning. Modify also the size of the hidden layer. In your experiment, use the data set

Inputs				Outputs		
0.0	0.4	0.9	1.0	0.3	0.2	0.8
0.6	0.2	0.7	0.0	0.6	0.3	0.7
1.0	0.2	0.9	1.0	0.9	1.0	0.8
0.0	0.6	0.9	0.3	0.0	0.0	0.8

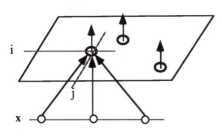

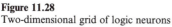

Figure 11.28
Two-dimensional grid of logic neurons

EXERCISE 11.8   Develop a complete learning algorithm for an OR/AND neuron implemented with the use of the product and probabilistic sum.

(a) Experiment with the training set from exercise 11.7.

(b) Add a momentum mechanism to the learning algorithm. Use the same data set. How does this mechanism affect the learning?

11.9.2 Self-Organization Mechanisms in Fuzzy Neural Networks

Consider learning in a network (Figueiredo, Gomide, and Pedrycz 1995) composed of an array of logic neurons situated on a two-dimensional grid (see figure 11.28). The proposed learning is unsupervised, because the patterns provided (the elements of the n-dimensional unit hypercube) $x_1, x_2, \ldots, x_N$ are unlabeled. The main idea governing this learning is that of self-organization. While one among the patterns, say x, is presented to the network, the winning neuron is determined (namely, the neuron producing the highest output). This particular neuron (the winner) takes control of all modifications of the neurons' connections. The resulting updates are carried out vigorously in neurons situated very close to the winning node; the changes become less visible among neurons positioned quite a distance from the winner.

The general scheme of unsupervised learning can be arranged in the following form (here $w_k(i, j)$ stands for the kth connection of the neuron situated at the (i, j)th coordinate of the grid):

repeat

Present all input data (e.g., going cyclically over all the patterns) to the network; declare neuron with the highest output a winner. The winning neuron is given preference to guide learning. Denote by (i^*, j^*) this neuron's coordinates. Denote by $\lambda(i, j)$ the distance between the (i, j)th node in the grid and the winning node.

Update the connections of the neurons where these modifications are dependent upon the computed distance

$$w_k(i, j; \text{iter} + 1) = (w_k(i, j; \text{iter}) + \lambda(i, j)x)/(1 + \lambda(i, j))$$

and increase the cycle counter;

until the number of learning cycles has exceeded the assumed limit.

The distance function is assumed to be in the form

$$\lambda(i, j) = \frac{\xi}{1 + \sqrt{(i^* - i)^2 + (j^* - j)^2}}$$

Note that if $i = i^*$ and $j = j^*$, $\lambda(i, j)$ assumes its maximal value equal to ξ.

11.10   Selected Aspects of Knowledge Representation in Fuzzy Neural Networks

Fuzzy neural networks are processing structures with an explicit form of knowledge representation; this is due to the individual neurons' well-defined semantics. In this section, we look at the detailed way in which these networks represent and process uncertainty. The Boolean and core neural networks induced based on the original network are used to facilitate further the ways of acquiring knowledge.

11.10.1   Representing and Processing Uncertainty

Let us begin by looking at a more detailed architecture of the fuzzy neural network as placed in the general context of fuzzy modeling (figure 11.29). The processing module is now realized as a fuzzy neural network. Because uncertainty representation resides primarily with the input interface composed of several linguistic labels and supported by the relevant transformation (matching) mechanisms, it becomes important that this representation be available to the network. Assuming that any input X is represented in the sense of the possibilities and necessities of the labels, we define the following inputs to the network:

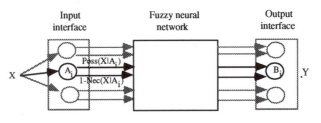

Figure 11.29
Representing uncertainty in a fuzzy neural network

$$x_i = \text{Poss}(X|A_i),$$

$$\bar{x}_i = 1 - \text{Nec}(X|A_i).$$

$i = 1, 2, \ldots, n$. Note, however, that in general, $x_i + \bar{x}_i \neq 1$. The two do not sum to 1 in particular when the possibility and necessity evaluations produce different outcomes (which usually happens when X is a non-numerical quantity). The network recognizes any departure from the unity constraint as a clear message about the uncertainty of the available information. To keep track of uncertainty propagation through the network, its output layer should consist of the outputs y_j as well as $\bar{y}_j$ regarded as the possibility and necessity values of the corresponding linguistic labels B_j forming the output interface. Hence,

$$y_j = \text{Poss}(Y|B_j),$$

$$\bar{y}_j = 1 - \text{Nec}(Y|B_j),$$

$j = 1, 2, \ldots, m$. Again, any violation of the equality $y_j + \bar{y}_j = 1$ comes as a straight signal that uncertainty exists at the network's output as a consequence of the uncertainty component associated with the input layer.

11.10.2   Induced Boolean and Core Neural Networks

An elicitation of the structure of the network can be enhanced by pruning some weaker connections of the neurons. Generally, in the OR neuron, one eliminates all the connections whose values are below a certain threshold $\lambda \in [0, 1]$. These connections are set to 0, and the values of the remaining ones are retained or elevated to 1. The opposite rule holds for the AND neuron: All connections with values above the threshold are set to 1. The neurons' threshold levels can be set up arbitrarily or may be subject to optimization.

Optimized pruning of the connections leads to a Boolean approximation of the fuzzy neural network. In optimized pruning, all of the network's connections are converted to either 0 or 1. Let $y = N(\mathbf{x}, \mathbf{w}, \mathbf{v}, \ldots)$ denote the neural network to be approximated, where $\mathbf{w}, \mathbf{v}, \ldots$ are collections of the connections (provided as matrices or vectors) between the successive layers. The idea of this approximation is to replace $N(\mathbf{x}, \mathbf{w}, \mathbf{v})$ by its Boolean counterpart, say $B(\mathbf{x}, \mathbf{w}_B, \mathbf{v}_B)$, in such a way that the Boolean network produces results as close as possible to those of the original network. The quality of approximation can be formally characterized by the performance index

$$\min_{\mathbf{w}_B, \mathbf{v}_B, \ldots} \sum_{\mathbf{x}} \|N(\mathbf{x}, \mathbf{w}, \mathbf{v}, \ldots) - B(\mathbf{x}, \mathbf{w}_B, \mathbf{v}_B, \ldots)\|,$$

where $\|\cdot\|$ stands for the distance function. The sum is taken over a certain collection of inputs X. The values of this performance index indicate how well the network can be represented by its optimal Boolean counterpart. The performance index is minimized with respect to the network's Boolean connections B while one attempts to approximate the network N over a set of inputs defined in X. More precisely, one can refer to the Boolean approximation of the network completed with respect to X. Obviously, different elements of X could result in fairly different approximations and, in sequel, different Boolean networks associated with the same fuzzy neural network. In particular, one can contemplate two specific families of inputs:

1. X is the same as the original training data set.

2. X covers the entire universe of discourse by including the elements randomly distributed in the input hypercube.

Obviously, some other options for X might be worth considering (figure 11.30).

The original multidimensional optimization task can be computationally demanding (particularly for larger fuzzy neural networks). Simplifying the strategy for building the Boolean network can greatly reduce the computational effort. The crux of this simplification is to reduce the search's dimensionality by selecting a uniform threshold strategy for all the AND and OR neurons. Let us introduce two threshold operations. The first, $I_\lambda(w)$, applies to all the OR neurons in the network and replaces the original connections by 0 or 1 depending upon their position with respect to the threshold λ:

$$T_\lambda(w) = \begin{cases} 0, & \text{if } w < \lambda \\ 1, & \text{if } w \geq \lambda, \end{cases}$$

$w \in [0, 1]$, $\lambda \in [0, 1]$. The second thresholding operation, $T_\mu(w)$, with a threshold value μ, is used for the AND neurons:

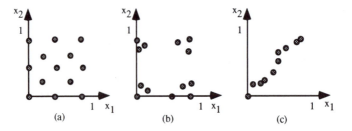

(a)   (b)   (c)

Figure 11.30
Examples of training data sets X: (a) uniform distribution in x_1-x_2 plane; (b) binary-biased (data centered around the unit square's vertices); (c) satisfying a certain functional constraint, $x_2 = f(x_1)$

$$T_\mu(w) = \begin{cases} 0, & \text{if } w \le \mu \\ 1, & \text{if } w > \mu. \end{cases}$$

With these threshold operations, the initial optimization task reduces to the following two-dimensional version:

$$\min_{\lambda,\mu} \sum_{\mathbf{x}} \| N(\mathbf{x}, \mathbf{w}, \mathbf{v}, \ldots) - B(\mathbf{x}, \mathbf{w}, \mathbf{v}, \ldots) \|,$$

which is computationally much more amenable than the previous one. Another feasible option for network induction might be to retain the neurons' most significant connections. In place of the above transformations, we can define less "drastic" modifications,

$$T_\lambda(w) = \begin{cases} 0, & \text{if } w < \lambda \\ w, & \text{if } w \ge \lambda, \end{cases}$$

$$T_\mu(w) = \begin{cases} w, & \text{if } w \le \mu \\ 1, & \text{if } w > \mu, \end{cases}$$

that preserve the values of the connections once they are recognized as essential under the assumed criteria. Figure 11.31 illustrates the thresholding operations.

11.11   Chapter Summary

Fuzzy sets and neural computing, organized together in the form of fuzzy neural networks, naturally embrace knowledge representation aspects along with significant learning features. The diversity of knowledge representation schemes fuzzy sets support helps accelerate network learning and facilitate further interpretation of the network obtained. Because of the network's initial semantics, its learning does not start from scratch

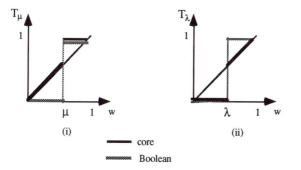

Figure 11.31
Core and Boolean thresholding operations in fuzzy neural networks: (a) AND neuron; (b) OR neuron

but rather focuses on further numerical refinement of the original configuration of the connections. In comparison to other classes of neural networks, the architectures under discussion here are heterogeneous. The exploitation of various types of neurons promotes the diversity necessary for downloading any prior domain knowledge and helps interpret the final network in the form of if-then statements. Fuzzy neural networks are useful in approximating multivariable input-output static and dynamic relationships. Interestingly, this form of approximation focuses on and reveals the logical nature of the dependences between variables and comes as an appealing analog of the Shannon theorem discussed in two-valued logic.

11.12 Problems

1. Derive a formula describing the fuzzy neural network with feedback constructed using referential neurons as shown in figure 11.32. (Assume that the referential neurons complete a disjunctive form of aggregation.) Treat the triangular norms as the product operation and probabilistic sum. Determine a steady state of the network as $k \to \infty$. Simulate the network in software; experiment with some other triangular norms; derive some general conclusions regarding this selection. How does the network's temporal response depend on the initial state $x(0)$?

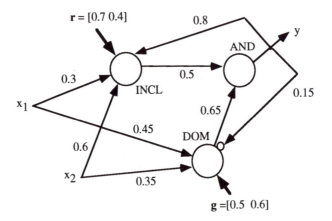

Figure 11.32
Example of fuzzy neural network for problem 1

2. The eigen fuzzy set A of a fuzzy relation $R : \mathbf{X} \times \mathbf{X} \to [0, 1]$, card $(\mathbf{X}) = n$, satisfies the relationship

$$A \circ R = A.$$

In the case of max-min composition, an algorithm exists that determines A in a finite number of steps. A generalization of this algorithm to deal with any s-t composition of A and R is not available. Formulate the problem of finding eigen fuzzy set(s) as a certain optimization task, propose an appropriate fuzzy neural network, and develop a complete learning algorithm. Note that an empty set $A = \emptyset$ is a trivial solution to the problem; exercise caution when initiating your learning scheme. Experiment with the network using different learning rates and initial conditions.

3.   Possibility and necessity measures are two vehicles commonly used to compare fuzzy quantities (sets) A and B; $\text{Poss}(A, B)$, $\text{Nec}(A, B)$. Assume that both A and B are defined in a finite universe of discourse (space) $\mathbf{X} = \{x_1, x_2, \ldots, x_n\}$.

(a) Propose a fuzzy neural network supporting computations of these measures. What are the connections of this network?

(b) Apply the structure developed in (a) to solve the following estimation problem: Given a collection of fuzzy sets B_k (fuzzy probes) and the corresponding possibility and necessity levels,

$$\lambda_k = \text{Poss}(A_k, B),$$

$$\mu_k = \text{Poss}(A_k, B),$$

estimate the membership function of B.

(c) What changes to the network in (b) are necessary to solve an estimation problem of the form

$$\lambda'_k = \text{Poss}(A, B_k),$$

$$\mu'_k = \text{Poss}(A, B_k),$$

where now B_k, λ_k and μ_k are given and A is to be determined.

4.   Consider the following verbal formulations of reliability problems:

(a) The system consists of five modules connected in series.

(b) The system consists of three modules put in parallel.

(c) The system includes two modules organized in parallel followed by two additional modules connected in series.

For each problem posed, propose the corresponding topologies of fuzzy neural networks and interpret the meaning of their connections and inputs.

5.   Fuzzy controllers are regarded as classic rule-based constructs. Considering an initial collection of the qualitative rules in the form

• If E_2 and DE_1, then U_1
• If E_4 and DE_2, then U_3
• If E_4 and DE_1, then U_1
• If E_2 and DE_2, then U_2

where E_i, DE_j, and U_l are fuzzy sets defined in the corresponding universes of discourse, $i = 1, 2, 3, 4$; $j = 1, 2$; $l = 1, 2, 3$; and the training set consisting of the activation levels of the linguistic terms standing in the rules:

E_1	E_2	E_3	E_4	DE_1	DE_2	U_1	U_2	U_3
1.0	0.3	0.1	0.0	1.0	0.2	0.0	0.1	1.0
0.7	0.6	0.1	0.0	1.0	0.0	0.0	0.7	0.5
1.0	0.1	0.1	0.0	0.1	0.9	1.0	0.1	0.0
1.0	0.2	0.1	0.0	0.0	1.0	0.2	1.0	0.0
0.0	0.2	0.6	1.0	1.0	0.2	1.0	0.2	0.0
0.0	0.0	0.1	1.0	0.1	1.0	1.0	0.4	0.0
0.7	1.0	0.3	0.0	1.0	0.2	0.3	1.0	0.4
0.0	0.2	1.0	0.2	0.0	1.0	0.0	0.2	1.0

Propose a relevant fuzzy neural network (logic processor) and complete its learning. Exploit various learning scenarios by changing learning rates and proceeding with different initial values for the connections. Once learning has been completed, simplify the network by eliminating the least significant connections and analyze the control rules obtained in this fashion. Derive the rules for the following conditions:

- if E_3 and DE_1, then ____.
- if E_1 and DE_2, then ____.
- if E_1 and DE_1, then ____.

References

Butnariu, D., and E. P. Klement. 1993. *Triangular Norm-Based Measures and Games with Fuzzy Coalitions.* Dordrecht, the Netherlands: Kluwer Academic Publishers.

Feldkamp, L. A., G. V. Puskorius, F. Yuan, and L. I. Davis, Jr. 1992. Architecture and training of a hybrid neural-fuzzy system. *Proc. 2nd Int. Conf. on Fuzzy Logic and Neural Networks,* Iizuka, Japan, 131–4.

Figueiredo, M., F. Gomide, and W. Pedrycz. 1995. Fuzzy neurons and networks: models and learning. *Proc. EUFIT-95* August 28–31, Aachen, Germany, I: 332–5.

Grossberg, S. 1988. *Neural Networks and Natural Intelligence.* Cambridge, MA: MIT Press.

Hecht-Nielsen, R. 1991. *Neurocomputing.* Reading, MA: Addison-Wesley.

Hirota, K., and W. Pedrycz. 1994. OR/AND neuron in modeling fuzzy set connectives. *IEEE Trans. on Fuzzy Systems* 2:151–61.

Kacprzyk, J. 1985. Group decision-making with a fuzzy majority via linguistic quantifiers (parts I and II) *Cybernetics and Systems* 16:119–29, 131–44.

Kacprzyk, J. 1986. Group decision making with a fuzzy linguistic majority. *Fuzzy Sets and Systems* 18:105–18.

Kolmogorov, A. N. 1957. On the representation of continuous functions of several variables by superposition of continuous functions of one variable and addition. *Doklady Akademii Nauk USSR,* 114:679–81.

Muroga, S. 1971. Threshold Logic and Its Applications. New York: J. Wiley.

Pedrycz, W. 1991. Direct and inverse problem in comparison of fuzzy data. *Fuzzy Sets and Systems* 34:223–36.

Pedrycz, W. 1991. Neurocomputations in relational systems. *IEEE Trans. on Pattern Analysis and Machine Intelligence* 13:289–96.

Pedrycz, W. 1993a. Fuzzy neural networks and neurocomputations. *Fuzzy Sets and Systems* 56:1–28.

Pedrycz, W. 1993b. *Fuzzy Control and Fuzzy Systems* (2d ed.) Taunton, NY: Research Studies Press/J. Wiley.

Pedrycz, W., and A. F. Rocha. 1993. Fuzzy-set based models of neurons and knowledge-based networks. *IEEE Trans. on Fuzzy Systems* 1:254–66.

Rocha, A. F. 1992. Neural Nets: *A Theory for Brain and Machine,* vol. 638 of Lecture Notes in Artificial Intelligence. Berlin: Springer-Verlag.

Schneeweiss, W. G. 1989. *Boolean Functions with Engineering Applications.* Berlin: Springer-Verlag.

Stoica, A. 1996. Synaptic and semantic operators for fuzzy neurons: Which t-norms to choose. *Proc. NAFIPS-96,* June 19–22, Berkeley, CA, 55–8.

Weisbuch, G. 1991. *Complex Systems Dynamics.* Redwood City, CA: Addison-Wesley.

Wolfram, S. 1994. *Cellular Automata and Complexity: Collected Papers.* Reading, MA: Addison-Wesley.

Yager, R. R. 1983. Quantifiers in the formulation of multiple objective decision functions. *Information Sciences* 31:107–39.

Zadeh, L. A. 1983. A computational approach to fuzzy quantifiers in natural languages. *Computers and Mathematics with Applications* 9:149–84.

Modeling evolutionary processes provides a basis for generating entities capable of problem solving and for creating intelligent behavior. Evolutionary computation means simulating natural evolution on a computer. This chapter deals with the issues related to evolutionary computation and their significance in assembling fuzzy systems. Emphasis is placed on genetic algorithms as a tool for system design. Evolution strategies, hybrid approaches, and cooperation schemes between fuzzy systems and genetic algorithms are discussed as well.

12.1   Evolution and Computing

Every living being is made of one or more cells, but beyond that, there are exceptions to just about every scheme in biological terms. However, certain characteristics of heredity and evolution seem to be universal or nearly so. In living beings, each cell contains a nucleus which, in turn, contains chromosomes. Chromosomes determine the hereditary characteristics whose determining factors are called genes. Genes are passed on when cells divide and offspring are parented through reproduction. During cell division, duplication and crossover occur, and occasionally the chromosome-copying process produces an altered gene different from the corresponding gene in the contributing parent. This is called mutation. From the molecular point of view, natural selection is enabled by the variation that follows from crossover and mutation enables natural selection. Crossover assembles existing genes into new combinations, whereas mutation produces new genes hitherto unseen in the chromosome contents. The fittest individuals, that is, those that emerge with the most favorable characteristics, tend to drive a population as a whole toward favorable characteristics. The fittest individuals appear as a consequence of competition and selection within an environment. These are the main issues behind Neo-Darwinism and are generally accepted among scientists. In other words, Neo-Darwinism states that the path of life is a result of a few processes acting on and within populations and species. The action of these processes—reproduction, mutation, competition, and selection—determines natural evolution.

In constructing a useful computational model of natural evolution, one should abstract evolution in terms of behavioral relationships between units of evolution instead of the mechanisms that produces these relationships. An important computational model for simulating certain characteristics of heredity and evolution is based on genetic algorithms (Holland 1975).

Genetic algorithms are essentially search algorithms inspired by the processes of natural evolution. They combine survival of the fittest among structures representing chromosomes with a randomized yet selected information exchange to form a search algorithm. Genetic algorithms are in the very realm of the evolutionary computation and use representation and processing schemes to model reproduction, mutation, competition, and selection.

Evolutionary computation (Fogel 1995) comprises several computational paradigms and includes evolutionary programming (Fogel, Owens, and Walsh 1966) and genetic programming (Koza 1993), evolution strategies (Rechenberg 1994), and classifier systems (Booker, Goldberg, and Holland 1989). Classifier systems and genetic programming may be viewed as special applications of genetic algorithms. Evolutionary programming and evolution strategies have been developed independently of genetic algorithm, but recently interactions among them are being pursued. In general terms, evolutionary computation gives rise to evolutionary algorithms, genetic algorithms and evolution strategies being among the most influential.

Evolution strategies also rely on computational models of natural evolution's collective nature. However, evolution strategies imitate the effect, not the appearance, of genetic information processing. Thus they can also be viewed as optimization procedures based on search, but they differ from genetic algorithms in many respects. For instance, the selection mechanism in evolution strategies is deterministic, and parameters controlling mutation are associated with each individual and are subjected to the same evolution operations as are the individuals of a population.

12.2 Genetic Algorithms

For our purposes, we may define a genetic algorithm as a method for solving optimization problems through a search procedure. By mimicking the natural evolution principle, genetic algorithms are able to evolve solutions for many problems, if the basic units, the chromosomes, are suitably encoded to represent problem solutions. For computational purposes, a chromosome is a list of elements called genes. For instance, a collection of binary digits may compose a string. In this case, the string represents a chromosome, the digits being the genes that assemble the chromosome. A chromosome is viewed as an individual to which a fitness value is assigned. The fitness value is associated with the objective function of the optimization problem to be solved and represents the selective facet of natural evolution. Thus, fitness provides a measure of how good a solution to the problem it is. The main operators genetic algorithms

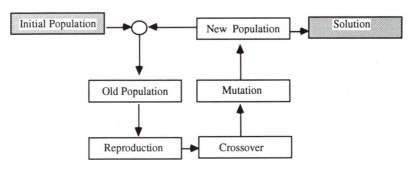

Figure 12.1
Main concept of a genetic algorithm

employ in modeling computational evolutionary processes are reproduction, crossover, and mutation.

Before getting into the computational details, we should emphasize that genetic algorithms are different from most optimization algorithms in various respects (Goldberg, 1989). More specifically, they

1. work with a coding of the decision variables, not with the decision variables themselves;

2. search from a population of individuals (coding points scattered in the decision space), not a single individual or, alternatively, a point in the decision space;

3. use objective function values only (no other information or knowledge is needed);

4. use probabilistic transition rules, not deterministic rules.

Thus, genetic algorithms evolve solutions by repeated application of the operators to populations which, after several generations, are expected to contain an individual representing the problem solution. Figure 12.1 summarizes the basic algorithm.

Before a genetic algorithm can be run, the following aspects should be determined: the representation scheme, the fitness measure, the parameters for controlling the algorithm, and the way to designate a solution and the criterion for terminating a run.

The representation scheme assumes that a potential solution to a problem is encoded to express each possible point in the search space as a list (string) of elements (characters). The list of elements forms a chromosome, viewed as an individual of a population. For instance, if our problem is to maximize a function of two variables $f(x, y)$, with x and y being integers in a certain interval, we might represent each variable by a 10-bit binary number. In this case, a chromosome would contain two genes and consist of 20 binary digits. In general, the representation scheme requires selecting the list length and the alphabet symbols and size.

A fitness measure must be devised for each problem to be solved. The fitness measure assigns a fitness value to each possible chromosome that is supposed to be proportional to the utility or usefulness of the individual the chromosome represents. For example, in optimization problems, a clearly appropriate fitness measure is the value of the objective function computed for a decoded chromosome. More generally, fitness can be a fuzzy function (see chapter 14). However, in many cases the fitness measure is not obvious as, for instance, in autonomous navigation problems in which obstacles must be avoided and targets must be reached.

The main parameters for controlling genetic algorithms include the population size, the number of generations to be run, and the probabilities P_r, P_c, P_m for the occurrence of reproduction, crossover, and mutation, respectively.

To designate a solution, a convergence criterion should be set. If the genetic algorithm has been correctly implemented, the population will evolve over successive generations such that the fitness of both the best and the average individual in each generation evolves toward the optimum. A gene is said to converge when most of the population shares the same value (the elements are the same). The population is said to converge when all genes converge (DeJong 1975). Therefore, convergence is a progression toward increasing uniformity. The termination criterion may be established as either when a solution is found or the maximum number of generations is achieved.

Once these steps for setting up the genetic algorithm have been completed, the algorithm can be run, and assuming fixed-length chromosomes, the following steps are executed:

Step 1: Create an initial population randomly.

Step 2: Until the termination criterion has been satisfied, perform the following:

Evaluate the fitness of each individual in the population.

Apply the genetic operators to individual chromosomes in the population, chosen with a probability based on fitness, to create a new population. That is:

Reproduction: Copy existing individual strings to the new population.

Crossover: Create two new chromosomes by crossing over randomly chosen sublists (substrings) from two existing chromosomes.

Mutation: Create new chromosomes from an existing one by randomly mutating the character in the list.

Step 3: Designate the best individual that appeared in any generation as the result. It may represent a solution or an approximate solution to the problem.

This is a basic version of genetic algorithm, and there are numerous variations of it. For instance, we may start with an initial population with only one chromosome, and as soon as a new population exceeding a maximum population size is generated, use a probabilistic selection based on fitness to keep the population at the maximum size. Mutation is often treated as an operation that can occur in sequence with either reproduction or crossover, so that a given individual might be mutated and reproduced or mutated and crossed over within a single generation. Also, the number of times a genetic operator is applied during one generation is often set to a given number for each generation, instead of determined probabilistically. Some versions includes ranking schemes to control the bias toward the best chromosome but also eliminate implicit bias introduced by inappropriate choice of the measurement scale. In addition, fitness, as defined so far, ignores diversity, which can be thought of as the degree to which chromosomes exhibit different genes. Accordingly, chromosomes tend to get wiped out if they score just a bit lower a chromosome closer to the best current chromosome. Even in large populations, the result is uniformity. On a large scale, however, unfit-looking individuals may survive quite well in certain niches that lie outside the view of other, relatively fit-looking individuals. Thus it can be as good to be different as it is to be fit, and this helps in handling local optimum points.

We next detail the genetic operators: reproduction, crossover and mutation.

12.2.1   Reproduction

Reproduction is a process in which individual chromosomes are copied according to their fitness. This means that chromosomes with higher fitness have higher probability of contributing one or more offspring to the next generation. The reproduction operator is typically implemented in a manner analogous to a biased roulette wheel where each current chromosome in the population has a roulette wheel slot sized in proportion to its fitness. Once a chromosome has been selected for reproduction, an exact copy of it is made and entered into a mating pool.

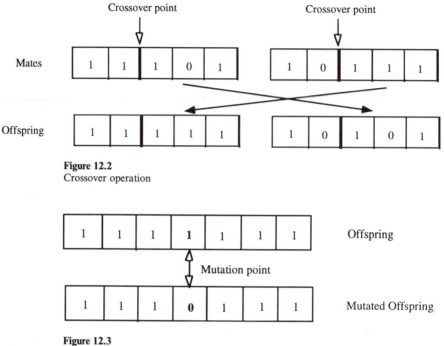

Figure 12.2
Crossover operation

Figure 12.3
Mutation operation

12.2.2 Crossover

The simplest procedure for crossing over may be divided into two steps. First, individuals possessing the newly reproduced chromosomes in the mating pool are mated randomly. Second, each pair of chromosomes undergoes crossing over by swapping all characters between positions selected at random, as illustrated in figure 12.2.

12.2.3 Mutation

In the basic genetic algorithm, mutation is applied to each chromosome after crossover. It randomly alters a gene with small probability. For instance, when binary coding is used, this simply means changing a 1 to 0 or vice versa, as shown in figure 12.3.

Reproduction according to fitness combined with crossover gives the genetic algorithms much of their processing power. This is intuitively appealing because we expect that, when combining good individuals, their offspring should have a high chance of better performance. Mutation provides a small amount of random search and helps ensure that no point in the search space has a null probability of being examined. By itself, mutation acts as a random walk through the search space.

12.3   Design of Fuzzy Rule–Based Systems with Genetic Algorithms

When designing fuzzy rule–based systems, major decisions must be made concerning determination of the input and output variables, the input and output space partition (granulation), the shape of membership functions, and the reasoning method. Once the target system is specified and the performance requirements and constraints are established, the input/output variables either are determined by the problem itself or should be determined under careful engineering analysis. However, very often the input/output space partition is not obvious, and usually decisions are made based on knowledge about the system, on experiments, or via a cut-and-try approach. This partition is critical because it is time consuming and may jeopardize the computational performance of the fuzzy systems to be designed once the partitions define the number of fuzzy rules to be processed.

Analogously, finding the most suitable membership functions is quite difficult, and commonly their shape is assumed a priori, for example, triangular, trapezoidal, and so forth, but their parameters still remain to be determined.

Determining the best reasoning method is still a challenge, and usually experience, analogy with previous solutions to similar problems, and simulation experiments provide insights for finding a good method.

Genetic algorithms provide a more systematic way to cope with the crucial problem of input/output space partition or, equivalently, the number of fuzzy rules. For this purpose, an appropriate parametrization of the fuzzy system as well as a suitable genetic encoding and a method for fitness evaluation should be decided upon. In the following, we first present a systematic method, devised by Takagi and Lee (1993) and Lee and Takagi (1993), for simultaneously determining the membership function parameters, the number of fuzzy rules, and rule-consequent parameters. Their approach uses binary chromosome coding. Following that, we provide an example in which a similar scheme with an alternative encoding method is used in designing a fuzzy rule–based system for a simple autonomous vehicle navigation problem.

There are various methods for parameterizing the shape of membership functions. For instance, in the Takagi and Lee method, the membership functions are restricted to being adjacent and fully overlapped. For instance, when a triangular shape is considered, this means that the center of one membership function serves as the left base point of the next. In addition, one membership function is constrained to having its center at the lower boundary of the input space. Thus, only $n - 1$

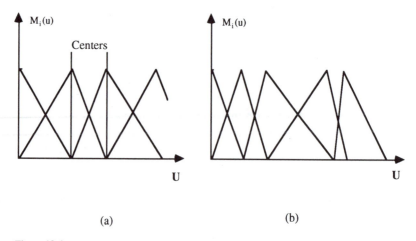

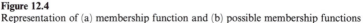

(a)   (b)

Figure 12.4
Representation of (a) membership function and (b) possible membership functions

membership functions' centers need to be specified, where n is the maximum number of partitions for a given dimension (figure 12.4).

In figure 12.4, $M_i(u)$ denotes generic membership functions defined in a space U that may be present in fuzzy rules of the type

If X is A,   then Y is B,

with variables and associated fuzzy sets defined in the corresponding space. Clearly, the maximum number of rules equals the number of all possible combinations of the input fuzzy sets defining the input spaces partition. Each possible rule may have as its consequent one output fuzzy set per output dimension, or it may have none, as specified by a rule identification number. For instance, an output variable whose domain is partitioned by three fuzzy sets could generate four possible rules for a given set of input variables; set one, set two, set three, or none of the three could be assigned to the rule consequent. When a set is not assigned to a rule, the rule is discarded.

The representation scheme for coding the system parameters into chromosomes is as follows. First, lists consisting of a set of parameters that represent membership functions' shape (the center, in the case of overlapped, triangular membership functions) and rule identification number, denoted by MFC and ORC, respectively, are defined. A chromosome is assembled as the concatenation of these lists, as illustrated in figure 12.5. Thus, a chromosome represents the entire fuzzy system. Once the coding scheme is set, a fitness measure is defined. Of course, this is dependent on the problem and dictated by system performance requirements and constraints. For instance, if a performance index is devised to

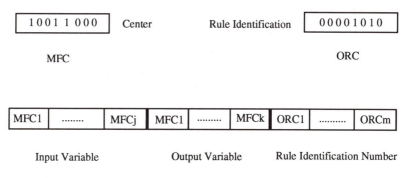

Figure 12.5
Fuzzy system as represented by a chromosome

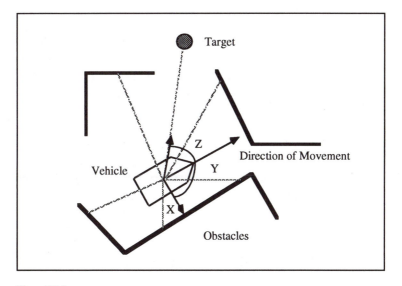

Figure 12.6
Autonomous vehicle in an environment

measure system effectiveness, it can be chosen as a fitness function. This is typical, for example, in control systems.

The type of rule being considered influences the scheme for representing fuzzy rule–based systems, which depends on the constraints imposed on the membership function shape and overlapping. (See problem 3 at the end of this chapter.) Next, we introduce an alternative scheme and an example to illustrate the many possibilities behind systematic design of fuzzy systems using genetic algorithm. (See also problem 6 at the end of the chapter for another interesting case.)

Consider an autonomous vehicle in a certain environment, as shown in figure 12.6, where X, Y, and Z denote the smallest distance from the vehicle to the obstacles, the angle between the movement direction and the closest obstacle, and the angle between the movement direction and the target, respectively. Ideally, the task is to develop a fuzzy rule–based navigator to control vehicle movement in achieving targets while avoiding collisions with obstacles. Assuming constant speed, the control variable is the driving direction at each moving step. We assume the navigator is a collection of if-then rules of the form

If X is A_i and Y is B_i and Z is C_i, then w is g_i,

where A_i, B_i, and C_i are fuzzy sets defined in the appropriate universes, and w is the control variable, assumed to be a real number; thus g_i is also a real number.

In this example, a chromosome is a navigator, that is, a collection of if-then rules, and thus a population comprises a number of possible navigators. As before, the problem is to find the fuzzy sets A_i, B_i, and C_i and the values of g_i. As a consequence, an appropriate partition of the input spaces should be found. This implicitly defines the number of rules.

The general encoding scheme is as follows (Oliveira, Figueiredo, and Gomide 1994). Let us denote as U_j an encoded partition of the space $\mathbf{U}_j$ associated with the jth input variable and U their concatenation. Thus, U represents all rule antecedents. Similarly, let S be the concatenation of real numbers g_i assigned as the corresponding rule consequents. More specifically, we have the situation depicted in figure 12.7, where u_{vj} is a real number representing the center of the membership function, and f_{vj} is a binary number representing the shape of the membership functions. For instance, if f_{vj} is such that

$$f_{vj} = \begin{cases} 0, & \text{if triangular} \\ 1, & \text{if trapezoidal,} \end{cases}$$

and a partition U_j has codes as shown in figure 12.8, then, the respective membership functions decodes as shown in figure 12.9. Whenever the centers fall outside the specified universes (input spaces), the corresponding rule that would be generated is discarded.

A fitness function for this example could be

$$f = p_1 n_t^2 + p_2 (d_{\min} - d_{tv})^{-1} + p_3 d_s^{-1}$$

where n_t is the number of targets reached by the vehicle, $d_{\min}$ is the minimum distance below which the vehicle is assumed to have reached a target or an obstacle, d_{tv} means either the distance between the vehicle and the target at the end of a simulation run or vehicle position when a collision occurs, and d_s is the length of the vehicle trajectory from initial

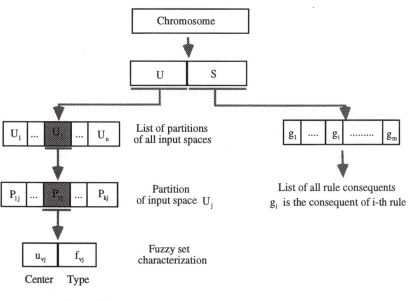

Figure 12.7
Example of navigators encoding

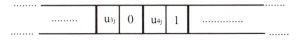

Figure 12.8
Input space partition encoding

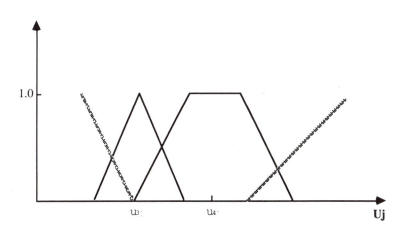

Figure 12.9
Example of input space partition

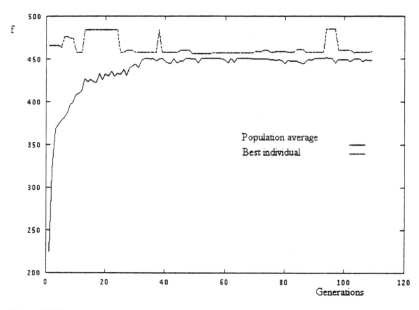

Figure 12.10
Fitness function values for obstacle avoidance behavior

to final position. The parameters p_i weight each term. In this particular example, they were set as $p_1 = 10.000$, and $p_2 = p_3 = 20$ if a simulation run terminated with a collision, and $p_2 = p_3 = 100$ otherwise. Each simulation run was limited to 10,000 steps.

The experimental results to be shown below were obtained as follows. First, a navigator was evolved as a target finder only. Second, another navigator was evolved for obstacle avoidance. In the last case, parameter $p_1 = 0$, and generations evolved as shown in figure 12.10. In the first case, $p_2 = p_3 = 0$, with the population evolving as in figure 12.11. Therefore, the rule base is composed of two sets of rules of the form

If X is A_i and Y is B_i, then y_{oi} is g_{oi} (obstacle avoidance behavior)

If Z is C_j, then y_{tj} is g_{tj} (target-seeking behavior).

Figures 12.12 and 12.13 illustrate the target-seeking and obstacle avoidance behaviors, respectively. In these figures, an external white-tipped arrow shows the current target, external black-tipped arrows the targets reached, and an internal black-tipped arrow the initial vehicle position. Figure 12.14 shows the corresponding membership functions generated.

The final control is a weighted combination of the two evolved behaviors and is determined as follows (Oliveira et al. 1995):

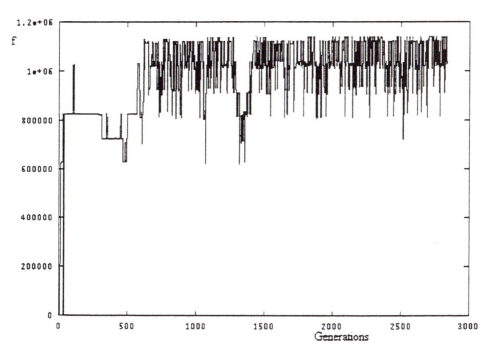

Figure 12.11
Fitness function values of the best individual for the target-seeking behavior

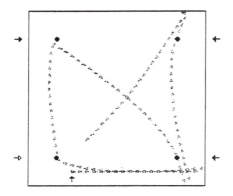

Figure 12.12
Target-seeking behavior

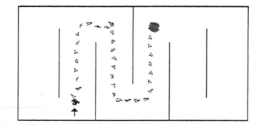

Figure 12.13
Obstacle avoidance behavior

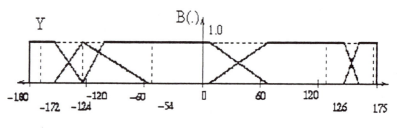

Position

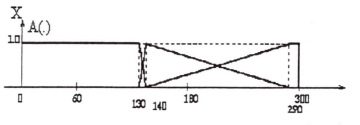

Angle

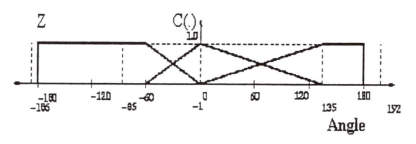

Angle

Figure 12.14
Membership functions evolved by the genetic algorithm

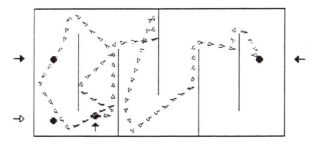

Figure 12.15
Combined behavior

$$y = y_o\left(1 - \frac{1}{1 + e^{6.2-d_o}}\right) + y_t\left(\frac{1}{1 + e^{6.2-d_o}}\right),$$

where y_o, y_y, and d_o are the the defuzzified obstacle and target navigator outputs, and the current smallest distance from the vehicle to the obstacles, respectively. Figure 12.15 shows the vehicle movement performance as a result of the combined behavior.

EXERCISE 12.1   The scheme just discussed provides a navigator that performs quite well in many environments different from the one shown in figure 12.15. However, in a few very specific situations, it may have difficulties! Can you give an example and explain why the difficulties arise?

12.4   Learning in Fuzzy Neural Networks with Genetic Algorithms

The range of applicability of genetic algorithm techniques can extend beyond the optimization schemes discussed so far. We next discuss the concept of the structural learning realized in a stratified scheme where each layer is responsible for a specific facet of network improvement: the higher the position of the stratum in the entire learning scheme, the more general the type of architectural changes to the optimized network. Because fuzzy neural networks are heterogeneous (composed of different types of neurons with distinct functional characteristics, as discussed in chapter 11), this layered approach can be strongly recommended. Figure 12.16 depicts the overall learning scheme (Pedrycz 1995).

At the highest (most general) level of the learning procedure, the network's overall architecture is established and coded in the form of a binary string (list). For instance, considering that only logical OR and AND neurons will contribute to the architecture, the coding may include the type of the individual neurons and a qualitative strength of the connection (e.g., positive and negative weights). We do not restrict ourselves

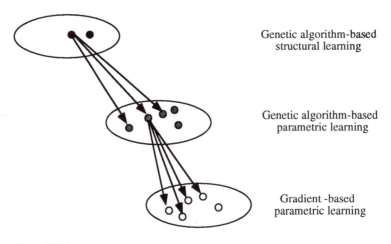

Genetic algorithm-based
structural learning

Genetic algorithm-based
parametric learning

Gradient -based
parametric learning

Figure 12.16
Stratified learning strategy

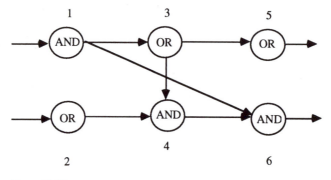

Figure 12.17
Example fuzzy neural network

to any particular arrangement of the neurons into layers—the connections
giving rise to this ordering in the network are also subject to learning.

For illustrative purposes, consider a small network in the configuration
shown in figure 12.17, where neurons 1 and 2 are the input processing
units; the outputs are associated with neurons 5 and 6. The coding of the
connections is determined at a qualitative level just by distinguishing
among positive (excitatory), negative (inhibitory), and no-connection
links between the neurons. All this structural information is organized in
a $(n \times n)$ binary matrix S, where n stands for the number of neurons
in the network. Each entry of S is coded with two bits. Additionally, an
n-dimensional vector T specifies the types of the neurons throughout the
network. The complete description of the example network is formalized

as

$$T = [1\ 0\ 0\ 1\ 0\ 1],$$

with the following coding: 1—AND neuron, 0—OR neuron, and

$$S = \begin{bmatrix} 00 & 00 & 01 & 00 & 00 & 10 \\ 00 & 00 & 00 & 10 & 00 & 00 \\ 00 & 00 & 00 & 01 & 01 & 00 \\ 00 & 00 & 00 & 00 & 00 & 01 \\ 00 & 00 & 00 & 00 & 00 & 00 \\ 00 & 00 & 00 & 00 & 00 & 00 \end{bmatrix},$$

where coding is defined as 00—no connection, 01—positive connection, 10—negative connection, and 11—unused.

EXERCISE 12.2   Provide a chromosome representation scheme to encapsulate the coding schemes used in the above example of fuzzy neural network structural design.

At the lower level of the stratified learning, the genetic algorithm optimizes parameters and tackles the process of modifying the values of the connections for the fixed architecture. The arrows pointing downward in figure 12.16 illustrate the direction in which the architectural information about the network is transmitted; the arrows pointing upward show the flow of information, including the values of the fitness function. Although the selection of the fitness function at the lower levels of learning may be straightforward (for example, of the form of the mean-squared error criterion), at the level of structural learning one should be sure to preserve some control over the size of the network, for example, by adding to the fitness function a term describing the network's size.

12.5   Evolution Strategies

The first efforts towards evolution strategies were due to Rechenberg (1973). As discussed at the beginning of this chapter, the idea was also to imitate the principles of evolution applied in optimization problems. As pointed out earlier, evolution strategies differ from genetic algorithms at the time they were conceived. The first applications used a simple mutation-selection scheme called two-membered evolution strategies. This scheme uses a population consisting of one parent individual—a real-valued vector—and one descendant, created by adding random numbers with normal distribution. The better of the two individuals

serves as the ancestor of the next generation. This scheme is denoted by
$(1 + 1)$-ES.

In contrast to its role in genetic algorithms, mutation plays a key role
in evolution strategies. To exemplify mutation, consider the problem of
assembling a $(m \times m)$ matrix M whose row, column and diagonal entries
add up to the same given number (Rechenberg 1994). Let M be

$$
M = \begin{bmatrix}
n_{1l} & .. & n_{1j} & .. & n_{1m} \\
. & .. & . & .. & . \\
n_{i1} & .. & n_{ij} & .. & n_{im} \\
. & .. & . & .. & . \\
n_{ml} & .. & n_{mj} & .. & n_{mm}
\end{bmatrix}.
$$

The problem may be stated as an optimization problem, that is, to find n_{ij},
$i = j = 1, \ldots, m$, such that the following objective function is minimized:

$$
Q = \sum_{i=1}^{m} \sum_{j=1}^{m} (n_{ij} - n)^2 + \sum_{j=1}^{m} \sum_{i=1}^{m} (n_{ij} - n)^2 + \sum_{i=j} (n_{ij} - n)^2
$$

$$
+ \sum_{i=1}^{m} (n_{i(m-i+1)} - n)^2.
$$

The mutation operation is as follows:

1. Select randomly a number from the square.
2. Change it by a small amount.
3. Find the number in the square that equals the changed number.
4. Exchange the first number with the second.

Rechenberg (1994) showed that this scheme successfully developed a
matrix for $m = 30$ and $n = 13{,}515$.

Generalizations and improvements were further pursued, giving rise to
the so-called $(\mu + \lambda)$-ES and (μ, λ)-ES. This notation means, in the first
case, that μ parents produce λ offspring that are reduced again to the μ
parents of the next generation. In $(\mu + \lambda)$-ES strategies, selection operates
on the joined set of parents and offsprings. Thus, parents survive until
they are surpassed by better offspring. Therefore, well-adapted individuals may survive in each generation. On the other hand, in (μ, λ)-ES
strategies, only the offspring undergo selection, that is, individuals are
limited to lifespans of one generation. The $(\mu + \lambda)$-ES and (μ, λ)-ES
strategies fit into the same formal framework, with the only difference
being the limited lifetime of individuals in (μ, λ)-ES. Therefore, only
(μ, λ)-ES is described here. Formally, (μ, λ)-ES is described by a tuple

(Back, Hoffmeister, and Schwefel 1994),

$$(\mu, \lambda)\text{-ES} = (P^o, \mu, \lambda; r, m, s; \delta\sigma; f, g, t),$$

where

- $P^o = (a_1^o, \ldots, a_\mu^o) \in I^\mu, I = \mathbf{R}^n \times \mathbf{R}^n$ (initial population)
- $\mu \in \mathbf{N}$ (number of parents)
- $\lambda \in \mathbf{N}, \lambda > \mu$ (number of offspring)
- $r : I^\mu \to I$ (recombination operator)
- $m : I \to I$ (mutation operator)
- $s : I^\lambda \to I^\mu$ (selection operator)
- $\delta\sigma \in \mathbf{R}$ (step-size meta-control)
- $f : \mathbf{R}^n \to \mathbf{R}$ (objective function)
- $g : \mathbf{R}^n \to \mathbf{R}^q$ (constraints)
- $t : I^\mu \to \{0, 1\}$ (termination criterion)

Each individual a_i^t in a population P^t is a pair $a_i^t = (x^t, \sigma^t) \in I$, where $x^t \in \mathbf{R}^n$ is the decision variable and σ^t the standard deviation of a vector of 0 mean, Gaussian distribution G of independent random numbers. Thus, the basic mutation mechanism produces offspring of the form

$$x^{\prime t} = x^t + G(\sigma^t).$$

Selection is performed by applying the selection operator as defined above for (μ, λ)-ES, whereas for $(\mu + \lambda)$-ES the selection operator is modified to $s : I^{\lambda + \mu} \to I^\mu$.

Because individuals in a population are coded as real-valued vectors, recombination includes methods that average parameter values at a given location, those that arbitrarily choose one parent's gene value or the other, and those that perform a weighted average of gene values. The major difference among the various evolution strategies results from the way parameter s is handled. For instance, if the action of the reproduction operator generates an individual a_i^t, that is, $a_i^t = r(P^t)$, thus we might have

$$m(a_i^t) = a_i^{\prime t} = (x^{\prime t}, \sigma^{\prime t}),$$

$$\sigma^{\prime t} = \sigma^t e^{G(\delta\sigma)},$$

$$x^{\prime t} = x^t + G(\sigma^{\prime t}),$$

which means that mutation occurs in both x^t and σ^t.

12.6   Hybrid and Cooperating Approaches

Classes of genetic algorithm have been developed obscuring the differences between them and evolution strategies. For instance, when using real numbers in coding chromosomes, genetic algorithms use crossover operators similar to those found in evolution strategies. Genetic algorithms may also encode mutation control parameters in their representation. However, differences still remain when selection is of concern.

Lee and Saloman (1995) proposed as a hybrid algorithm a selection mechanism that shares ideas from genetic algorithms and evolution strategies. The main idea is as follows. Mating among parents chosen to reproduce is uniformly distributed in an evolution strategy. In genetic algorithms, mating probabilities are frequently based on fitness. In the hybrid approach proposed, the selection scheme uses the multielitist selection of evolution strategies and then performs rank-based selection among the elite members for mating. An application example concerning the design of a fuzzy rule–based controller for the classic cart-pole control problem illustrates the usefulness of their approach.

An interesting cooperation approach between fuzzy systems and genetic algorithms has also been explored. The idea here is to use fuzzy rule-based systems to dynamically control the genetic algorithm's parameters, such as population size, crossover and mutation rates. For instance, Lee and Takagi (1993) introduce the idea of dynamic parametric genetic algorithms. In their approach, a fuzzy system is used to provide genetic algorithm parameter settings given the genetic algorithm performance measures or current parameter settings as input. For instance, the average fitness and best fitness ratio, current population size, or current mutation rate could be used as inputs. Rules in a fuzzy rule base and fuzzy reasoning provide, based on the inputs, a decision about parameter settings. For example, rules of the following form may compose a rule base (Takagi 1993):

- If average fitness/best fitness is big, then increase population size.
- If worst fitness/average fitness is small, then decrease population size.
- If mutation is small and population is small, then increase population size.

A fuzzy rule base and the corresponding membership functions that dynamically change the population size and crossover and mutation rates is provided in Lee and Takagi (1993).

12.7   Chapter Summary

Evolutionary computation has been discussed in this chapter as a promising and effective paradigm for developing fuzzy systems. The key facets of evolutionary computation were reviewed and examples of their use in designing fuzzy rule–based systems and fuzzy neural networks provided. In addition, this chapter has discussed the symbiosis between genetic algorithms and evolution strategies when the two are combined into hybrid systems, and the role of fuzzy systems in guiding genetic algorithms' parameter settings.

12.8   Problems

1. Suppose you are to find the maximum of a function $f(x, y)$, using the basic genetic algorithm, where $f(x, y)$ is defined by the following table:

9	2	3	4	5	6	5	4	3	2
8	3	4	5	6	7	6	5	4	3
7	4	5	6	7	8	7	6	5	4
6	5	6	7	8	9	8	7	6	5
5	6	7	8	9	10	9	8	7	6
4	5	6	7	8	9	8	7	6	5
3	4	5	6	7	8	7	6	5	4
2	3	4	5	6	7	6	5	4	3
1	2	3	4	5	6	5	4	3	2
↑ y/x →	1	2	3	4	5	6	7	8	9

Develop a representation scheme for chromosomes with integer genes. Will the basic genetic algorithm perform well using reproduction and crossover operations only? Does the mutation operator play any role in finding the maximum of $f(x, y)$ via the basic algorithm? Explain your answers.

2. Consider the same situation as in problem 1. How would you represent chromosomes using binary coding of the genes? Will the basic genetic algorithm work well with reproduction and crossover operations only? What is the role of mutation in this case? Explain and compare your answers against those of problem 1.

3. Assume you are to design a fuzzy rule–based system with the following rule structure,

    If X is A and Y is B, then $Z = w_0 + w_1 X + w_2 Y$,

    where A and B have adjacent and fully overlapped triangular membership functions, as shown in figure 12.18, and ws are integers. Develop a representation of a chromosome, using binary coding, to evolve a rule base by means of a genetic algorithm. If you were to evolve a rule base to act as a controller for the classic inverted pendulum (cart-pole) problem, how would you assemble a fitness function?

4. Given the fuzzy neural network depicted in figure 12.19, obtain the binary coding scheme that represents the network for structural learning. If more than two types of neurons were present, what modifications should be made to the coding scheme?

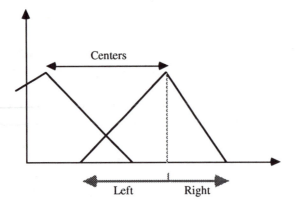

Figure 12.18
Membership functions for problem 3

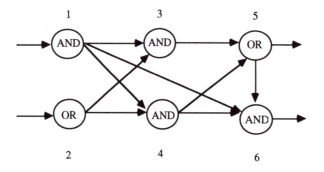

Figure 12.19
Fuzzy neural network and its structural learning

5.   Formulate a stratified learning strategy under the framework of evolution strategies.

6.   To encode fuzzy rules in a genetic string, the input and output variables and their corresponding values can be numbered. The pair (nX, nA) corresponds to the statement "X is A," where nX and nA are integer numbers coding the variable name and its value, respectively. For instance, we may have the following case:

Variable name (Code)	Values (Code)
$X(1)$	*negative*(1), *zero*(2), *positive*(3)
$Y(2)$	*small*(1), *medium*(2), *large*(3)
$Z(3)$	*left*(1), *center*(2), *right*(3)

Thus the rule "If X is *zero* and Y is *large* then Z is *left*" can be represented by the string (1,2) (2,3) (3,1), where each pair forms a gene. Verify that the individual genes of a fuzzy rule can be arranged in any order because their meaning is independent of their position in the sequence. What are the advantages and drawbacks of such a coding scheme?

References

Back, T., F. Hoffmeister, and H. Schwefel. 1994. *A Survey of Evolution Strategies*. Department of Computer Science Technical Report, University of Dortmund, Germany.

Booker, L. B., D. E. Goldberg, and J. H. Holland. 1989. Classifier systems and genetic algorithms. *Artificial Intelligence* 40:235–82.

DeJong, K. 1975. The Analysis and Behavior of a Class of Genetic Adaptive Systems. PhD diss., University of Michigan.

Fogel, D. B. 1995. *Evolutionary Computation*. New York: IEEE Press.

Fogel, L. J., A. J. Owens, and M. J. Walsh. 1966. *Artificial Intelligence Through Simulated Evolution*. New York: John Wiley.

Holland, J. H. 1975. *Adaptation in Natural and Artificial Systems*. Ann Arbor; University of Michigan Press.

Koza, J. R. 1993. *Genetic Programming*. Cambridge, MA: MIT Press.

Lee, M., and R. Saloman. 1995. Hybrid evolutionary algorithms for fuzzy system design. *Proc. of the 6th IFSA World Congress*. São Paulo, Brazil, 269–72.

Lee, M., and H. Takagi. 1993. Dynamic control of genetic algorithms using fuzzy logic techniques. *Proc. of the 5th International Conference on Genetic Algorithms*, Urbana-Champaign, IL 76–83.

Oliveira, M., M. Figueiredo, and F. Gomide. 1994. A neurofuzzy approach to autonomous control. *Proc. of the 3rd International Conference on Fuzzy Logic, Neural Nets and Soft Computing*, Iizuka, Japan, 597–8.

Oliveira, M., M. Figueiredo, F. Gomide, and L. Romero. 1995. Neurofuzzy navigation control and the neural group selection. *Proc. of the 6th IFSA World Congress*, São Paulo, Brazil, 73–6.

Pedrycz, W. 1995. Genetic algorithms for learning in fuzzy relational structures. *Fuzzy Sets and Systems* 69:37–52.

Rechenberg, I. 1994. Evolution strategy. In *Computational Intelligence: Imitating Life*, eds. J. M. Zurada, R. J. Marks, C. J. Robinson, 147–59. New York: IEEE Press.

Rechenberg, I. 1973. *Evolutionsstrategie: Optimierung Technischer Systeme nach Prinzipien der Biologischen Evolution*. Stuttgart: Frommann Holzboog.

Takagi, H. 1993. Fusion techniques of fuzzy systems and neural networks, and fuzzy systems and genetic algorithms. *Proc. of SPIE Technical Conference on Applications of Fuzzy Logic Technology*, Boston, MA (vol. 2061), 402–13.

Takagi, H., and M. Lee. 1993. Neural networks and genetic algorithms approaches to auto-design of fuzzy systems. *Proc. of Fuzzy Logic in Artificial Intelligence*, Linz, Austria.

13 Fuzzy Modeling

13.1 Fuzzy Models: Beyond Numerical Computations

Fuzzy modeling (Zadeh 1965; Bezdek 1993; Pedrycz 1993a; Pedrycz 1995) undoubtedly becomes vital whenever any application of fuzzy sets is anticipated. The advancements in modeling translate directly into more advanced methodology, better understanding of fundamental concepts of fuzzy sets, and design practices of user-friendly models. Because mathematical and experiential components obviously coexist in fuzzy sets, the development of a sound modeling methodology becomes a must. With the current array of fuzzy models, it is important to pose several generic questions getting to the technical core of the issue:

• What makes fuzzy models different from other widely used classes of models characterized by numerical mappings?

• What commonalities are involved?

• Which part of the existing modeling methodology should be preserved as universal to all the available modeling techniques?

• How can the fuzzy models be optimized and verified/validated?

Briefly speaking, fuzzy models are modeling constructs featuring two main properties:

• They operate at the level of linguistic terms (fuzzy sets); similarly all system dependencies can be portrayed in the same linguistic format,

• They represent and process uncertainty.

Relative to this definition, two features need to be emphasized:

1. Because the linguistic labels can easily be modified, the way the elements of the universe of discourse are delimited and encapsulated into manageable chunks of information (fuzzy sets) will greatly affect the processing level at which the model operates. Subsequently, such collections of fuzzy sets should be chosen carefully according to the specifications the potential user provides. In fact, the designer/user can easily modify the cognitive perspective from which the model is developed and utilized. As shown in figure 13.1, the same system may induce a variety of fuzzy models depending on the granularity of the fuzzy sets used as an interface between the particular model and the modeling environment. The linguistic labels' granularity constitutes a viable source of the available modeling flexibility, embracing a significant range of information granularity. Moreover, the fuzzy models can be arranged into a hierarchy of submodels focusing on specific aspects of information processing. In

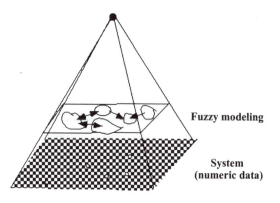

Figure 13.1
Fuzzy system modeling—hierarchical information processing

particular, this structural organization could also give rise to so-called
pyramidal or layered fuzzy models (figure 13.1), with the lowest layer
being the most specific and embracing the processing at the level of
numeric computing.

2. As the interrelationships among the linguistic labels move from individual numeric values to the more abstract set-theoretic level, the ensuing
dependencies should be elucidated primarily at the level of logic-oriented
relations.

13.2   Main Phases of System Modeling

It is instructive to outline the main conceptual phases encountered in the
development of any model, regardless of the specific framework in which
it is constructed. In its very basic scheme, such a scheme comprises several
essential phases:

1. Preprocessing: This phase involves a specification of input and output
variables and incorporation of all relevant domain knowledge, in particular at the level of an initial data analysis (including any sort of preliminary filtering and elimination of eventual outliers).

2. Parameter estimation: At this stage, the pertinent parameters of the
model are determined making use of a variety of optimization techniques
(including, in particular, gradient-based algorithms).

3. Model verification: The model is verified with respect to available data
by analyzing how well it represents (captures) the data. This feature can
be evaluated in terms of some carefully selected performance index (e.g.,

sum of squared errors). For instance, statistical models are usually verified with the aid of parametric and nonparametric tests.

4. Model validation: This phase is far more complicated and challenging than the verification endeavors undertaken in the previous step. The intent here is to ensure that the model is valid, meaning that it helps support the particular task for which it has been developed and meets the user's expectations. In such a context, there are no universal models. A model that is superb for short-term prediction may eventually fail miserably when applied to any long-term forecasting. Similarly, the level of granularity at which the model was developed may not meet a different set of requirements, thus making it invalid in another context.

Quite often the two last phases are referred to together as a VV process (model verification and validation).

13.3   Fundamental Design Objectives in System Modeling

It is important to emphasize that any model (including, of course, a fuzzy model) is constructed (designed) with some fundamental objectives in mind. One strives to accomplish at least two dominant goals in system modeling:

• Approximation abilities: The model should be able to capture the data by minimizing a certain performance index (e.g., a sum of squared errors).

• Interpretation abilities: The fundamental guiding instrument, known as Occam's razor principle, states that a simple model is always preferred over a more complex one if both approximate (explain) the data to the same degree. The available data always carry some residual component (which could be symbolically expressed as data = fit + residual) that blurs interpretation abilities. This means that the simpler, less accurate model will generalize better to new data.

These two fundamental modeling thrusts are somewhat in conflict. Very high approximation abilities could easily trigger simple memorization of data. The danger of so-called model overfitting becomes particularly high if one tends to admit models of a very high dimensionality of the parameter space (that is, models rich in parameters). This happens, for example, to polynomial models in which the polynomials used are of a high order as well as to significantly oversized neural networks. Evidently, not only does the learning itself become longer and somewhat inefficient, but the model's behavior can be highly surprising beyond the range of the

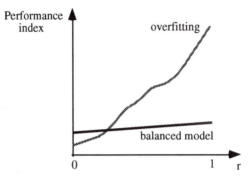

Figure 13.2
Performance index as a function of *r* for well-balanced and overfitted models

training data used in its development. This phenomenon is also known as overfitting effect: The model gets much more complicated that it was supposed to be. As a consequence, it primarily interpolates the data, implying that the model's generalization abilities are practically nonexistent. Subsequently, one gets very distinct values of the performance index it assumes on the training set and testing set. Consider a mixture of data coming from both training and testing sets and denote by *r* a proportion of the test data. Obviously *r* ranges from 0 to 1; $r = 0$ means that we are dealing with the training data only whereas $r = 1$ states that only testing data are considered. Figure 13.2 illustrates the performance index characterizing a model with strong overfitting as opposed to a well-balanced one.

More specifically, the reader can refer to figure 13.3, in which an excessively high-order polynomial has been applied to a fairly "smooth" collection of data. The effort to approximate all the data points, including several evident outliers, has resulted the model's behaving somewhat unpredictably and undesirably, figure 13.3b. Note that the higher-order polynomials have not visibly improved the resulting model but rather produced a vast number of minute ripples in the nonlinear form of the model itself.

13.4   General Topology of Fuzzy Models

A general architecture of the fuzzy model as portrayed in figure 13.4, is composed of three principal modules: a fuzzy encoder, a processing module, and a fuzzy decoder. Fundamentally, fuzzy sets are instrumental in forming an interface between the computational part of this hybrid and an application (modeling) environment. They allow us to look at the

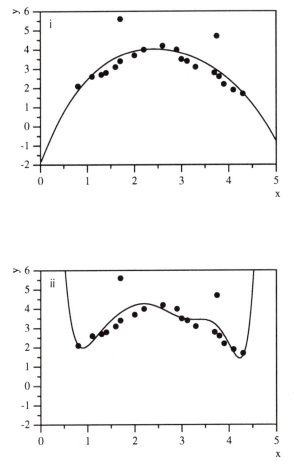

Figure 13.3
Data set and its polynomial approximations: (i) second-order polynomial; (ii) eighth-order polynomial

environment (data) from the most relevant perspective by assuming an appropriate level of information granularity (Zadeh 1979; Pedrycz 1992). By modifying the form of fuzzy sets (linguistic terms) used in the construction of the fuzzy encoder and decoder as well as increasing or decreasing their number, we preprocess data before the computational module uses them. The processing module can vary significantly depending upon the particular problem at hand. Several interesting implementation options can be envisioned:

• One can think of the processing module as a collection of rules encapsulated in the form of fuzzy neural networks (Pedrycz 1993b). These are essential in the design of rule-based systems.

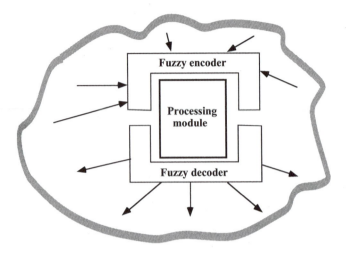

Figure 13.4
General topology of a fuzzy model

• The computational module can be viewed as a linear or nonlinear discriminant function. Here we are interested in building linguistic classifiers.

Quite often, the fuzzy model's processing module is logic based, and the resulting structure is very much knowledge-transparent. This feature is very helpful in striking an appropriate balance between prescriptive and descriptive aspects of system modeling (figure 13.5). The prescriptive facet of modeling allows us to download initially available qualitative domain knowledge and set up a structural canvas of the model. Making use of the available data and accommodating them within the previously assumed structure, we shape its remaining numerical details (the model's parameters). As a straightforward example of the prescriptive-descriptive duality of fuzzy models, one can refer to the topology of a fuzzy neural network and its connections. The prescriptive part of the model concerns the network's structure. The network's connections handle the descriptive aspects efficiently.

13.5   Compatibility of Encoding and Decoding Modules

Information compatibility between the mechanisms of encoding and decoding exploited by practically any fuzzy model is vitally important in fuzzy modeling. We may also view this problem as the design of a lossless communication channel. What we require (figure 13.6) can then be posed as follows:

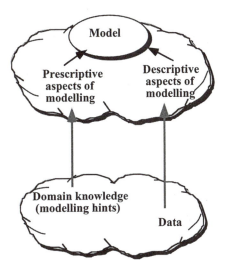

Figure 13.5
Prescriptive and descriptive aspects of fuzzy models and modeling

Figure 13.6
Encoding and decoding mechanisms: the development of a lossless fuzzy communication
channel

$$\text{Decode}(\text{Encode}(X)) = X,$$

where X is any piece of input information (a numeric datum, fuzzy set,
numeric interval, etc.). To clarify the role of the lossless communication
channel, let us identify its functional role. Consider the identification of
the fuzzy model. As usual, each data pair, say (X, Y), is encoded in the
internal format (x, y), as shown in figure 13.7(a), and as such is used to
guide the optimization activities (parametric enhancements to the model).
Once the processing module has been developed, it is verified by examining the mapping properties achieved for X. This is completed in the
configuration as depicted in figure 13.7(c). Note that by making the communication channel lossless, we assure an ideal interaction of the fuzzy
model with the modeling environment, ensuring that the optimization at
the level of the processing module is really meaningful and that the ensuing decoding will not generate any additional error, as shown in figure
13.7(c). We elaborate on this issue in section 13.7.

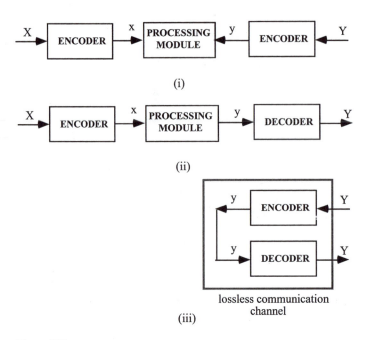

Figure 13.7
Encoding and decoding schemes in the optimization of the processing module: (i) a general structure: transformation of (X, Y) into an internal format (x, y); (ii) mapping the results of the processing module through the decoder; (iii) communication of the processing module with the modeling environment

13.6   Classes of Fuzzy Models

We can enumerate several general types of fuzzy models. The proposed taxonomy depends on the level of structural dependencies envisioned among the system's variables and captured by the specific model. One can enlist models from several commonly used classes, such as

- tabular representations,
- fuzzy grammars,
- fuzzy-relational equations,
- fuzzy neural networks,
- rule-based models,
- local regression models,
- fuzzy regression models.

The above categories are arranged in order of increasing level of structural dependencies. Without attempting to cover all the minute details and study specific identification procedures, let us elaborate briefly on the models' main features. Furthermore, because previous chapters have already covered some of these models in more depth, we concentrate on a few selected categories.

13.6.1   Tabular Format of the Fuzzy Model

Fuzzy models in tabular format are the least structured among those under discussion, because they capture basic relationships between the assumed linguistic landmarks (labels) of the system's variables. For instance, a first-order discrete-time dynamic fuzzy model with a single control variable (u) and state variable (x) (both of which are linguistically quantified) can be conveniently described in the tabular form

$$
\begin{array}{c|ccc}
\diagdown x(k) & B_1 & B_2 & B_3 \\
u(k) & & & \\
\hline
A_1 & B_3 & B_1 & B_3 \\
A_2 & B_2 & B_2 & B_1 \\
A_3 & B_2 & B_1 & B_1 \\
\end{array}
$$
$$x(k+1)$$

where A_1, A_2, A_3 and B_1, B_2, B_3 are linguistic labels associated with the corresponding variables. In fact, this table can be converted into a series of rules (conditional statements):

If $u(k)$ is A_i and $x(k)$ is B_j, then $x(k+1)$ is B_1,

$i, j, 1 = 1, 2, 3$. In contrast to rule-based systems, tabular systems impose no particular inference scheme that makes the overall model much more rigid and specialized. In fact, the tabular form of the fuzzy model is the most "readable" among all the types of fuzzy models, but its operational abilities are very much limited, with no provisions, for example, for computing $x(k+1)$ for $u(k)$ and $x(k)$ given. As outlined in Zadeh 1965, the table's entries can also be linguistic, involve linguistic probabilities, or be a mixture of the two, as exemplified in the following table:

$$
\begin{array}{c|ccc}
\diagdown x(k) & B_1 & B_2 & B_3 \\
u(k) & & & \\
\hline
A_1 & \begin{matrix}B_3\\ very\ likely\end{matrix} & B_1 & \begin{matrix}B_3\\ possible\end{matrix} \\
A_2 & B_2 & \begin{matrix}B_2\\ highly\ probable\end{matrix} & \begin{matrix}B_1\\ possible\end{matrix} \\
A_3 & B_2 & B_1 & \begin{matrix}B_1\\ unlikely\end{matrix} \\
\end{array}
$$
$$x(k+1)$$

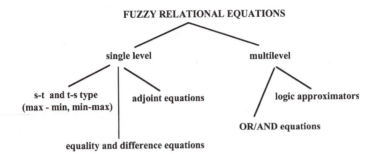

FUZZY RELATIONAL EQUATIONS

single level   multilevel

s-t and t-s type   adjoint equations   logic approximators
(max - min, min-max)

                                              OR/AND equations

equality and difference equations

Figure 13.8
Taxonomy of fuzzy-relational equations

13.6.2   Fuzzy-Relation Equations

Fuzzy-relational equations, originally introduced to the area of fuzzy sets in Sanchez 1976, are quite commonly used in applications where both representation capabilities and computational abilities are of a vital interest. Fuzzy-relational equations have been pursued vigorously at both the theoretical and the application levels (Di Nola et al. 1989; Pedrycz 1990; Pedrycz 1991). Figure 13.8 provides a concise taxonomy of fuzzy-relational equations. Fuzzy-relational models express dependencies between a system's variables in terms of fuzzy *relations* rather than *functions*, a genuine generalization in comparison with function-oriented models.

The theory of fuzzy-relational equations provides us with a vast number of analytic solutions; their structures have been investigated in depth. Approximate solutions are also available (Pedrycz 1993a). Interestingly, analytical solutions are not available for general forms of composition operators (e.g., s-t composition t- and s-norms). Similarly, multilevel fuzzy-relational equations are also out of reach of analytical methods (except for a few quite specific cases).

13.6.3   Fuzzy Grammars

Fuzzy grammars and fuzzy languages (Lee and Zadeh 1969; Mizumoto, Toyoda, and Tanaka 1972; Kandel 1982; Santos 1974) are fuzzy symbol–oriented formalisms that can be readily used in the description of various systems. They are of particular interest in characterizing time series and developing signal classifiers. In a formal setting, a fuzzy grammar is defined as a quadruple

$$G = (V_N, V_T, P, \sigma)$$

where

V_T is a vocabulary of terminal symbols (alphabet),

V_N is a set of nonterminal symbols ($V_N \cap V_T = \emptyset$),

P is a list of production (rewrite) rules,

σ is an initial symbol included in V_N.

Quite often the nonterminal symbols are denoted using upper-case letters while the lower-case letters are used to identify terminal symbols. The elements of P (production or rewrite rules) are of the form

$$a \xrightarrow{\beta} b$$

where a and b are two strings belonging to $(V_T \cup V_N)^+$ and $(V_T \cup V_N)^*$, respectively. The coefficient β quantified in the unit interval stands for the strength of the rule. A^+ denotes a set of strings over A of length greater than or equal to one. Similarly, A^* is union of A^+ and $\{\varepsilon\}$ where ε stands for an empty string. A string $a_1 a a_2$ and a production $a \xrightarrow{\beta} b$ yield the replacement of a by b. We write this down as

$$a_1 a a_2 \Rightarrow^\beta b_1 b b_2$$

where the symbol $\Rightarrow$ denotes a derivation. The multistep derivation is written down as a sequence

$$\alpha_1 \Rightarrow^{\beta_1} \alpha_2, \alpha_2 \Rightarrow^{\beta_2} \alpha_3, \ldots, \alpha_{n-1} \Rightarrow^{\beta_{n-1}} \alpha_n \quad (n \geq 1)$$

The strength of the individual production rules used determine the strength of this derivation. In particular, it can be computed as

$$\beta = \text{AND}(\beta_1, \beta_2, \ldots, \beta_{n-1})$$

A fuzzy grammar generates a fuzzy language $L(G)$, namely, all strings ($\mathbf{x}$) in V_T^* that can be derived from σ. The strength of derivation of $\mathbf{x}$ from G is a membership degree of $\mathbf{x}$ in $L(G)$ and is computed as

$$\text{Degree of } \mathbf{x} \text{ in } L(G) = \max \min \|\mathscr{D}(\mathbf{x})\|$$

where $\mathscr{D}$ is a derivation of σ from $\mathbf{x}$. The strength of such a derivation is equal to the weakest derivation used (the lowest β). Subsequently, the maximum is taken over all possible derivations.

EXAMPLE 13.1   Let us consider a fuzzy grammar

$$\langle V_N, V_T, P, \sigma \rangle$$

where

$$V_N = \{A, B\}$$
$$V_T = \{a, b\}$$
$$P:$$

$$\sigma \to^{0.3} AB$$
$$A \to^{0.4} aA$$
$$A \to^{0.7} a$$
$$B \to^{0.6} bB$$
$$B \to^{0.9} b$$

Backus-Naur form allows us to combine all production rules with the same left-hand side. Thus we get

P:

$$\sigma \to^{0.3} AB$$
$$A \to^{0.4,0.7} aA|a$$
$$B \to^{0.6,0.9} bB|b$$

(Note that the strengths of the rules are shown as a vector of numbers in the unit interval). Figure 13.9 shows the so-called derivation tree of $\mathbf{x} = aaabb$. (One can also write this string as $\mathbf{x} = a^3 b^2$.)

Let us compute the degree to which $\mathbf{x}$ belongs to $L(G)$. Proceeding with σ and moving down the tree one gets

$$\text{deriv}_1: \quad \sigma \Rightarrow^{0.3} AB \Rightarrow^{0.4} aAB \Rightarrow^{0.4} aaAB \Rightarrow^{0.7} aaaB$$

$$\Rightarrow^{0.6} aaabB \Rightarrow^{0.9} aaabb$$

The total membership degree of this derivation is computed by aggregating the corresponding strengths of derivations. Take, for instance, minimum. This implies

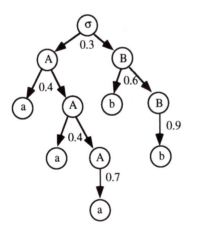

Figure 13.9
Derivation tree of *aaabb*

$\min(0.3, 0.4, 0.4, 0.7, 0.6, 0.9) = 0.3$

Some other derivations are

deriv$_2$: $\sigma \Rightarrow^{0.3} AB \Rightarrow^{0.6} AbB \Rightarrow^{0.4} aAbB \Rightarrow^{0.4} aaaB$

$\Rightarrow^{0.6} aaAbB \Rightarrow^{0.9} aaAbb \Rightarrow^{0.7} aaabb$

deriv$_3$: $\sigma \Rightarrow^{0.3} AB \Rightarrow^{0.4} aAB \Rightarrow^{0.6} aAbBB \Rightarrow^{0.9} aAbb$

$\Rightarrow^{0.4} aaAbb \Rightarrow^{0.7} aaabb$

deriv$_4$: $\sigma \Rightarrow^{0.3} AB \Rightarrow^{0.4} aAB \Rightarrow^{0.4} aaAB \Rightarrow^{0.6} aaAbB$

$\Rightarrow^{0.7} aaabB \Rightarrow^{0.9} aaabb$

All produce the same derivation tree and lead as well to the same degree of membership of derivation. The grammars that exhibit unique derivation trees are unambiguous. Similarly, ambiguous grammars emerge if any derivation of **x** yields distinct derivation trees. This justifies the use of the maximum in the entire computation of the strength level taken over all possible derivations.

EXAMPLE 13.2   We look into the preprocessing phase, during which one starts with a signal as it emerges at the level of physical fabric of discrete samples and ends up as a sequence of symbols. The terminal symbols $(a, b, \ldots)$ are used to describe a slope between two successive samples of the signal (figure 13.10). The nonterminal symbols are denoted by $A, B, C,$ and D; these are used to build up longer segments of the signals they emerge at the symbolic level. The terminal symbol describes a signal belonging to a certain class of signals, say

σ : quasi periodic signal.

The production rules assume the following general format:

$T_i \rightarrow^{\beta_{ijk}} t_j T_k.$

For instance,

$A \rightarrow^{0.8, 0.7, 0.3, 0.1, 0.05} a|aA|bA|cA|dA.$

Although the source of fuzziness resides within the production rules, another important point producing fuzzy sets comes at the encoding level. Observe that in the previous studies, mapping the ranges (intervals) of angles generates the terminal symbols, as figure 13.10(b) shows. This encoding can be done linguistically; see figure 13.10(c). Observe that the fuzziness comes now both at the level of the terminal symbols (which are essentially fuzzy sets) and the production rules themselves. Computations

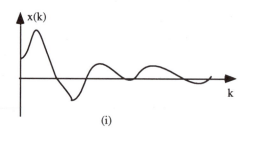

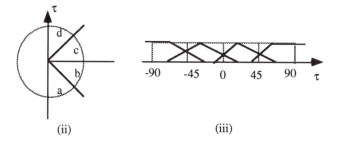

Figure 13.10
Syntactic modeling: (i) temporal signal; (ii) encoding; and (iii) encoding with fuzzy sets

of the strength of derivation need to take these two components into account.

13.6.4 Local Fuzzy Models

The idea underlying local fuzzy models concerns a construction of a model as a collection of submodels whose relevance is restricted to some region in the space of the input variables (input space):

If $\mathbf{x}$ is A_1, then $y = f_1(\mathbf{x})$.

If $\mathbf{x}$ is A_2, then $y = f_2(\mathbf{x})$.

$$\vdots$$

If $\mathbf{x}$ is A_c, then $y = f_c(\mathbf{x})$.

In general, A_1, A_2, and A_c are viewed as fuzzy relations (assuming that $\mathbf{x}$ is a vector) defined in the input space.

The tessellation of the input space depends upon the form and a distribution of A_is; figure 13.11 provides some illustrative examples. Note that the distribution of A_i implies a specific partitioning of the input space. Furthermore, in some regions of $\mathbf{X}$, the granularity of the relation is higher, inducing a fine-grained coverage of the space.

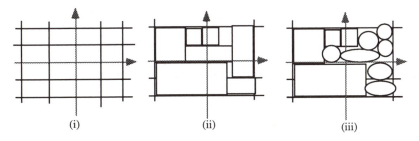

Figure 13.11
Examples of tessellation of the input space: (i) uniform coverage; (ii) and (iii) nonuniform coverage

As the fuzzy relations overlap, each local model becomes invoked to some extent as implied by the level of matching between $\mathbf{x}$ and A_i. The commonly encountered formula (Takagi and Sugeno 1985; Sugeno and Yasukawa 1993) governing the behaviours of the model reads as

$$y = \frac{\sum_{i=1}^{c} A_i(\mathbf{x}) f_i(\mathbf{x})}{\sum_{i=1}^{c} A_i(\mathbf{x})}.$$

We can rewrite in a slightly different way by introducing the weight coefficient w_i:

$$w_i = \frac{A_i(\mathbf{x}) f_i(\mathbf{x})}{\sum_{i=1}^{c} A_i(\mathbf{x})}.$$

Hence y becomes a linear combination of f_is,

$$y = \sum_{i=1}^{c} w_i(\mathbf{x}) f_i(\mathbf{x}).$$

Specifically, if these fuzzy relations form a so-called fuzzy partition, meaning that

$$\sum_{i=1}^{c} A_i(\mathbf{x}) = 1,$$

for each $\mathbf{x}$ in $\mathbf{X}$, then y is a convex combination of f_is. The form of the local models (f_i) has not been specified so far. The diversity of such models can be substantial. In particular, one can discuss

• linear models, where the local functions are linear,

$$f_i(\mathbf{x}) = \mathbf{a}_i^T \mathbf{x},$$

and $\mathbf{a}_i$ stands for a vector of the parameters;

- nonlinear models, especially polynomial models, whose local relation-ships are governed by the expression

$$f_i(x) = \sum_{j=1}^{n} a_{ij}x_j + \sum_{j=1}^{n} b_{ij}x_j^2 + \sum_{j=1}^{n} c_{ij}x_j^3 \ldots \text{etc.}$$

One can also anticipate harmonic models governed by the equations

$$f_i(\mathbf{x}) = a_0 \sin(\omega_i^T \mathbf{x}) + b_0 \cos(\omega_i^T \mathbf{x}).$$

In addition to static relationships, we can introduce local dynamic models. In this class of models, the right-hand side of the modeling rule can comprise differential or difference equations, namely,

If $\mathbf{x}$ is A_i, then $\dot{\mathbf{x}} = f_i(\mathbf{x}, \mathbf{a})$,

or

if $\mathbf{x}$ is A_i, then $\mathbf{x}(k + 1) = B\mathbf{x}(k) + C\mathbf{u}(k)$,

where k represents a discrete moment in time.

Even though the generic form of the entire model arises as a linear combination of the linear local models, the entire model can exhibit a highly nonlinear form, assuming that the A_is themselves are kept non-linear. On the other hand, if the A_is are linear, then the overall model resulting from the above form of aggregation is piecewise linear.

EXAMPLE 13.3   To illustrate the model's qualitative performance of the model, let us discuss three modeling rules of the form

If x is A_i, then $y = a_i x + b_i$,

and consider triangular fuzzy sets A_i with 1/2 overlap between successive terms (figure 13.12). To pursue a detailed analysis, let us consider the segment of $\mathbf{R}$ spread between −4 and 3. The calculations of the output of the model are carried out as follows. First, note that the weight coefficients are equal to

$$w_i(x) = A_i(x),$$

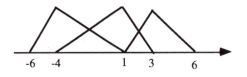

Figure 13.12
Triangular fuzzy sets defined in the input space

thus $w_i(x)$ (which can be regarded as *activation* profiles of the local models) are identical with the membership functions defined in the input space. Next, combine the constant functions associated with the individual fuzzy sets in the input space, namely -2, 1.5, and 3.2, respectively. Finally, obtain a linear model—in this sense the smooth aggregation of the local models increases the order of the polynomial of the local models—here from order 0 to 1. The observed increase in dimensionality is primarily caused by the smooth switching between the submodels. Figure 13.13 illustrates the model's ensuing input-output characteristics. The changes in overlap among the linguistic terms imply a qualitatively different performance for the overall model.

In general, one can show that if f_i are polynomials of order r, then the aggregation of such local submodels produces a polynomial whose order is $r + 1$. Figure 13.14 illustrates the aggregation effect for f_i as a first and a second-order polynomial. Thus in general one can state that with the piecewise linear partition of the input space and the polynomial type of the local models, the entire model is again polynomial, yet of an increased order. Let us complete detailed calculations. Because the corresponding triangular fuzzy sets overlap at the $1/2$ level, this simplifies the model. We derive

$$y = \frac{\sum_{i=1}^{c} A_i(\mathbf{x})f_i(\mathbf{x})}{\sum_{i=1}^{c} A_i(\mathbf{x})} = \sum_{i=1}^{c} A_i(\mathbf{x})f_i(\mathbf{x}),$$

where x is in $[-4, 3]$. Furthermore the local models are given as

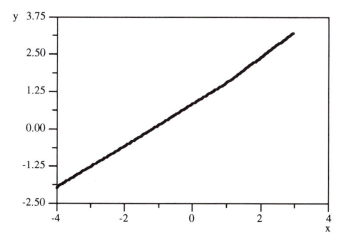

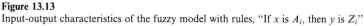

Figure 13.13
Input-output characteristics of the fuzzy model with rules, "If x is A_i, then y is Z_i"

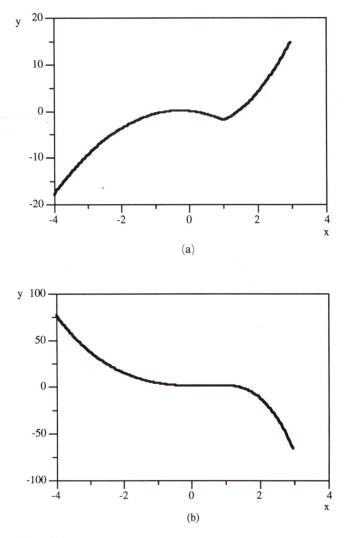

Figure 13.14
Input-output characteristics of the fuzzy model for polynomial types of local models: (a) linear local models: $y = 5x + 2$, $y = -1.5x - 0.5$, $y = 4x + 3.2$; (b) quadratic local models— $y = 5x^2 + 1.4x + 2$, $y = 1.5x^2 - 0.5x - 0.25$, $y = -8x^2 - 0.4x + 3.2$

$$L_1(x) = 5x + 2,$$

$$L_2(x) = -1.5x - 0.5,$$

$$L_1(x) = 4x + 3.2.$$

Consider any x situated in $[-4, 1]$. Here only two local models become activated,

$$A_1(x) = \frac{1 - x}{5},$$

$$A_2(x) = \frac{x + 5}{5},$$

$$A_3(x) = 0.$$

Subsequently we derive

$$y = L_1(x)A_1(x) + L_2(x)A_2(x)$$
$$= \frac{(5x + 2)(1 - x)}{5} + \frac{(-1.5x - 0.5)(x + 4)}{5} = \frac{-7.5x^2 - 3.5x - 2}{5}$$

Let us derive detailed parameter estimation formulas for the local models. Here we assume that each model is given as

$$y_i = \mathbf{a}_i^T \mathbf{x},$$

where $\mathbf{x}$ and $\mathbf{a}_i$ and n-dimensional vectors of inputs and parameters, respectively. As usual, we assume that the optimization of the model is carried out in the presence of a finite data set that the model must approximate:

$$(\mathbf{x}_1, y_1)(\mathbf{x}_2, y_2) \ldots (\mathbf{x}_N, y_N).$$

Introduce a notation w_{ik} to denote the value of the weight coefficient assumed for $\mathbf{x}_k$:

$$w_{ik} = \frac{A_i(\mathbf{x}_k)}{\sum_{i=1}^{c} A_i(\mathbf{x}_k)}.$$

Then the output of the model for x_k is expressed as a weighted sum:

$$\sum_{i=1}^{c} w_{ik} \mathbf{x}_k^T \mathbf{a}_i = w_{ik} \mathbf{x}_k^T \mathbf{a}_1 + w_{2k} \mathbf{x}_k^T \mathbf{a}_2 + \cdots + w_{ck} \mathbf{x}_k^T \mathbf{a}_c.$$

Furthermore, let us introduce the notation that allows us to rewrite the problem in a concise matrix form. Set

$$\mathbf{z}_{ik} = w_{ik}\mathbf{x}_k.$$

The output of the model then reads as

$$[\mathbf{z}_{1k}^T \mathbf{z}_{2k}^T \dots \mathbf{z}_{ck}^T]\begin{bmatrix} \mathbf{a}_1 \\ \mathbf{a}_2 \\ \vdots \\ \mathbf{a}_c \end{bmatrix}.$$

Denote by A a $(c \times n)$ vector of the parameters of the entire fuzzy model:

$$A = \begin{bmatrix} \mathbf{a}_1 \\ \mathbf{a}_2 \\ \vdots \\ \mathbf{a}_c \end{bmatrix}.$$

Similarly, we arrange all input data in the matrix form having N rows and $c \times n$ columns:

$$Z = \begin{bmatrix} \mathbf{z}_{11}^T & \mathbf{z}_{21}^T & \cdots & \mathbf{z}_{c1}^T \\ \mathbf{z}_{12}^T & \mathbf{z}_{22}^T & \cdots & \mathbf{z}_{c2}^T \\ & & \vdots & \\ \mathbf{z}_{1N}^T & \mathbf{z}_{2N}^T & \cdots & \mathbf{z}_{cN}^T \end{bmatrix}.$$

The output data points $(y_1, y_2, \dots, y_N)$ generate the N-dimensional vector of outputs

$$\mathbf{y} = \begin{bmatrix} y_1 \\ y_2 \\ \vdots \\ y_N \end{bmatrix}.$$

Assuming a standard form of the squared error, we conceive the optimization problem as follows:

$$\underset{A}{\text{Min}} \, \|\mathbf{y} - ZA\|^2,$$

where

$$\|\mathbf{y} - ZA\|^2 = (\mathbf{y} - ZA)^T(\mathbf{y} - ZA)$$

is the model's squared error. The solution to the problem is instantaneous and emerges in the following concise form:

$$A_{\text{opt}} = (Z^T Z)^{-1} Z^T \mathbf{y}$$

or equivalently,

$$A_{\text{opt}} = Z^{\#} \mathbf{y}.$$

where $Z^{\#}$ is often referred to as a pseudoinverse of Z. The optimization's main computational step is an inverse of $Z^T Z$. If the number of the local models (c) or the number of inputs (n) increases, then the inverse of $Z^T Z$ could become more cumbersome. Note that the entire problem becomes linear with respect to the parameters (but not necessarily linear with respect to the inputs).

The derived formulas are fairly general and can be applied to other types of local models. Consider, for instance, a collection of second-order local models described as second-order polynomials,

$$y_i = \mathbf{a}_i^T \mathbf{x} + \mathbf{b}_i^T \mathbf{x}',$$

where

$$\mathbf{x}'^T = [x_1^2, x_2^2, \ldots, x_n^2].$$

The same optimization formula as derived before also applies here. The problem itself needs some slight notational refinements. The overall model's parameters are arranged in a single vector,

$$\tilde{\mathbf{a}} = \begin{bmatrix} a_1 \\ b_1 \\ \ldots \\ a_2 \\ b_2 \\ \ldots \\ a_c \\ b_c \end{bmatrix},$$

whose dimensions are $2 \times c \times n$. Similarly, Z reads as

$$Z = \begin{bmatrix} \mathbf{z}_{11}^T (\mathbf{z}_{11}')^T & \cdots & \mathbf{z}_{c1}^T (\mathbf{z}_{c1}')^T \\ \mathbf{z}_{12}^T (\mathbf{z}_{12}')^T & \cdots & \mathbf{z}_{c2}^T (\mathbf{z}_{c2}')^T \\ \vdots & & \vdots \\ \mathbf{z}_{1N}^T (\mathbf{z}_{1N}')^T & \cdots & \mathbf{z}_{cN}^T (\mathbf{z}_{cN}')^T \end{bmatrix},$$

and $\dim(Z) = N \times 2 \times c \times n$, where

$$\mathbf{z}_{ik} = w_{ik} \mathbf{x}_k,$$

$$\mathbf{z}_{ik}' = w_{ik} \mathbf{x}_k'.$$

Obviously the increase in the order of the local model has a primordial impact on the dimensionality of the problem to be solved.

13.7   Verification and Validation of Fuzzy Models

As with other modeling frameworks, model verification and validation emerges very specifically as an issue. Owing to the nonnumerical character of information processed therein and the variable level of information granularity, new facets of model verification and validation appear that need to be carefully addressed. Generally speaking, by model verification we mean all procedures aimed at characterizing the model's descriptive (approximative) capabilities when it is confronted with available data. Two basic scenarios of model verification are usually envisioned:

• Using training data: The model's performance is quantified by considering the same data as was used for its construction.

• Using testing data: The model's is evaluated using a collection of data different from the one used originally in its development. In general, this approach is more demanding and realistic than the first.

The above taxonomy is commonly employed in pattern recognition in testing various pattern classifiers.

The other aspect of model verification specific to fuzzy models concerns the level at which the verification activities are carried out. We consider two main approaches: internal model verification, and external model verification. These two levels are tightly linked with the way in which the model was developed. With respect to the internal level of model verification, as depicted in figure 13.10(b), the performance index operates at the level of the encoded fuzzy data. In fact, this index evaluates the processing module's preformance. First, the target value, *target*, is encoded, yielding an element in the unit hypercube, whose result is denoted by **target**, which in turn is compared with the result provided by the processing module. To emphasize this, we can also refer to the resulting error as an internal performance index. The verification occurring at the external level is aimed at evaluating the fuzzy model by comparing the model's external (mainly numerical) outcomes with the available target value. (See figure 13.15.)

The results of internal and external verification can differ significantly. In general, they lead to the inequality

$Q(\text{internal}) < Q(\text{external})$.

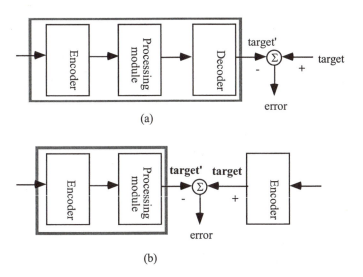

Figure 13.15
External and internal model verification

The verification completed internally quite often becomes overly optimistic. Differences between the performance indices are due to the improperly implemented mechanisms of encoding and decoding. The verification at both levels returns the same result, $Q(\text{internal}) = Q(\text{external})$, once the mechanisms of encoding and decoding used in the model are equipped with an ideal (lossless) communication channel. Only under these circumstances can the optimization of the processing module be completed at either level because the tandem decoder-encoder behaves in a translucent way. If the channel introduces some losses, the level at which the model is optimized does matter considerably. Let us discuss the case in which the encoding/decoding mechanisms are fixed (given). When confining ourselves to the internal format of model optimization, we have no control over losses of information caused by the communication channel and essentially give up on compensating for this undesired effect. The situation can be alleviated to some extent by taking into account the decoder's characteristics and optimizing the processing module in this setting.

Taking a classic gradient-based optimization method governed by the performance index

$$Q = (y - \text{target})^2,$$

we obtain (see figure 13.16)

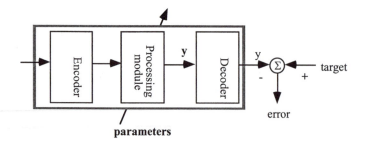

parameters

Figure 13.16
Processing module optimization involving decoder

$$\frac{\partial Q}{\partial \mathbf{param}} = 2(y - \text{target})\frac{\partial y}{\partial \mathbf{y}}\frac{\partial \mathbf{y}}{\partial \mathbf{param}}.$$

Clearly, the optimization of the model involves the characteristics of the decoder.

$$\frac{\partial y}{\partial \mathbf{y}} = \frac{\partial}{\partial \mathbf{y}}(\text{Decoder}(\mathbf{y}))$$

Further, assuring computational transparency of the decoding mechanism makes use of the optimization procedure feasible. Computational transparency now becomes a synonym for computational feasibility of the respective gradient in the last expression. In this sense, some decoding methods used today (e.g., center of area or height method) are definitely lacking.

13.7.1 Verification Algorithms for Fuzzy Models

In the following, we elaborate on how to quantify fuzzy models' performance. More specifically, let us consider an output of the processing module. The already optimized module (after identification) returns vectors $\mathbf{y}(1), \mathbf{y}(2), \ldots, \mathbf{y}(N)$. The corresponding target vectors are denoted by $\mathbf{target}(1), \mathbf{target}(2), \ldots, \mathbf{target}(N)$. The elements of these two series of data are placed in the n-dimensional unit hypercube. Interestingly, the processing module's output can be regarded as a type 2 fuzzy set (refer to chapter 2). For instance, if the codebook consists of three linguistic labels, *small, medium* and *large*, and $\mathbf{y} = [0.9\ 0.4\ 0.1]$, it can be explicitly interpreted as $[0.9/small\ 0.4/medium\ 0.1/large]$.

The ideal processing module satisfies the system of equalities

$$\mathbf{y}(k) = \mathbf{target}(k).$$

Practically speaking, the above equalities can be satisfied approximately. The model's performance depends on the differences between $\mathbf{y}(k)$ and

target(k). The greater the similarity between these two vectors, the better the model (more precisely, the better its mapping properties). Computing the similarities leads to an interesting notion concerning confidence intervals. To some extent the computations play the role of confidence intervals encountered in statistical models, but their origin is totally different. First, we express the similarity level between the data and the model's output with the aid of the equality index (see section 2.6.2):

$$\gamma_i = (\mathbf{y} \equiv \mathbf{target})_i = \frac{1}{N} \sum_{k}^{N} (y_i(k) \equiv target_i(k)).$$

Here i denotes the ith coordinate of the unit hypercube. The above computations are repeated for each coordinate of the hypercube. We end up with a series of indices $\gamma_1, \gamma_2, \ldots, \gamma_N$. The higher the value of the equality index, the better the mapping properties of the model at the given coordinate. The equality indexes give rise to so-called fuzzy confidence intervals, whose essence is the following. Assume that the processing module returns y_j for a certain input. Associated the jth coordinate is the equality index γ_j. Determine all possible membership values (y) such that

$$y_j \equiv y \geq \gamma_j.$$

The resulting set (interval in $[0,1]$),

$$\Gamma(y_j) = \{y \mid y_j \equiv y \geq \gamma_j\},$$

is referred to as a (fuzzy) confidence interval associated with y_j and implied by the confidence level. The solution to the problem (confidence interval) depends on a specific form of the equality index (that is, an implication function specified therein). A few essential properties hold:

1. If $\gamma_j = 1$, then the confidence interval reduces to a single point (membership value):

$$\Gamma(y_j) = \{(y_j)\}.$$

2. If $\gamma_j > \gamma_j'$, the ensuing confidence interval satisfies the inclusion

$$\Gamma(y_j) \subset \Gamma'(y_j),$$

namely, the lower values of the equality index broaden the confidence interval. For such low values of the equality index, the induced confidence interval can easily point to a limited precision (and usefulness) of the fuzzy model. Even though the processing module always returns a single membership value, this needs to be interpreted properly, as the associated confidence interval can be fairly broad. In this sense, the width of the confidence interval can make the designer aware of the model's deficiencies and prompt eventual enhancements of it.

The output of the processing module equipped with the confidence intervals comes in the format

$$[[y_1^-, y_1^+]/A_1 \;\; [y_2^-, y_2^+]/A_2 \;\; \cdots \;\; [y_n^-, y_n^+]/A_n].$$

In other words, the resulting structure is a (interval-valued) type 2 fuzzy set.

An important question arises in turn about measuring the fuzzy model's relevancy based on the associated equality indexes. This problem can be addressed in many possible ways. Here we exploit one that involves the notion of nonlinear summarization (aggregation) offered by fuzzy measures and fuzzy integrals. As discussed in chapter 9, this structure embraces two key components:

1. fuzzy measures (more specifically, λ-fuzzy measures), g_λ,
2. a function to be integrated, **h**

These two components need to be interpreted and constructed to capture the meaning of the verification procedure. The fuzzy measure is developed based on the equality indexes computed for each coordinate of the n-dimensional unit hypercube. The fuzzy measure (g_λ) over any subset of the coordinates (K) is interpreted as the extent to which one can evaluate the model's quality given K. The fuzzy measure characterizes the fuzzy model's *global* mapping properties, which pertain to the entire data set. Now the same fuzzy model invoked by a particular input returns **y** instead of the given target, **target**. The discrepancies (differences) between **y** and **target** describe some *local* behavior of the model. Denote by **h** the values of the resulting equality index,

$$\mathbf{h} = \mathbf{y} \equiv \mathbf{target}.$$

For another input, the same model returns $\mathbf{y'}$ instead of $\mathbf{target'}$; this implies the equality index

$$\mathbf{h'} = \mathbf{y'} \equiv \mathbf{target'}.$$

Quite commonly, **h** and $\mathbf{h'}$ can be incomparable, meaning that neither $\mathbf{h} > \mathbf{h'}$ nor $\mathbf{h'} > \mathbf{h}$. Here a problem arises: how can we compare the relevancy of these two results? The solution offered comes in the form of the fuzzy integral

$$\mathbf{I} = \int \mathbf{h} \, \square \, \mathbf{g}_\lambda$$

The higher the value of the integral, the more relevant the results the fuzzy model delivers. Then **h** and **h′** are compared indirectly: by stating that **h** dominates **h′**, we denote the situation

$$\int \mathbf{h} \,\square\, \mathbf{g}_\lambda \geq \int \mathbf{h}' \,\square\, \mathbf{g}_\lambda$$

13.7.2   Validation of Fuzzy Models

The essence of validation activities is more subjective and task-oriented. The entire issue revolves around evaluating the model's usefulness when placed in a certain application. We can set up two main criteria:

• the fuzzy model's usefulness in representing and solving specific classes of problems for which it has been designed: For instance, we may think of the model's suitability in handling control tasks or dealing with the problems of single or multistep prediction.

• the fuzzy model's ability to cope with information of different granularity: Quite commonly, the model's validity is articulated by looking at its performance in the presence of numerical data. It is also of interest to study the model's performance when exposed to nonnumerical quantities. Here the question of uncertainty management becomes crucial. It is intuitively compelling to expect that if the data become uncertain (linguistic), so should the results of modeling. One should not anticipate that the fuzzy model would absorb the uncertainty and react as if no uncertainty were associated with the input. How well the fuzzy model lives up to these expectations depends very much on the design of the model itself. In a nutshell, the effect of uncertainty propagation should also be calibrated, making use of carefully selected data, including those of a nonnumerical character and exhibiting enough diversity of information granularity.

13.8   Chapter Summary

We have highlighted the main features of fuzzy models and fuzzy modeling. We have also proposed a general conceptual scheme of modeling emphasizing information processing at several different levels of information granularity. The essential role of fuzzy encoding and decoding has been clarified in depth. As the paradigm of nonnumerical modeling is confined at the level of linguistic terms, new opportunities and new challenges arise. Although some elements of system modeling encountered in numerical models are totally valid in this new environment, there are new items totally specific to this framework. Several trends in fuzzy modeling can be anticipated. One involves the development and utilization of fuzzy models in the presence of genuine fuzzy data. The dominant tendency today is to design and exploit fuzzy models for numerical data. The essential generalization of data format can easily open up new application avenues; the general architecture outlined in this study is universal enough to accommodate this generalization.

13.9 Problems

1. Probabilistic models are concerned with modeling systems in the presence of probabilistic information. In a simple conceptual scenario, a single input-single output probabilistic model is governed by relationship

$$\mathbf{P}_y = \mathbf{P}_{y|x}\mathbf{P}_x,$$

where $\mathbf{P}_y$ and $\mathbf{P}_x$ are the probabilities defined in a finite numeric universes of discourse $\mathbf{X}$ and $\mathbf{Y}$, $\text{card}(\mathbf{X}) = n$, $\text{card}(\mathbf{Y}) = m$, and $\mathbf{P}_{y|x}$ denotes a matrix of conditional probabilites. More specifically

$$\mathbf{P}_y = [p(y_1)\ p(y_2)\ \cdots\ p(y_m)],$$

$$\mathbf{P}_x = [p(x_1)\ p(x_2)\ \cdots\ p(x_n)],$$

$$\mathbf{P}_{y|x} = \begin{bmatrix} p(y_1|x_1) & p(y_1|x_2) & \cdots & p(y_1|x_n) \\ p(y_2|x_1) & p(y_2|x_2) & \cdots & p(y_2|x_n) \\ \vdots & \vdots & & \vdots \\ p(y_m|x_1) & p(y_m|x_2) & \cdots & p(y_m|x_n) \end{bmatrix}.$$

In view of the above matrix relationship, we get

$$p(y_j) = \sum_{i=1}^{n} p(y_j|x_i)p(x_i),$$

$j = 1, 2, \ldots, m$. Furthermore, the fundamental probabilistic relationships are straightforward:

$$\sum_{j=1}^{m} p(y_j) = 1, \quad \sum_{i=1}^{n} p(y_i) = 1, \quad \sum_{i=1}^{n} p(y_j|x_i) = 1, \quad \text{for all } j = 1, 2, \ldots, m.$$

Even though, from a conceptual point of view, the above probabilistic framework is very distinct from the modelig framework estabished by fuzzy sets, it has a lot of common with the modeling framework when it comes to its utilization. Identify the decoding and encoding module and describe briefly their main functional properties.

2. Consider a fuzzy grammar

$$\langle V_N, V_T, P, \sigma \rangle$$

where

$V_N = \{A, B\}$
$V_T = \{a, b\}$
P:

$\quad \sigma \to bA \,|\, aB \ (0.4, 0.5)$
$\quad A \to a \,|\, a\sigma \,|\, bAA \ (0.9, 0.1, 0.6)$
$\quad B \to b \,|\, b\sigma \,|\, aBB \ (0.8, 0.2, 0.4).$

Provide a derivation tree for $\mathbf{x} = aabb$ and compute the degree of this derivative (that is, the degree of membership of x in the language induced by G). Repeat the calculations for the strings $aaabbb$ and $aaaabbbb$. Can you observe any regularity? Comment on the membership degrees obtained.

3.   Consider the collection of data shown in figure 13.17. Construct a rule-based model with three rules and triangular fuzzy sets with 1/2 overlap and second-order polynomials in the conclusion. Compare the results with a global approximation completed with the aid of third-order plynomials. Draw conclusions as to the performance of these two types of models (namely, global and local).

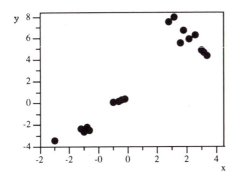

Figure 13.17
Experimental input-output data for problem 4

4.   This problem directs us to the issue of the fuzzy model's approximation-interpolation capabilites. Assume that we want the model to fit at some points of the output, say $g_1, g_2, \ldots, g_c$, for some specific input $i_1, i_2, \ldots, i_c$, that is

$$f(i_j) = g_j,$$

$j = 1, 2, \ldots, c$, while approximating some other points $\{x_k, y_k\}$, $k = 1, 2, \ldots, N$. The partitioning of the input space is to cope with the approximation-interpolation form of the problem. Let us study local models of the form

If x is A_i, then $y = a_i(x - m_i) + g_i$,

$i = 1, 2, \ldots, c$, where A_i are triangular membership functions with an 1/2 overlap between two successive linguistic terms and centered on m_i.

(a) Show that this type of model satisfies the required interpolation constrains.

(b) Derive optimization formulas to estimate the remaining parameters (slopes) of the local models.

(c) Using the results of analysis completed in (b), construct the fuzzy model for the data from the previous example. Elaborate on how you can determine m_i and g_i. What is a reasonable number of local submodels (c) to be used in this model.

(d) Build the fuzzy model for the data in (c) but using the local models not associated with the clusters of data, namely,

If x is A_i then $y = a_i x + b_i$.

Compare the results of modeling with those produced by the approximation-interpolation format of the models studied in (c).

5.   Along the lines of fuzzy models' generalization-approximation abilities, elaborate on maintaininig this balance for the local fuzzy models. As an illustration, use the small data set in figure 13.18.

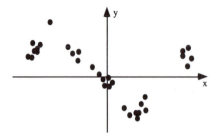

Figure 13.18
Experimental input-output data for problem 5

6.   Let us discuss linear local fuzy models and Gaussian membership functions

$$y = a_i x + b_i,$$

$$A_i(x) = \exp\left(\frac{(x - m_i)^2}{\sigma_i^2}\right),$$

$i = 1, 2, \ldots, c$. Derive complete formulas for gradient-based optimization of the entire models by modifying the membership functions' parameters as well as those of the linear functions. Assume that the training is aimed at minimizing the standard squared errors between the model's output (y_k) and the target value coming from the learning set,

$$Q = \sum_{k=1}^{N} (y_k - \text{target}_k)^2$$

The finite family of the data is arranged in a collection of pairs (x_k, target_k), $k = 1, 2, \ldots, N$.

(a) Using the data from problem 4, run the algorithm. Elaborate on how to initialize the model's parameters.

(b) Discuss the option of having two different learning rates, the first applying to the linear functions' parameters (α_1) and the other used to update the membership functions' parameters (α_2). What is the relationship between α_1 and α_2?

(c) If the overall model does not perform well (the performance index Q exceeds acceptable values), one is usually tempted to increase the number of the local models. Obviously, with too many submodels, the entire model may not generalize. The idea is to make each fuzzy set in the input space meaningful by keeping these fuzzy sets distinct enough. If no provisions to this effect are made, at the end of the optimization process, some of the fuzzy sets in the input space could easily become almost identical (overlap). To prevent this from happening, the previous performance index is augmented by a so-called regularization component R whose role is to keep the fuzzy sets distinct,

$$Q = \sum_{k=1}^{N} (y_k - \text{target}_k)^2 + R.$$

Explain how the regularization factor can be implemented.

References

Bezdek, J. C. 1993. Fuzzy models—What are they, and why? *IEEE Trans. on Fuzzy Systems* 1:1–6.

Di Nola, A., S. Sessa, W. Pedrycz, and E. Sanchez. 1989. *Fuzzy Relational Equations and Their Applications in Knowledge Engineering*. Dordrecht, Netherlands: Kluwer Academic Press.

Kandel, A. 1982. *Fuzzy Techniques in Pattern Recognition*. New York: J. Wiley.

Lee, E. T., and L. A. Zadeh. 1969. Note on fuzzy languages. *Information Sciences* 1:421–34.

Mizumoto, M., J. Toyoda, and K. Tanaka. 1972. General formulation of formal grammars. *Information Sciences* 4:87–100.

Pedrycz, W. 1990. Processing in relation structures: fuzzy relational equations. *Fuzzy Sets and Systems* 40:77–106.

Pedrycz, W. 1991. Neurocomputations in relational systems. *IEEE Trans. on Pattern Analysis and Machine Intelligence* 13:289–96.

Pedrycz, W. 1992. Selected issues of frame of knowledge representation realized by means of linguistic labels. *Int. J. of Intelligent Systems* 7:155–70.

Pedrycz, W. 1993a. Fuzzy Control and Fuzzy Systems. 2d ext. ed. Taunton, NY: Research Studies Press/J. Wiley.

Pedrycz, W. 1993b. Fuzzy neural networks and neurocomputations. *Fuzzy Sets and Systems* 56:1–28.

Pedrycz, W. 1995. Fuzzy Sets Engineering. Boca Raton, FL: CRC Press.

Sanchez, E. 1976. Resolution of composite fuzzy relation equations. *Information and Control* 30:38–47.

Santos, E. S. 1974. Context-free fuzzy languages. *Information and Control* 26:1–11.

Sugeno, M., and T. Yasukawa. 1993. A fuzzy-logic-based approach to qualitative modeling. *IEEE Trans. on Fuzzy Systems* 1:7–31.

Takagi, T., and M. Sugeno. 1985. Fuzzy identification of systems and its application to modeling and control. *IEEE Trans. on Systems, Man, and Cybernetics* 15:116–32.

Zadeh, L. A. 1965. Fuzzy sets and systems. *Proc. Symp. Syst. Theory*, Polytechnic Institute of Brooklyn, 29–37.

Zadeh, L. A. 1979. Fuzzy sets and information granularity. In *Advances in Fuzzy Set Theory and Applications*, eds. M. M. Gupta, R. K. Ragade, and R. R. Yager, 3–18. Amsterdam: North Holland.

III PROBLEM SOLVING WITH FUZZY SETS

14 Methodology

Fuzzy set theory has become of the utmost relevance in most engineering areas. In a broad sense, engineering applications involve goals to be achieved under technical, economical, and social constraints. Steps must be taken to develop solutions in situations where the criteria for choice are not always as certain as we would wish them to be. This chapter addresses a number of methodologies that have been proven useful in solving a diversity of engineering problems. After introducing an overview of the essential issues of analysis and design, the keys for engineering problem solving, the chapter proceeds by presenting techniques for decision making. These include control and optimization methodologies.

14.1 Analysis and Design

Typically, engineering tasks concern activities involved in translating an idea or technical need into the detailed specification from which a product, process, system, or solution can be derived. Having an idea in mind about a required outcome, the engineer must be able to originate systems and to predict how these systems will fulfill objectives. The result is a description of a system that satisfies a set of performance requirements and constraints.

Engineering tasks can be considered to incorporate two main activities: analysis and design. Analysis concerns mainly defining and understanding problems to produce a clear and explicit statement of goals, constraints, and desired system behavior characteristics. Design involves producing feasible solutions, then judging and selecting among alternatives. A cycle is then induced in which the solution is revised and improved by reexamining the analysis. Therefore, analysis and design form the basis of a framework for planning, organizing and evolving design solutions, as figure 14.1 summarizes.

Given that knowledge is an important ingredient for engineering problem solving, its acquisition, representation, organization, and manipulation is at the heart of analysis and design tasks. In an engineering framework, knowledge is usually considered within a wider perspective to include basic engineering knowledge, procedures, experience, and expert knowledge. It can be represented in different forms, ranging from mental images in one's thoughts, spoken or written words in some language, and graphical or other pictures to detailed character strings or collections of memory positions in a computer system. Any choice of representation depends on the type of problem to be solved, the solution procedure available, and the level at which the problem is being attacked.

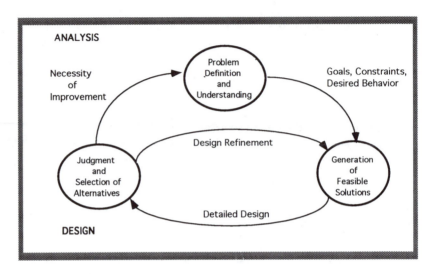

Figure 14.1
Analysis and design cycle

The organization of knowledge is key to efficient problem solving. It is essential that the appropriate item be easy to locate and retrieve. Otherwise, much time and effort will be wasted in searching for and developing solutions. This applies both to higher-level tasks such as problem modeling, requirement analysis, and project organization, and to lower-level tasks such as coding and field testing.

Knowledge can come from various sources, such as textbooks, reports, technical articles, experts, and the like. To be useful, knowledge must be accurate, presented at the right level, complete in the sense that all essential elements are included, free of inconsistencies as much as possible, and so on.

In the field of artificial intelligence, for example, experience in building dozens of knowledge-based systems has shown that eliciting facts, heuristics, procedures and relevant ingredients is a slow, error-prone process. Thus, knowledge acquisition refers to the ability in engineering to elicit from the sources coherent and consistent information for decision making. It is not a passive intuition about the problem environment, nor is it a self-coherent and self-consistent agglomeration of items. It is a particular distinction, selection, and organization of concepts derived from the knowledge sources that brings the engineer a solution to the problem.

Figure 14.2 illustrates a more detailed step-by-step analysis and design cycle. These activities can be given a more specific character by exploiting relevant concepts of fuzzy sets. In particular, by assigning fuzzy sets to the linguistic labels, we are dealing with a numerical concept calibration

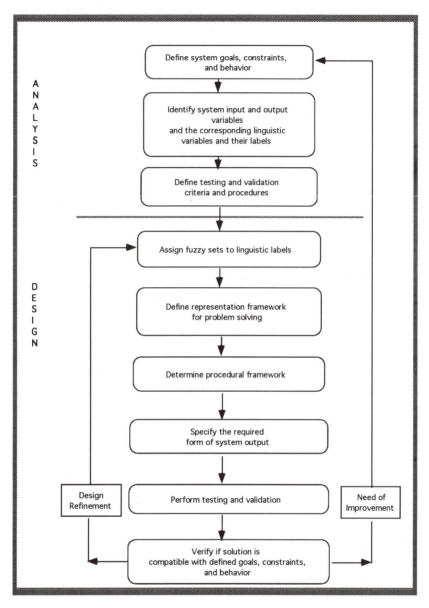

Figure 14.2
Step-by-step analysis and design

through associating respective membership functions. The representation framework for defining problem solving may involve ideas of knowledge representation, including rules and frames. The desired format of output information can be achieved by exploiting various schemes of decoding as well as using mechanisms of linguistic approximation. Finally, testing and validation are accomplished through system simulation and system prototyping.

In actual, real-world applications, it is very unlikely that only a single issue will predominate. In general, the nucleus methodology is completed, with a number of complementary tasks required to assemble the final solution. For instance, interfaces with system environment may be needed, and sensors and actuators may be used. The overall system may need to be supervised to avoid pitfalls or failures in both software and hardware. In other cases, man-machine interfaces may be essential in providing information, with a focus on user and problem-solving needs. Appropriate interfaces must be developed to convey the essential information and to avoid cognitive overload. Thus, system integration at the methodological, computational and physical levels must be carefully planned and pursued.

Another important point concerns the need of methodological adaptations when a real problem is to be tackled. Often, after the main problem-solving paradigm is sought, careful analysis must be carried out to fit the basic methodological framework to the actual problem needs. This is at the very heart of engineering tasks. Therefore, the theoretical, computational, and methodological background composing a sound technical knowledge must always be associated with engineering judgment and insight.

Domain knowledge plays a strong role in developing good designs. Domains must be carefully and patiently explored because in general, they indicate the route to be pursued during design decisions and solution trade-offs.

14.2 Fuzzy Controllers and Fuzzy Control

Fuzzy controllers commonly control engineering artifacts that take advantage of fuzzy sets and rule-based systems in particular.

14.2.1 Knowledge Acquisition in a Simple Control Problem

To gain better insight into how control rules reflect domain knowledge about a problem to be solved, we study a class of dynamic systems whose time response to a unit step function is of so-called S shape (figure 14.3).

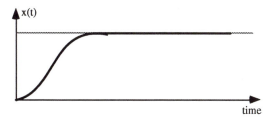

Figure 14.3
Dynamic system with S-shape unit step response

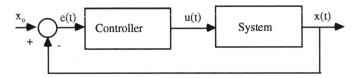

Figure 14.4
Generic structure of a closed-loop control system

In other words, when the system is exposed to a unit step input, its time response is S-like. The system is dynamic, since x depends on its history (namely, the previous values). Usually the class of dynamic systems is represented in the form of differential (or difference) equations of first or higher order,

$$\frac{\partial x(t)}{\partial t} = g(x, u),$$

$$\frac{\partial^2 x(t)}{\partial t^2} = h(x, u),$$

$$\ldots$$

$$\frac{\partial^n x}{\partial t^n} = f(x, u).$$

Given now a point of reference (setpoint) x_0, we are interested in applying such control action u that the system achieves this setpoint smoothly and as quickly as possible. Imagine that we will assume the role of such a controller and form a closed-loop control structure (figure 14.4). What information is required to take relevant action? After some thought, we may consider error and change of error as two essential pieces of information that definitely should guide our decisions. Error is defined as the difference between the setpoint and current output x,

$$e(t) = x_0 - x(t).$$

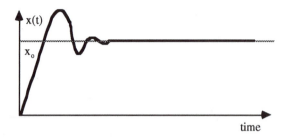

Figure 14.5
Typical response of a closed-loop control system

Change of error reflects a trend in the reported values of error and is defined as the difference between two successive values of error,

$$de(t) = e(t) - e(t'),$$

where $t' < t$ and $t' \rightarrow t$. (More precisely, we can talk about the derivative of error, $\partial e / \partial t$).

When control actions are applied in the closed-loop structure, the behavior shown in figure 14.5 would be a typical system response. Because of the system's dynamics, one cannot expect it to reach the setpoint immediately. In fact, there is always a compromise: too strong control actions lead to a faster response but yield substantial overshoots and oscillations.

Now let us think in terms of rules. Looking at the control actions previously taken and monitoring the current value of error and its change, we ask ourselves how control actions should next be changed to get the desired performance from the system. It depends. Based on our intuition we can summarize our experience. For instance, if error is around 0 and change of error is around 0, it is intuitively justifiable to maintain the previous control. This means that the change in control should be around 0. We have developed the following control rule:

If error is *zero* and change of error is *zero*, then change of control is *zero*.

This rule looks straightforward. What about control if error assumes *negative big* values while change of error is around 0? This is an example of the system exhibiting overshoot, as depicted in figure 14.6(a). A possible control action would be to reduce control significantly; that is, change of control should be *negative big* or *negative medium* to avoid eventual oscillations. The rule reads now as

If error is *negative big* and change of error is *zero*, then change of control is *negative big*.

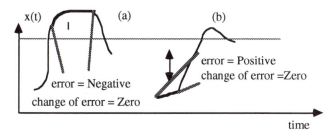

Figure 14.6
Derivation of control rules

Table 14.1
Linguistic control protocol

DE\E	NB	NM	NS	ZE	PS	PM	PB
NB	NB				NM		
NM		NM		NM			
NS				NS	ZE		PM
ZE	NB	NM	NS	ZE	PS	PM	PB
PS	NM		ZE	PS			
PM				PM		PM	
PB			PM	PB			PM

If error is *positive big* and change of error is around 0, we are very far from the setpoint, as shown in figure 14.6(b), and the change of control should be *positive big* (or *positive medium*). This becomes formalized as the rule

If error is *positive big* and change of error is *zero*, then change of control is *positive big*.

Performing similar reasoning about control, we come up with a protocol of control rules, summarized in table 14.1. The rules in this protocol assume the form

If error is E_i and change of error is DE_i, then change of control is U_i,

$i = 1, 2, \ldots, N$. The control protocol is built using a finite (usually very small) number of symbols (linguistic terms) defined for all the system's variables. For instance, the labels of the linguistic terms (term sets) include

- For error and change of error: NS, NM, NB, ZE, PB, PM, PB.
- For change of control: NS, NM, NB, ZE, PS, PM, PB.

This description of control actions obviously does not involve numbers but takes place at the level of words (more precisely, linguistic terms). In both the space of error and the change of error, we are concerned with a few linguistic landmarks. Similarly, our perception of required control is focused at the same level of specificity, calling for the relevant linguistic notions of control.

While the control protocol (rule base) elucidated operates at the level of linguistic terms, the system under control requires a single numerical control signal. Similarly, it is common for the status of the system to be characterized by numerical measurements of error and change of error. To cope with all these visible information incompatibilities, in addition to the rules, we have to develop interfaces between the protocol and the control environment by deciding upon relevant encoding and decoding schemes. Furthermore, to make the rules fully operational, we need to incorporate a certain mechanism of inference or mapping from the space of error and change of error to the space of control. All these elements together generate the concept known as a fuzzy control (frequently referred to as a fuzzy logic controller; we do not fully subscribe to this notation as there is no underlying logical construct exploited in the controller).

Everything we have stated about control rules has concerned the conceptual (virtual) level at which control is perceived: the rules reflect our perception on how to solve the control problem. The fuzzy controller evolves as a realization of this protocol. More precisely: The same *virtual* presentation of the control protocol could lead to many (and sometimes quite distinct) *physical* realizations known as fuzzy controllers. These variants depend on different ways in which the encoding and decoding are worked out as well as the mapping mechanism applied.

There are several ways of eliciting knowledge. We elaborate on two main classes:

1. Using commonsense knowledge: Knowledge of this type is quite readily available and can be easily structured into a more formal framework. This is the case in the present study: Practically everybody can propose the control protocol, because the rules obviously appeal to our intuition. Two points should be raised:

• Rules can be developed for relatively small problems involving very few variables. As the number of variables increases, so does our effort to come up with meaningful and transparent rules. Under these circumstances, we usually end up with control protocols that are incomplete or inconsistent (including conflicting rules).

• Rules that comply somewhat with our intuition on how the control actions should be taken easily apply to systems. The protocol we have developed pertain to a class of systems that fortunately is broad enough to be of practical relevance. The same qualitative rules will not be completely relevant when considering another class of systems, like those with delay in system variables, say

$$\frac{\partial x}{\partial t} = g(x(t - \tau), u(t)),$$

(where τ stands for the delay time) or non-minimal phase systems. This is because systems of these classes do not respond to control entirely as expected. For instance, in any system with delay, the control affects the system after a certain time (caused by so-called transportation delay). This makes the commonsense error–control relationships somewhat less apparent. The control of non-minimal phase systems is even more intricate, because the common reversed behavior of system output in a time period is commonly reversed under a control action that we would expect to drive the system output in the desired direction.

2. Acquiring knowledge through interactive computer simulations: Interactive computer simulations provide a powerful and flexible way of acquiring knowledge. Through a series of such simulations, the designer (user) plays with the system simulated via a detailed model and learns or tests the control protocols. After a series of such experiments, the control knowledge becomes highly refined and can be transformed into a formal structure of rules.

The developer becomes fully aware of eventual specificities of control that should be incorporated into the control protocol. Because the simulation capabilities are to a large extent unlimited as well and the designer, supported by diverse mechanisms of virtual reality (VR), can easily and effectively exercise his/her control skills, this approach to knowledge acquisition will very likely gain in importance. The VR environment can also accommodate collecting and organizing experimental data.

14.2.2   Construction of the Fuzzy Controller: Algorithmic Aspects

In this section, we elaborate on the main components that assemble a fuzzy controller and describe the characteristic mechanisms involved in its construction as well as how they function. Figure 14.7 depicts a fuzzy controller's basic structure. Note that a fuzzy controller resembles in its general architecture a fuzzy model. Recall that, in general terms, a fuzzy model comprises an input interface, a processing module and an output interface. In this view of a fuzzy controller, the encoding and decoding mechanisms embody the input and output interface, whereas the rules and inference are constituents of the processing module.

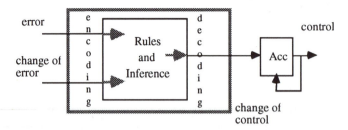

Figure 14.7
Basic structure of a fuzzy controller

14.2.2.1   Rules and Rule Base

The fuzzy controller's underlying structure comes from its rule-based organization. Its key premise is that control knowledge is available to specify a control strategy represented by a collection of if-then rules. Thus the control strategy is structured into control protocols linking the system's current state with the corresponding control action. In figure 14.7, the state is represented by the error and change of error. The fuzzy controller in essence comes up with interpolation or approximation mechanisms to provide a control decision for any system state. Actually, Castro (1995) has developed the theoretical machinery showing its universal approximation capabilities.

Putting this in a broader context, fuzzy controller design is concerned with the calculus of fuzzy rules, and its computing procedures are governed by rule-based computations, discussed in chapter 10. To summarize briefly the technical details of the basic construction of the fuzzy controller, as frequently encountered in the current literature, let us concentrate on a collection of control rules (the rule base) in the form

If X is A_i, then U is $B_i, i = 1, \ldots, N$,

where X is a linguistic variable expressing the state of the system and U is a linguistic variable representing the control actions. A_i and B_i are normal fuzzy sets in $\mathbf{X}$ and $\mathbf{U}$, respectively. Based on the discussion in chapter 10, we see that one way to construct a fuzzy controller is to aggregate the N rules into the form of a single fuzzy relation in $\mathbf{X} \times \mathbf{U}$,

$$R = \bigcup_{i=1}^{N} A_i \times B_i,$$

or, alternatively in terms of membership functions,

$$R(x, u) = \bigvee_{i=1}^{N} R_i(x, u),$$

which could be regarded as a (associative) memory of the controller.

14.2.2.2   Inference

Let A be an input datum entering the controller for which we must determine the corresponding control action, say B. The sup-t composition provides a general procedure for computing B, as seen previously, but let us concentrate on the sup-min composition, a widely accepted basic mapping mechanism between the state space and the control space, that is,

$$B = A \circ R.$$

We thus derive the following:

$$B(u) = \sup_x [A(x) \wedge R(x, u)] = \sup_x \left[A(x) \wedge \bigvee_{i=1}^{N} R_i(x, u) \right].$$

If we assume that the min t-norm assembles the relations R_i, then the following holds:

$$B(u) = \bigvee_{i=1}^{N} \left\{ \sup_x [A(x) \wedge R_i(x, u)] \right\} = \bigvee_{i=1}^{N} \left\{ \sup_x [A(x) \wedge A_i(x) \wedge B_i(u)] \right\}.$$

If we denote $\sup_x [A(x) \wedge A_i(x)] = \text{Poss}(A, A_i) = \lambda_i$, it is clear that

$$B(u) = \bigvee_{i=1}^{N} \{\lambda_i \wedge B_i(u)\}.$$

So B becomes a union of the B_is previously weighted via the activation level of the respective rule λ_i. Alternativelly, $B(u)$ may be found using the scaled method in its additive form, namely

$$B(u) = \sum_{i=1}^{N} [\lambda_i \wedge B_i(u)].$$

Max aggregation ignores overlap in fuzzy sets, whereas sum combination adds overlap. When input changes slightly, the additive output also changes slightly. The max-combined output may ignore small input changes, since for large sets of rules most change occurs in the overlap regions of the fuzzy sets (Kosko 1992). Further choices of rule interpretation and aggregation are possible, as discussed in section 10.4, including alternative additive forms of the fuzzy controller (Mizumoto 1994).

14.2.2.3   Encoding

At this point, the task of encoding becomes transparent (which is why its description has purposely been postponed until now). Note that the computation of λ_i involves A and A_i only, If A is only a single point $A(x_0)$ representing, for example, a sensor datum x_0, then the computations of λ_i are straightforward—the activation level is nothing more than a membership functions at value of the x_0. If the sensor's readings are inherently biased, it is reasonable to treat the measurements as fuzzy sets around x_0 (This gives rise to a model of a fuzzy sensor.) subsequently, the activation level is viewed as a possibility measure of A taken with respect to A_i.

14.2.2.4   Decoding

Finally, the inferred fuzzy set of control is to be converted into a single numeric value, a control signal, say $b \in \mathbf{U}$, to drive the system under control. This decoding task, known as defuzzification, can be accomplished in many different ways. Because the set-based information needs to be transformed into a specific pointwise format, there is no unique way of carrying out this task (Yager and Filev 1994). The most common alternatives are reviewed below (assuming a discrete universe $\mathbf{U}$):

- Mean of maxima:

$$b = \begin{cases} \max_{u \in \mathbf{U}} B(u), & \text{if } B(u) \text{ is unimodal} \\ \sum_{j=1}^{L} \dfrac{u_j}{L}, & \text{otherwise,} \end{cases}$$

where u_j is the support value at which $B(u)$ reaches the maximum, and L is the number of such support values.

- Center of gravity:

$$b = \frac{\sum_{j=1}^{n} B(u_j)\lambda_j}{\sum_{j=1}^{n} B(u_j)},$$

where n is the number of discretization levels of $\mathbf{U}$.

14.2.2.5   Fuzzy Controllers and PI and PD Controllers

The velocity-type PI (proportional and integral) controller (Astrom and Wittenmark 1984) is described as follows:

$$du(k) = k_1 e(k-1) + k_2\, de(k),$$

where k is the discrete time, and

$de(k) = e(k) - e(k-1).$

We show that the controller can be rewritten in an equivalent form with explicitly defined proportional and integral parts. We have

$u(1) = k_1 e(0) + k_2 [e(1) - e(0)] + u(0),$

$u(2) = k_1 [e(1) + e(0)] + k_2 [e(2) - e(0)] + u(0).$

Iterating over discrete time moments, we get

$$u(k) = k_1 \sum_{j=0}^{k-1} e(j) + k_2 [e(k) - e(0)] + u(0).$$

It is now obvious that k_2 is the proportional term (k_p) and k_1 plays the role of the integral gain, k_I. Comparing the format of the fuzzy controller's rules with that of the above structure, we conclude that the fuzzy controller (or more precisely, its numerical manifestation) may constitute an analog of the PI controller. On the other hand, if the total control output is computed in the form

$u(k) = k_1 e(k) + k_2 de(k),$

we obtain an anolog of the PD controller.

14.2.3 Possibility-Necessity Computations in the Fuzzy Controller

The need to exploit both possibility and necessity computations in fuzzy controllers stems from the fact that the inputs of the controller can be nonnumerical values. In that case, the possibility and necessity values are usually not equal. We can take advantage of this difference and use it as a signal informing the fuzzy controller about the input uncertainty. Subsequently, the fuzzy controller can release this information along with the suggested control values. As we indicated in section 2.6.3, the computation of possibilities is not sufficient to support this idea.

The proposed fuzzy controller becomes a tandem of two parallel parts, the first of which realizes standard possibilistic calculations and the second of which performs necessity computations. Here the basic formula reads

$$\bar{B}(u) = \bigwedge_{i=1}^{N} [1 - \text{Nec}(A_i, A) \vee \bar{B}_i(u).$$

To illustrate the effect produced by both types of computations, we study an illustrative example involving a single rule, "If A_i, then B_i," where B_i is described by a triangular membership function, as shown in figure 14.8. If the input information is numerical, then the possibility and

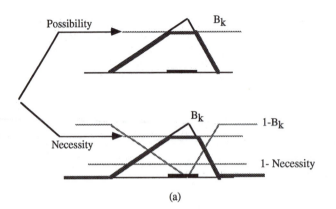

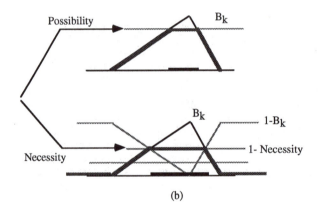

Figure 14.8
Intervals of possible and necessary control in the case of (a) numerical information and (b)
linguistic information

necessity values are equal, (panel (a) of figure 14.8). Following the standard computations, we get

1. For possibility computations: $B(u) = \text{Poss}(A, A_i) \wedge B_i(u)$.

2. For necessity-oriented calculations: $\bar{B}(u) = 1 - \text{Nec}(A_i, A) \vee \bar{B}_i(u)$.

As figure 14.8(a) shows, the indicated intervals of control values that
could be referred to as ranges of possible control and necessary control
(because they identify numerical control that is either possible or necessary) fully coincide.

    When the input is nonnumerical, say, a normal fuzzy set, then the pos-
sibility and necessity values tend to differ; the fuzzier the input, the more

evident the gap between these two values. As sketched in figure 14.8(b), we end up with two substantially distinct ranges of possible control and necessary control.

14.2.4   Fuzzy Hebbian Learning

To clearly reveal the analogy between Hebbian learning, which is encountered quite often in associative neural networks (associative memories) (Pedrycz 1995) and the fuzzy controller, several specific formulas may be worth emphasizing. In any associative memory, one is concerned with the mechanisms of an efficient storage and recall. The data (items) to be handled are organized as a finite family of input-output pairs (associations),

$\mathbf{x}_1, \mathbf{y}_1$

$\mathbf{x}_2, \mathbf{y}_2$

$\vdots$

$\mathbf{x}_N, \mathbf{y}_N$

Here we confine ourselves to the pairs $\mathbf{x}_i, \mathbf{y}_i$ lying in the unit hypercubes, $[0, 1]^n$ and $[0, 1]^m$, respectively. The associative memory problem is basically formulated as follows: Store the associations and propose a recall mechanism capable of retrieving the data without any error. Let $\mathbf{A}$ be an association architecture. The recall scheme $\mathbf{R}$ is called error-free if, when acting upon $\mathbf{A}$, it satisfies the recall formula

$\mathbf{R}(\mathbf{x}, \mathbf{A}) = \mathbf{y}$

for any $\mathbf{x} = \mathbf{x}_1, \mathbf{x}_2, \ldots, \mathbf{x}_N$; thus $\mathbf{y}$ becomes $\mathbf{y}_1, \mathbf{y}_2, \ldots, \mathbf{y}_N$, respectively.

In the following, we outline a generic form of the transformation scheme. The functional concept of the controller can be split into two phases:

1. Summarization (storing):   At this stage, the data (associations) are aggregated (summarized) into the format of a single relation (rather than a function).

2. Recall: Here, for any new data (input variables), the associated output (control) is recalled.

Quite often the fuzzy controller is referred to a fuzzy logic controller. The construct we are about to discuss, at least, does not strongly justify the use of any logical notions. It is perhaps prudent to emphasize that the widely used architecture of the controller has much more in common with the associative memories and should be viewed from this perspective (cf Kosko 1992).

The general scheme described above can be realized in various ways (by coming up with the different manifestations of the summarization and association steps). The standard way exploits fuzzy Hebbian learning and recall. (This becomes an essence of the processing module.) Before discussing that, let us summarize some basic facts about associative memories (cf. Kosko 1992). In this setting, the association structure A is very often achieved as a correlational convolution of the items to be stored, producing a correlation matrix $\mathbf{R}$:

$$A : \mathbf{R} = \sum_{i=1}^{N} \mathbf{y}_i \mathbf{x}_i^T .$$

This form of aggregation is known as Hebbian learning.[1] The recall scheme is also a linear, producing the recall vector

$$\mathbf{y} = \mathbf{R}\mathbf{x}$$

for a given item $\mathbf{x}$. The fundamental issue of error-free recall can be addressed by analyzing the character of the associations to be stored in the memory. If the inputs of the associations are mutually orthonormal, namely,

$$\mathbf{x}_i^T \mathbf{x}_j = \delta(i,j) = \begin{cases} 1, & \text{if } i = j \\ 0, & \text{if } i \neq j, \end{cases}$$

then the recall is error-free. To validate this claim, simply set $\mathbf{x} = \mathbf{x}_{k_o}$; the detailed computations reveal that

$$\mathbf{y} = \sum_{i=1}^{N} \mathbf{y}_i (\mathbf{x}_i)^T \mathbf{x}_{k_o} = \sum_{i \neq k_o}^{N} \mathbf{y}_i [(\mathbf{x}_i)^T \mathbf{x}_{k_o}] + \mathbf{y}_{k_o} [(\mathbf{x}_{k_o})^T \mathbf{x}_{k_o}] = \mathbf{y}_{k_o},$$

since the first term of the expression vanishes because of the orthonormality condition. Note also that if the orthonormality requirement does not hold, then $\mathbf{y} \neq \mathbf{y}_{k_o}$; the first term forms a crosstalk component

$$\sum_{i \neq k_o}^{N} \mathbf{y}_i [(\mathbf{x}_i)^T \mathbf{x}_{k_o}]$$

that is responsible for any distortions in the recall process. Two observations arise:

1. The correlational associative memory is a conceptually simple and quite attractive construct, since no extensive learning is involved; essentially all the associations are combined in the form of a single correlational matrix built within a single pass through the data set,

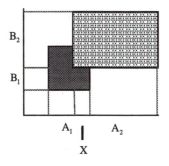

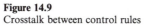

Figure 14.9
Crosstalk between control rules

2. A limited capacity offsets the conceptual and computational simplicity. (Again, one can easily figure out that only N different associations can be stored without error.) There is also no special mechanism for treating some of the associations more selectively, even knowing that they might be more reliable than the others.

We have concentrated on this scheme of the associative memory because the fuzzy controller forms a straightforward analog of this construct; the only technical difference lies in the different operations used in the associative memory and fuzzy controller. Similarly, the strengths and weaknesses of the structure discussed also occur in the fuzzy controller.

Let us focus on the input-output associations, (A_k, B_k), with A_k, B_k viewed as fuzzy sets defined in the corresponding finite universes of discourse, namely $A_k \in [0, 1]^n$, $B_k \in [0, 1]^m$. Figure 14.9 illustrates the crosstalk between the rules that occurs because of Hebbian learning and the architecture introduced. For clarity of presentation, the rules include overlapping sets (as opposed to fuzzy sets). The figure illustrates that in the regions where they overlap, the mapping generates a sum of B_1 and B_2.

The fuzzy controller has the following fundamental information processing properties:

1. Boundary conditions:

• If X is an empty set, $X = \varnothing$, so is B, $B = \varnothing$;

• $X = \mathbf{X}$ implies that the fuzzy set of control is given as $B = \bigcup_{k=1}^{N} B_i$, which constitutes a union of all the control actions forming the control protocol.

2. Monotonicity:

• If $X \subset X'$, then $B \subset B'$, so B' is less specific than what is produced by B.

• In particular, the second boundary condition outlined above becomes a clear-cut illustration of the monotonicity aspect.

3. Crosstalk:

• Crosstalk leads to the containment rather than equality condition between the recalled items, namely,

If $X = A_{k_o}$, then $B \supset B_{k_o}$.

This relationship is a manifestation of the crosstalk phenomenon (refer to figure 14.9).

14.2.5. Design Considerations and Controller Adjustments

14.2.5.1 Relational Partition of the Input Space

The linguistic terms occurring in the control rules discussed so far constitute a regular grid in two- or multiple-dimensional space (panel (a) of figure 14.10). Sometimes it is possible to end up with a far more irregular grid (partition) of the space (panel (b) of figure 14.10). Partitions of this type are of interest if some underlying domain knowledge about control policy allows us to classify some regions in the Cartesian product of the spaces of error and change of error. Notice, however, that now the antecedents of the rules include relations rather than straightforward Cartesian products of the corresponding linguistic terms (fuzzy sets). For instance, the rule shown in figure 14.11 can be read as

If error and change of error are *approximately* equal, then control is B_1,

or equivalently

If relation (error, change of error) is R_1, then control is B_1,

where R_1 denotes the fuzzy relation "*approximately* equal." The relation itself can be defined in many different ways, with one possibility specified as

$R_1(e, de) = \exp[-(e - de)^2/3]$.

The relational rules are more general, and thus the total number of these rules can be substantially reduced in comparison to the previous format. In fact, a relational rule may embrace a series of rules of the form

If E_i and DE_i, then B_1,

$i = 1, 2, \ldots, N$, as shown in figure 14.11.
The only essential requirement is that the domain knowledge about control be provided in this format. Sometimes one encounters control protocols that involve both types of rules.

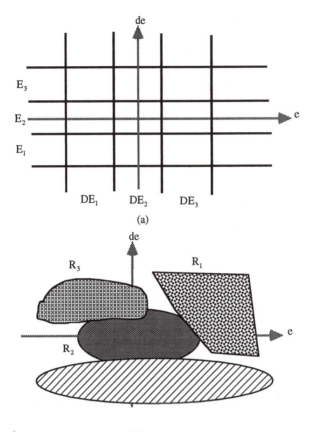

Figure 14.10
Error-change of error space with (a) regular and (b) irregular partitions

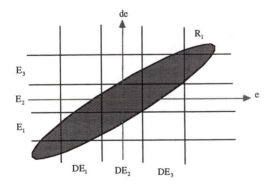

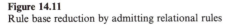

Figure 14.11
Rule base reduction by admitting relational rules

14.2.5.2   Controller Adjustments

The underlying premise of controller adjustments is that control rules are universal in that they convey qualitative hints about control activities (control policy) deemed applicable to a broad class of dynamic systems. This *universality* of rules (which holds, at least, at the symbolic level) implies that when optimizing the fuzzy controller, one should leave the rules unchanged or should resort to changes only if other attempts to improve the fuzzy controller's performance have failed. This by no means implies that the universes of discourse are universal: They could be subject to essential changes. The methods discussed below hinge on this principle. In a nutshell, accepting that the rules are qualitatively correct, we modify the spaces in which the basic linguistic terms are defined. This mode of design emphasizes the fact that the semantics (membership functions) of the linguistic terms are very much context-dependent. There is no universal meaning of such terms like "*small* error"; the definition can vary during the process of controlling the system. When close to a setpoint, an error is regarded as small when being far from the setpoint now becomes quite big as we intend to reduce its value even further. The main approaches are discussed below.

Scaling Factors
Scaling factors (gain factors or gains) are used quite often in the design of fuzzy controllers. They apply to the fuzzy controller's inputs and output and scale the original signals, say, $e' = k_e e$, $de' = k_{de} de$, $u' = k_u u$. Depending on the values of these factors, the signals are amplified or reduced. Usually these gain factors are tuned but afterwards left unchanged.

Context-Dependent Adjustment of the Universe of Discourse
In comparison to the previous method, in which the signals are modified, another approach involves changing the universe of discourse. (Note that the gains have modified the data, leaving the universe fixed.) Furthermore, the changes applied in this approach are continuous depending upon the current datum being furnished to the fuzzy controller. Let the universe of discourse be formed as an interval distributed around 0, say, $\mathbf{X} = [m_-, m_+]$. (This interval need not be symmetrical.) Depending upon the recent data available, the universe can be modified (generalized-stretched or specialized-squeezed) to "accommodate" this data fully. Figure 14.12 illustrates two cases. In figure 14.12(a), the data go far beyond the original universe of discourse and begin activating only two fuzzy sets at the bounds, making all intermediate linguistic terms totally useless. Figure 14.12(b) illustrates another example in which the linguistic labels are underutilized; the data activate only a single fuzzy set, leaving

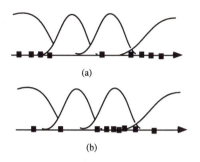

(a)

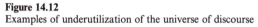

(b)

Figure 14.12
Examples of underutilization of the universe of discourse

the rest idle. The obvious way of alleviating this shortcoming would be to stretch the universe in the first case and squeeze it in the other. The effect of this dynamic adjustment relies on the current datum or a short series of several data points reported in the past. In the method to be presented, we assume that the zero point is left unchanged; this implies that the upper and lower bound will be adjusted separately. Let x be the input datum. If $x > 0$, we distinguish two situations:

1. If $x < m_+$, then

$$m_+(\text{new}) = \alpha m_+(\text{old}) + (1 - \alpha)x,$$

where α describes the contraction rate; the higher the value of α, the stronger the contraction effect, $\alpha \in [0, 1]$.

2. If $x > m_+$, then the universe must be stretched. We now get

$$m_+(\text{new}) = \alpha x + (1 - \alpha)m_+(\text{old}).$$

Here α plays the role of an expansion rate.

For the lower bound of the universe of discourse, the following formulas hold. (It is assumed that $x < 0$).

1. If $x > m_-$, then

$$m_-(\text{new}) = \alpha m_-(\text{old}) + (1 - \alpha)x.$$

2. If $x < m_-$, then

$$m_-(\text{new}) = \alpha x + (1 - \alpha)m_-(\text{old}).$$

Windowing Effect
The generic idea behind the windowing effect is that if the error and change of error are hovering around 0, we switch to a more detailed

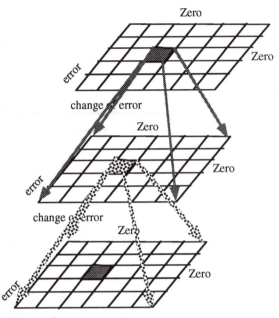

Zero

Zero

error

change of error

Zero

Zero

error

change of error

Zero

Zero

error

change of error

Figure 14.13
Windowing effect in a fuzzy controller undergoing multiresolution control

(precise) bank of control rules (figure 14.13). The process of switching can consist of several stages, each of which involves operating on fuzzy sets of increasing specificity. This achieves an effect of multiresolution, focusing on ore minute control details.

14.2.6 Fuzzy Logic Controller

Referring to the inferences realized in fuzzy logic, the mapping $\eta_i : [0,1] \rightarrow [0,1]$, expressed by

$$\eta_i(r) = \sup_{p,q \in [0,1]} [\tau_i(p) \wedge v_i(q) \wedge R(p,q,r)],$$

can be interpreted as a fuzzy truth value associated with the ith control rule. This expression is a fuzzy-relational equation to be solved with respect to v_i,

$$\eta_i = \tau_i \circ v_i \circ R.$$

If the solution set exists, the maximal truth value of the conclusion is

$$\hat{v}_i = (\tau_i \circ R) \, \varphi \, \eta_i.$$

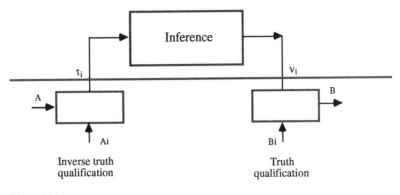

Figure 14.14
General structure of the fuzzy logic controller

This solution is a key ingredient of the inference step that operates in the fuzzy logic controller at the level of fuzzy truth values. The encoding and decoding tasks perform truth valuation and return fuzzy truth values (fuzzy sets in [0,1]) rather than single elements of the unit interval.

Consider the following statement including condition A_i and the current controller input A:

(controller input is A_i) is $\tau_i \equiv$ controller input is A,

with $\tau : [0, 1] \rightarrow [0, 1]$ sought as a linguistic truth value. Among the labels than can be used to describe τ_i are *true, very true, more or less true, false,* and so forth. For instance,

(controller input is *positive big*) is *true* $\equiv$ controller input is *positive big*.

(input datum is *positive small*) is *false* $\equiv$ controller input is *positive big*.

Truth qualification entails linking A and A_i via a linguistic truth value τ_i,

$$A(x) = \tau_i(A_i(x)),$$

for $x \in \mathbf{X}$, and the inverse truth qualification of τ_i for A and A_i provided,

$$\tau_i(v) = \sup_{x:A_i(x)=v} A(x).$$

The general fuzzy logic controller structure (figure 14.14) consists of three modules, two of which perform linguistic truth qualification and the third of which completes the logical inference in the space of fuzzy truth values. Following the inference step producing $\hat{v}_i$, truth qualification applies now at the control level, providing

control is $B'_i \equiv$ (control is B_i) is $\hat{v}_i$,

that is,

$$B_i'(u) = \hat{v}_i(B_i(u)).$$

Since inference involves a collection of N rules, the aggregation of the conclusions resulting from the individual rules is completed by intersecting B_is,

$$B = \bigcap_{i=i}^{N} B_i'.$$

14.3 Mathematical Programming and Fuzzy Optimization

The concept of optimization is rooted in the analysis of many decision problems. Usually, one approaches a decision problem involving a number of interrelated variables by focusing on a single objective developed to quantify performance and measure decision quality. This single objective is maximized (or minimized, depending on the purpose) subject to constraints that limit the choice of decision variable values. For instance, single objectives may represent profit or loss in business applications, speed or physical dimensions in an engineering problem, social welfare in social systems, best parameters of a controller or a neural network, and so on.

In real world problems, rare is the situation in which it is possible to represent fully all the complexities of variable interaction, constraints, and appropriate objectives. Thus, in general an optimization model is an approximation. Such a model should capture the problem's essential elements, and good judgment is required in interpreting the results to obtain meaningful and useful solutions.

14.3.1 General Optimization Problems

A general optimization problem, often referred to as a mathematical programing problem, can be concisely stated as follows.

minimize $f(\mathbf{x})$

subject to $h_i(\mathbf{x}) = 0, \quad i = 1, \ldots, m$

$\qquad\quad g_j(\mathbf{x}) \le 0, \quad j = 1, \ldots, r$

$\qquad\quad \mathbf{x} \in S,$

where $\mathbf{x}$ is an n-dimensional vector of decision variables, $\mathbf{x} = (x_1, x_2, \ldots, x_n)$; f, h_i, $i = 1, \ldots, m$, and g_j, $j = 1, \ldots, r$ are real-valued functions of $\mathbf{x}$; and S is a subset of the n-dimensional space. The function

f is the objective function, whereas the equations, inequalities, and the set restrictions are constraints. In general, the functions involved are nonlinear, and the model above becomes a nonlinear optimization model.

In the particular case where all the functions involved are linear and S takes the form of intervals, the optimization problem becomes a linear programming problem and has the general form

minimize $\mathbf{c}^T \mathbf{x}$

subject to $\mathbf{A}\mathbf{x} = \mathbf{0}$

$\qquad \mathbf{x} \geq \mathbf{0}$,

because, if inequalities are present, we may transform them into equalities by introducing slack or surplus variables. Here $\mathbf{c}$ is an n-dimensional vector and, as before, $\mathbf{x}$ is also an n-dimensional vector. $\mathbf{A}$ is usually an $n \times m$ matrix with real entries, where m encompasses the eventual r linear inequalities that may be present.

When a nonlinear optimization has no constraints, it becomes an unconstrained problem, and techniques based on extensions of the well-known calculus methods to find extrema (maxima and minima) of functions is used. If the function has a unique decision variable, at optimum decision point x^*, the derivative of f with respect to x, evaluated at x^*, is clearly null. In the multivariable case, this condition is replaced by $\nabla f(\mathbf{x}^*) = \mathbf{0}$, that is, the gradient of f with respect to the decision variables should be 0. Since most of the time this reduces to a difficult system of nonlinear equations to be solved, the algorithms to find $\mathbf{x}^*$ are iterative.

For instance, the steepest descent method is a basic scheme defined by the algorithm

$\mathbf{x}(k+1) = \mathbf{x}(k) - \alpha_k \nabla f(\mathbf{x})$,

where α_k is a nonnegative scalar minimizing $f(\mathbf{x}(k) - \alpha_k \nabla f(\mathbf{x}))$. In other words, from the current point $\mathbf{x}(k)$, we search along the direction of the negative gradient to a minimal point on this line; this minimal point is taken to be the next $\mathbf{x}(k+1)$. Figure 14.15 illustrates the gradient descent algorithm for the unconstrained problem in which we are asked to minimize $f(x_1, x_2) = (x_1 - 1)^2 + 0.5(x_2 - 1)^2$.

The steepest descent method is a gradient scheme from which a diversity of more sophisticated algorithms exploring the characteristics of the problem are derived. Refer to Gill, Murray, and Wright (1981) for comprehensive and practical approaches to constrained and unconstrained optimization of nonlinear models. Note that gradient methods are key not only in a broad class of decision-making problems, but also in very specific cases such as neural network learning algorithms, controller

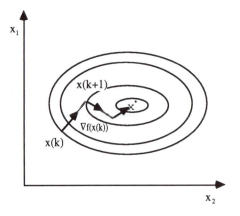

Figure 14.15
Gradient-based optimization algorithm

parameter optimization, and so on. Clearly, these methods assume f to be differentiable. If this is not the case, nondifferentiable methods should be pursued, genetic algorithms being of use in many situations. Unless some specific properties, such as convexity, are fulfilled, there is no a priori guarantee that the algorithms will find global optimal solutions. The function in figure 14.15 is convex, and therefore the algorithm converges to the optimal point $\mathbf{x}^* = (1, 1)$.

Linear optimization problems are inherently constrained, and several well-known methods are available for their solution, the simplex method and its variations being among the most popular. Essentially, the simplex method is a specific elimination method for solving a set of linear equations. (See Luenberger 1973 for detailed coverage.) However, low-dimensional problems solved graphically can provide insight into the method's characteristics. For instance, consider the linear optimization problem

minimize $2x_1 + x_2$

subject to $3x_1 + x_2 \le 4$

$$4x_1 + 3x_2 \ge 6$$

$$x_1 + 2x_2 = 3$$

$$x_1, x_2 \ge 0.$$

This problem can be easily put in the standard form involving only equality constraints by adding slack and surplus variables in the first and second inequalities, respectively. This is important for using the simplex algorithm. But in this case, figure 14.16 offers a graphical visualization of the solution process. The optimal decision is at a corner of the feasible set

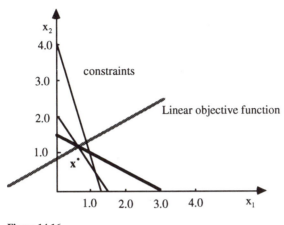

Figure 14.16
Linear programming example

(regions established by the inequalities and equality), and the value of the objective function at $\mathbf{x}^* = (x_1^*, x_2^*)$ is 2.5.

So far we have considered only classes of continuous optimization problems, that is, problems in which the decision variables may range over a certain continuous domain. Many situations involve decisions among a set of integer values, and sometimes in a set $S = \{0, 1\}$. These situations involve integer programming problems, for the solution of which we must rely on special search strategies and heuristics, the branch-and-bound being among the most popular (Hillier and Lieberman 1990). Clearly, mixed programming problems are also common, and in this case methods for continuous and integer programming solutions are combined to develop solutions.

Finally, realistic optimization problems usually involve not just one single objective function but a number of objectives that should be pursued. Such a situation creates a multiobjective optimization problem. Care should be taken in handling this class of problem, because is quite unusual to find a unique solution that looks optimal under all given objective functions. Several strategies may be employed, the nondominant approach being of most use because it generates several feasible candidate solutions, and the decision maker trades off among them to select the most preferable (Haimes et al. 1990). For instance, assume we have the following multiobjective problem:

$$\text{minimize} \begin{bmatrix} f_1(\mathbf{x}) \\ f_2(\mathbf{x}) \end{bmatrix}$$

subject to $\mathbf{x} \geq 0$,

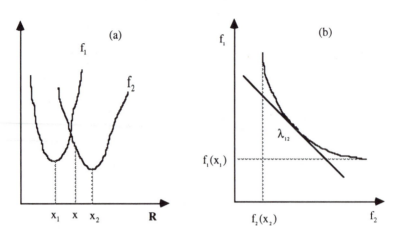

Figure 14.17
Multiobjective optimization

where $\mathbf{x} \in \mathbf{R}^n$. If f_1 and f_2 are strictly convex functions, they may look in the decision and function spaces as shown in figure 14.17(a) and 14.17(b), respectively, for $n = 1$. From figure 14.7(b), we easily recognize that computing

$$\lambda_{12} = -\frac{\partial f_1}{\partial f_2}$$

provides information about how to trade off. Note that λ_{12} can be obtained by solving a family of standard ε-constrained nonlinear optimization problems of the form

minimize $f_1(\mathbf{x})$

subject to $f_2(\mathbf{x}) \le \varepsilon$,

where λ_{12} is the Lagrange multiplier associated with the ε-constraint.

14.3.2   Fuzzy Optimization Problems

Bellman and Zadeh (1970) introduced the basis of most fuzzy optimization problems, in which both objectives and constraints in an ill-defined situation are represented by fuzzy sets. In the case presented, a collection of N fuzzy objective functions F_k, $k = 1, \ldots, N$, and M constraints C_1, $l = 1, \ldots, M$, defined in the decision space $\mathbf{X}$, are assumed to be given. A decision is defined by its membership function

$$D(\mathbf{x}) = F_k(\mathbf{x}) * C_1(\mathbf{x}), \quad k = 1, \ldots, N \text{ and } l = 1, \ldots, M,$$

where $\mathbf{x} \in \mathbf{X}$, and * denotes an appropriate aggregation operator. In the original formulation, this operator was taken as the minimum operator. Since this approach does not discriminate objectives and constraints, it is often referred to as the symmetrical optimization method.

If there exists a subset $X \subseteq \mathbf{X}$ for which $D(\mathbf{x})$ reaches its maximum, then X is called the maximizing decision (if the optimization is viewed as a maximization problem). If $\mathbf{x} \in \mathbf{X}$ is to be found which simultaneously satisfies objectives and constraints with the highest possible degree of membership, we must solve a conventional, equivalent nonlinear (generally) optimization problem:

maximize λ

subject to $F_k(\mathbf{x}) \geq \lambda, \quad k = 1, \ldots, N$

$\qquad\quad C_1(\mathbf{x}) \geq \lambda, \quad 1 = 1, \ldots, M$

$\qquad\quad \lambda \in (0, 1]$

$\qquad\quad \mathbf{x} \in \mathbf{X}.$

EXAMPLE 14.1   Suppose your task is to determine a product price to maximize a company profit that is a squared function of the prices but limited to four units for marketing reasons. If we denote by x the prices, we may easily formulate the following nonlinear optimization problem

maximize $f(x) = x^2$

subject to $0 \leq x \leq 4$.

Clearly, the optimal, precise decision is $x = 4$. But suppose you have been told that, in accomplishing your task, competitors' price is around 3, but the company owner wants as much profit as possible. In this case, the optimization problem formulation above must cope with these directives, and a fuzzy optimization model becomes necessary. Therefore we may define the problem as follows. Assign a fuzzy set to the objective function defining its membership function as follows:

$$F(x) = \frac{f(x) - I}{S - I},$$

where $I = \inf_{x \in \mathbf{X}} f(x)$ and $S = \sup_{x \in \mathbf{X}} f(x)$. In the example, $I = 0$ and $S = 16$. Thus

$$F(x) = \frac{1}{16}x^2, \quad \text{and} \quad F(x) \in [0, 1].$$

Let *around three* be described by a fuzzy set whose membership function is triangular and defined by $C(2, 3, 4)$. Therefore, the decision set is as shown in figure 14.18 (with $^* = \wedge$). In this case, a unique maximizing

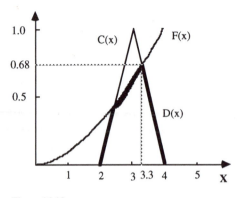

Figure 14.18
Fuzzy sets for example 14.1

decision exists, $x^* = 3.3$. This intuitive and geometrical solution can also be derived easily from the associated conventional nonlinear optimization problem

maximize λ

subject to $\frac{1}{16} x^2 \geq \lambda$,

$$x - 2 \geq \lambda, \quad \text{if } x \in [0, 3)$$
$$-x + 4 \geq \lambda, \quad \text{if } x \in (3, 4]$$
$$\lambda \in (0, 1], \quad x \geq 0,$$

whose solution $(x^*, \lambda^*) = (3.3, 0.68)$ can be determined graphically, as depicted in figure 14.19. Note that here $\mathbf{X} = \{x \mid x \in \mathbf{R} \text{ and } x \geq 0\}$.

Before proceeding, note that the equivalent nonlinear optimization problem may be restated as a minimization problem as follows. Let $\lambda = 1 - \theta$. Thus we get

minimize θ

subject to $F_k(\mathbf{x}) \geq 1 - \theta, \quad k = 1, \ldots, N$
$$C_l(\mathbf{x}) \geq 1 - \theta, \quad l = 1, \ldots, M$$
$$\theta \in (0, 1]$$
$$\mathbf{x} \in \mathbf{X}.$$

This idea can be easily understood geometrically by looking at figure 14.19.

14.3.3   Fuzzy Linear Programming

Fuzzy linear programming is a family of optimization problems in which the optimization model parameters are not well defined. This means that

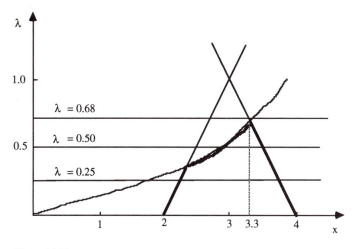

Figure 14.19
Equivalent problem solution for example 14.1

the objective function and constraint coefficients are not exactly known and that some of the inequalities involved may also be subject to unsharp boundaries. Precise values are not known, and approximations can be specified more comfortably than point values. Depending on the information available to construct a optimization model, several classes of problems arise: fuzzy constraints, fuzzy objective functions or goals, and fuzzy coefficients in the objective function, the constraints, or in both (Verdegay 1995). In the following, we provide a number of approaches for solving some of the above problems. For an extensive overview and more detailed approaches, see Verdegay (1995) and Delgado et al. (1994).

14.3.3.1   Fuzzy Objective Functions and Fuzzy Constraints

The optimization model associated with a linear programming problem in which the objective and constraints are fuzzy is as follows:

$$\text{maximize}_f \; \mathbf{c}^T \mathbf{x}$$
$$\text{subject to } \mathbf{A}\mathbf{x} \leq_f \mathbf{b}$$
$$\mathbf{x} \geq \mathbf{0},$$

where maximize_f means that the objective function is fuzzy and that the inequality $\leq_f$ is to be understood as essentially smaller than or equal to. Let us denote the ith constraint associated with $\mathbf{A}\mathbf{x} \leq_f \mathbf{b}$ by $(\mathbf{A}\mathbf{x})_i \leq_f b_i$, $i = 1, \ldots, m$. In this case, it is assumed that $\mathbf{c}$ and $\mathbf{A}$, as well as an aspiration level b_o and maximum tolerances p_o for the objective function and p_i for the constraints, are given. Thus the problem becomes symetric, because the aim is to find $\mathbf{x}$ such that (Zimmermann 1993)

$$\mathbf{c}^T \mathbf{x} \geq_f b_o$$
$$\mathbf{A}\mathbf{x} \leq_f \mathbf{b}$$
$$\mathbf{x} \geq 0$$

with membership functions given by

$$F(\mathbf{x}) = \begin{cases} 1, & \text{if } \mathbf{c}^T \mathbf{x} > b_o \\ 1 - \dfrac{b_o - \mathbf{c}^T \mathbf{x}}{p_o}, & \text{if } b_o - p_o \leq \mathbf{c}^T \mathbf{x} \leq b_o \\ 0, & \text{if } \mathbf{c}^T \mathbf{x} < b_o - p_o, \end{cases}$$

$$C_i(\mathbf{x}) = \begin{cases} 1, & \text{if } (\mathbf{A}\mathbf{x})_i < b_i \\ 1 - \dfrac{(\mathbf{A}\mathbf{x})_i - b_i}{p_i}, & \text{if } b_i \leq (\mathbf{A}\mathbf{x})_i \leq b_i + p_i \\ 0, & \text{if } (\mathbf{A}\mathbf{x})_i > b_i + p_i \end{cases}$$

Therefore, according to the previous section, it can be translated into the following equivalent problem

minimize θ

subject to $\mathbf{c}^T \mathbf{x} \geq b_o - \theta p_o$

$$(\mathbf{A}\mathbf{x})_i \leq b_i + \theta p_i, \quad i = 1, \ldots, m$$
$$\theta \in (0, 1]$$
$$\mathbf{x} \geq \mathbf{0}.$$

Note that this is a standard linear programing problem that can be solved by any appropriate or available computer code.

14.3.3.2   Fuzzy Constraints with Fuzzy Coefficients

Given that no fuzzy objective function is considered, this class of linear programming model assumes fuzziness both in, adherence to the constraints and in the coefficients involved in the constraints. Formally, the model can be written as

maximize $\mathbf{c}^T \mathbf{x}$

subject to $\mathbf{A}_f \mathbf{x} \leq_f \mathbf{b}_f$

$$\mathbf{x} \geq \mathbf{0},$$

where $\mathbf{A}_f$ and $\mathbf{b}_f$ denote a matrix and a vector, respectively, whose entries are fuzzy numbers. Delgado, Verdegay, and Vila (1989) have developed a general solution strategy assuming the existence of a ranking relation $\leq_f$ among fuzzy numbers. The ranking relation is required only to preserve ranking under multiplication by a positive scalar. Thus, if $\mathbf{t}_f$ is a vector of fuzzy numbers expressing the allowed violation of the constraint, the problem considered becomes

maximize $\mathbf{c}^T\mathbf{x}$

subject to $\mathbf{A}_f\mathbf{x} \leq_f \mathbf{b}_f + (1 - \alpha)\mathbf{t}_f$

$\qquad \mathbf{x} \geq \mathbf{0}$

$\qquad \alpha \in (0, 1]$.

In particular, we may assume fuzzy numbers of LR type (e.g., triangular fuzzy numbers). The equivalent conventional linear programming problem associated depends, in this case, on the choice of the ranking relation. Examples of ranking relations include, for triangular fuzzy numbers, the following. Let $a_f = (\underline{a}, a^o, \bar{a})$ and $b_f = (\underline{b}, b^o, \bar{b})$. Thus we have

$a_f \leq_f^1 b \leftrightarrow a^o \leq b^o$,

$a_f \leq_f^2 b \leftrightarrow \bar{a} \leq \underline{b}$.

Hence, the following auxiliary problems can be defined for each $\leq_f^1$ and $\leq_f^2$, and $t_f = (t_1, t_2, \ldots, t_m)$ with $t_j = (\underline{t}_j, t_j^o, \bar{t}_j)$, $j = 1, 2, \ldots, m$

maximize $\mathbf{c}^T\mathbf{x}$

subject to $\sum_{i=1}^{n} a_{ij}^o x_i \leq b_j^o + (1 - \alpha)t_j^o, \quad j = 1, \ldots, m$

$\qquad \mathbf{x} \geq \mathbf{0}$

$\qquad \alpha \in (0, 1]$

and

maximize $\mathbf{c}^T\mathbf{x}$

subject to $\sum_{i=1}^{n} \bar{a}_{ij} x_i \leq \underline{b}_j + (1 - \alpha)\underline{t}_j, \quad j = 1, \ldots, m$

$\qquad \mathbf{x} \geq \mathbf{0}$

$\qquad \alpha \in (0, 1]$,

respectively. Note that the smaller α is, the lesser the fulfillment degree of the constraints. For alternative ranking relations and associated auxiliary problems, see Campos and Verdegay (1989).

14.3.3.3   Fuzzy Coefficients in the Objective Function

We now examine an optimization model in which the constraints are not fuzzy, but the coefficients of the objective function are fuzzy numbers characterized by membership functions C_i whose universe is the set $\mathbf{R}$ of real numbers, $i = 1, \ldots, n$. The approach to be discussed follows the method presented in Verdegay (1995). Formally, the problem is

maximize $\mathbf{c}_f^T \mathbf{x}$

subject to $\mathbf{A}\mathbf{x} \leq \mathbf{b}$

$\qquad \mathbf{x} \geq \mathbf{0}$,

where $\mathbf{c}_f$ denotes the n-dimensional vector whose entries are fuzzy numbers with membership functions given by

$$C_i(v) = \begin{cases} 0, & \text{if } \bar{c}_i \leq v \text{ or } v \leq \underline{c}_i \\ \underline{H}(v), & \text{if } \underline{c}_i \leq v \leq c_i \\ \overline{H}(v) & \text{if } c_i \leq v \leq \bar{c}_i, \end{cases}$$

where $[\bar{c}_i, \underline{c}_i]$ is the support of C_i, and $\underline{H}_i$ and $\overline{H}_i$ are strictly increasing and decreasing continuous functions, respectively, such that

$$\underline{H}_i(c_i) = \overline{H}_i(c_i) = 1.$$

Thus, considering the $(1 - \alpha)$-cuts of each objective function coefficient, $\alpha \in (0, 1]$, $\forall v \in \mathbf{R}$ and $i = 1, \ldots, n$,

$$C_i(v) \geq 1 - \alpha \leftrightarrow \underline{H}_i^{-1}(1 - \alpha) \leq v \leq \overline{H}_i^{-1}(1 - \alpha),$$

and letting $\phi_i(1 - \alpha) = \underline{H}_i^{-1}(1 - \alpha)$ and $\psi_i(1 - \alpha) = \overline{H}_i^{-1}(1 - \alpha)$, we get

$$\phi_i(1 - \alpha) \leq v \leq \psi_i(1 - \alpha).$$

A fuzzy solution for the problem can be found from the parametric solution of the following multiobjective linear programming model:

$$\text{maximize} \begin{bmatrix} \mathbf{c}^1 \mathbf{x} \\ \vdots \\ \mathbf{c}^p \mathbf{x} \end{bmatrix}$$

subject to $\mathbf{A}\mathbf{x} \leq \mathbf{b}$

$\qquad \mathbf{x} \geq \mathbf{0}$

$\qquad \mathbf{c}^k \in \mathbf{E}(1 - \alpha), \quad k = 1, \ldots, p = 2^n$

$\qquad \alpha \in (0, 1],$

where $\mathbf{E}(1 - \alpha)$ is the set of vectors of $\mathbf{R}^n$ whose components are either on the upper bound $\psi_i(1 - \alpha)$, or on the lower bound $\phi_i(1 - \alpha)$, of the respective $(1 - \alpha)$-cuts. That is, $\forall i, i = 1, \ldots, n$,

$$\mathbf{c} = (c_1, \ldots, c_n) \in \mathbf{E}(1 - \alpha) \leftrightarrow c_i = \phi_i(1 - \alpha) \text{ or } \psi_i(1 - \alpha).$$

To find a parametric solution, from which a fuzzy solution can be obtained, one can use any classical multiobjective linear programming methodology.

14.4   Chapter Summary

The essential steps in performing analysis and design tasks as well as the principal issues in applying fuzzy set methodologies in engineering problems have been discussed. In addition, this chapter has introduced techniques for important classes of decision-making problems, namely, control and optimization problems.

14.5   Problems

1.   Consider the numerical input-output characteristics of the fuzzy controller

$$du = f(e, de),$$

with the rules defined in table 14.1, whose linguistic terms have Gaussian membership functions with the parameters μ and σ. Approximate the characteristics by the PI velocity controller

$$du^* = k_1 e + k_2\, de.$$

Hint: Define the performance index

$$\sum_e \sum_{de} [f(e, de) - k_1 e - k_2\, de]^2$$

and minimize it with respect to the parameters of the controller.

2.   Reconsider problem 1 by constructing several separate PI controllers for the indicated portions of the control protocol. Compare the results and comment on differences in the quality of approximation.

3.   Recalling that the necessity of the input A taken with respect to the condition of the rule A_k is computed as

$$\mathrm{Nec}(A_k, A) = \inf_{x \in X} [(1 - A(x)) \vee A_k(x)],$$

show that the fuzzy controller based on the calculus of possibilities can be expressed in the equivalent format

$$\bar{B}(u) = \bigwedge_{k=1}^{N} [\mathrm{Nec}(A_k, \bar{A}) \vee \bar{B}_k(u)].$$

4.   Consider a fuzzy controller composed of the following two rules:

If E is A_1 and DE is B_1, then U is C_1,

If E is A_2 and DE is B_2, then U is C_2,

where the membership functions are triangular shaped and defined by $A_1(2, 5, 8)$, $A_2(3, 6, 9)$, $B_1(5, 8, 11)$, $B_2(4, 7, 10)$, $C_1(1, 4, 7)$ and $C_2(3, 6, 9)$. Suppose that sensor readings provide exact numerical measurements $e = 4$ and $de = 8$. Assume the sup-min rule of inference and interpret the connective *and* as the min operator and the rule relations as conjunctions via the min t-norm. Find the fuzzy controls suggested by each rule. Determine the control signal that should be applied to a process considering both the mean-of-maxima and center-of-gravity. Sketch the successive steps you performed to get your results.

5.   Repeat problem 4, but compute the control signals via a fuzzy logic controller, Compare your results with those derived in the problem 4. Assume the truth values of the rules to be *true* and *very true*, respectively.

6.   Solve graphically the following nonlinear optimization problem:

minimize $(x_1 - 0.5)^2 + (x_2 - 0.5)^2$

subject to $x_1 + x_2 = 1$

$$x_1, x_2 \geq 0.$$

Find its solution analytically. Following that, use the gradient-based method; assume the problem to be constraint free.

7.   Determine the optimal solution to the following linear optimization problem:

maximize $3x_1 + x_2 + 3x_3$

subject to $2x_1 + x_2 + x_3 \leq 2$

$$x_1 + 2x_2 + 3x_3 \leq 5$$

$$2x_1 + 2x_2 + x_3 \leq 3$$

$$x_1, x_2, x_3 \geq 0.$$

Find the value of the objective function at the optimal decision.

8.   Solve the following scalar fuzzy optimization problem graphically:

maximize$_f$ x

subject to $x \leq_f 1$

$$x \geq 0,$$

where $\leq_f$ may be interpreted as *around*. Assume triangular membership function $A(0.5, 1, 2)$ to describe the idea of *around*.

9.   Assume the following fuzzy linear programming problem with fuzzy constraints and fuzzy coefficients:

maximize $5x_1 + 6x_2$

subject to $3_f x_1 + 4_f x_2 \leq_f 18_f$

$$2_f x_1 + 1_f x_2 \leq 7_f$$

$$x_1, x_2 \geq 0$$

where $3_f = (3, 2, 4), 4_f = (4, 2.5, 5.5), 18_f = (18, 16, 19), 2_f = (2, 1, 3), 1_f = (1, 0.5, 2),$ $7_f = (7, 6, 9),$ and with tolerances given by $t_{1f} = (3, 2.5, 3.5)$ and $t_{2f} = (1, 0.5, 1.5).$ Solve the problem using the $\leq_f^1$ and $\leq_f^2$ ranking relations. Experiment and compare your solutions, trying several values of α. Interpret the results graphically.

10.   Does a fuzzy linear programming problem always induce a conventional linear model for its solution? Support your argument with examples.

Note

1. The essence of Hebbian learning is that the strength of each synapse is changed proportionally to the product of the activations of the pre- and postsynaptic neurons.

References

Astrom, K., and B. Wittenmark. 1984. Computer Controlled Systems: Theory and Design. Englewood Cliffs, NJ: Prentice Hall.

Bellman, R., and L. Zadeh. 1970. Decision-making in a fuzzy environment. *Management Science* 17:141–64.

Campos, L., and J. Verdegay. 1989. Linear programming problems and ranking of fuzzy numbers. *Fuzzy Sets and Systems* 32:1–11.

Castro, J. 1995. Fuzzy logic controllers are universal approximators. *IEEE Trans. on Systems, Man, and Cybernetics* 25(4):629–35.

Delgado, M., J. Kacprzyk, J. Verdegay, and M. Vila. 1994. *Fuzzy Optimization: Recent Advances*. Heidelberg, Germany: Physica-Verlag.

Gill, P., W. Murray, and M. Wright. 1981. *Practical Optimization*. New York: Academic Press.

Haimes, Y., K. Tarvainen, T. Shima, and J. Thadathil. 1990. *Hierarchical Multiobjective Analysis of Large-Scale Systems*. New York: Hemisphere Publishing.

Hillier, F., and G. Lieberman. 1990. *Introduction to Operations Research*. New York: McGraw-Hill.

Kosko, B. 1992. *Neural Networks and Fuzzy Systems: A Dynamical Systems Approach to Machine Intelligence*. Englewood Cliffs, NJ: Prentice Hall.

Luenberger, D. 1973. *Introduction to Linear and Nonlinear Programming*. Reading, MA: Addison-Wesley.

Mizumoto, M. 1994. Fuzzy controls under product-sum-gravity methods and new fuzzy control methods. In *Fuzzy Control Systems*, eds. A. Kandel and G. Langholz, 275–94. Boca Raton, FL: CRC Press.

Pedrycz, W. 1995. *Fuzzy Sets Engineering*. Boca Raton, FL: CRC Press.

Pedrycz, W. 1993. *Fuzzy Control and Fuzzy Systems*, 2d ed. New York: RSP Press.

Verdegay, J. 1995. Fuzzy optimization: models, methods and perspectives. Sixth IFSA World Congress, July 22, 1995, vol. 1, pp. 21–28. São Paulo, Brazil.

Yager, R., and D. Filev. 1994. *Essentials of Fuzzy Modeling and Control*. New York: Wiley Interscience.

Zimmermann, H. 1993. *Fuzzy Set Theory and Its Applications*, 2d ed. Dordrecht, the Netherlands: Kluwer Academic Publishers.

15 Case Studies

In the previous chapters we have presented the fundamentals, computational models, and methodologies for the analysis and design of fuzzy systems. In this chapter, we reconsider the analysis and design issue, but now with the tools in hand. Several characteristic applications are presented in something of a case-study format to illustrate the use of the methodologies and tools we have introduced for fuzzy system analysis and design.

15.1 Traffic Intersection Control

Most urban traffic network links intersect frequently, leading to conflicts between traffic flows. Since the operation of intersections may often be a critical factor in determining the overall capacity and performance of an urban network, traffic engineers are continuously faced with the problem of controlling flow at intersections to improve performance (Wohl and Martin 1967).

Traffic lights are used to control vehicle flows through most cities' intersections. The cycle length of a signal is the time period required for one complete sequence of signals at a given intersection. The cycle length is normally divided into a number of phases, each phase being a part of the time cycle allocated to one or more traffic or pedestrian movements. The green phase is a particular state that provides a green light (right of way) for a particular direction. The green time represents the time duration of the green phase. Also, for each phase there is time lost (reaction time) while getting the waiting line of traffic moving again.

Often the control strategy consists of changing the green time (and consequently the cycle length) as a function of the incoming traffic such that the vehicle share the intersection more efficiently. This problem is complicated by the fact that each intersection has unique characteristics of physical layout, vehicle flow rates, turning movements, pedestrian movements, and so forth. Mathematical models exist for describing the traffic flow through an intersection for a given flow density, but the exact description is nonlinear, and it is difficult to take into account fluctuations in the density of the traffic flows that approach the intersection from different directions. As the traffic flows in one direction, the vehicles are forced to wait to proceed in another direction. The longer the waiting time, the heavier the load that must pass through the intersection at a later time.

Clearly, the state of an intersection may be characterized by the vehicle arrivals and queue length, whereas the control decision is the the green time. Intersection parameters include its physical layout and dimensions and the number of lanes per approach.

The policies used in traffic control systems can be classified into two main categories (Al-Khalili 1985):

1. Fixed-time systems: In a fixed-time system, traffic plans are generated off-line and applied on-line. This type of control consists of a set of control plans computed from average measured traffic flows. The controller can store a number of these plans, and the appropriate plan is applied at different times of the day as needed. This policy depends on periodic traffic counts to keep the system updated.

2. On-line systems: In an on-line system, traffic plans are generated on-line and applied directly for traffic control. The control strategy is computed according to actual traffic flows measured in real time, and the controller is continuously updated to choose the new strategy. This policy requires sensing devices to measure the traffic flow. The fuzzy traffic controller presented here belongs to this category.

Both methods have advantages and disadvantages (Robertson 1979), but a common objective: to minimize the average vehicle delay the intersection causes and average queue length at each of its points, as proposed by Webster (1958).

15.1.1   Fuzzy Traffic Controller

The pioneer work in using fuzzy sets for traffic light control, by Pappis and Mamdani (1977), considers a single intersection of two one-way streets. The results obtained from the implementation of the fuzzy logic controller were compared with those obtained using to a conventional, effective vehicle-actuated controller. Considering the average delay of vehicles as the performance criterion, Pappis and Mamdani showed that using a fuzzy logic controller results in better performance.

Figure 15.1 depicts a block diagram of the fuzzy traffic controller (FTC) discussed here and highlights its main components:

1. Sensing devices: a set of two inductive loops spaced by a distance d (one set per lane); their disposition allows vehicle detection as well as speed estimation

2. Estimator: computes each vehicle's speed and the time it takes it to cross the intersection, especially at the end of the green phase; provides estimates of the arrival rate and queue length at each approach

3. Fuzzy controller: determines the green time according to the traffic condition

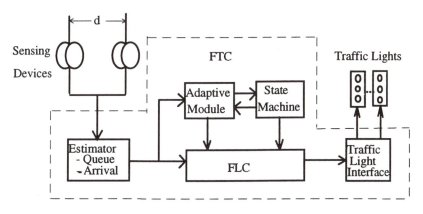

Figure 15.1
Block diagram of fuzzy traffic controller

4. State machine: controls the sequence of states the FTC should pass through

5. Adaptive module: changes the fuzzy controller settings to adjust its performance

6. traffic light interface: provides the circuitry for turning the lights on and off according to the fuzzy controller's decisions.

The following discussion describes the fuzzy controller, the state machine, and the adaptive module, emphasizing their role in achieving acceptable levels of vehicle delay and average queue lengths.

15.1.2   Fuzzy Controller

To perform the intersection control task, the controller must be provided with information about the traffic coming in on the approach currently in the green phase and the vehicle queue that has formed in the remaining approaches. Based on this information, it decides whether to extend the current green time. If it decides not to do so, the traffic controller state must change to allow traffic from another approach to flow. Therefore, the fuzzy controller must have the following inputs:

1. Arrival rate of vehicles for the approach in the green phase: defined over a universe of discourse **A**, whose generic element is $a \in \mathbf{A}$

2. length of queue of vehicles for the approaches in the red phase: defined on a universe of discourse **Q**, with elements $q \in \mathbf{Q}$.

The output variable is the extension of the current green phase, which the controller can increase up to a previously defined maximum value.

Extending the green phase for a particular approach also changes the state of the traffic controller. The extension is defined on a universe of discourse **E**, with elements $e \in$ **E**. The universes **A**, **Q** and **E** are assumed to be discrete.

The input and output variables have associated the linguistic variables *arrival*, *queue*, and *extension*, respectively, which assume linguistic values in the following term sets:

- Arrival: {*almost_none, few, many, too_many*}
- Queue: {*very_small, small, medium, long*}
- Extension: {*very_short, short, medium, long*}

These linguistic values define the fuzzy partition (granulation) of the input and output spaces. Figure 15.2 shows the membership functions of the fuzzy sets Q, A, and E (the meaning of the linguistic terms) associated with the linguistic labels *queue*, *arrival*, and *extension*, the variables that compose antecedent and consequent rules. An initial set of fuzzy control rules were derived based on expert knowledge, control engineering knowledge, and an intersection simulation model, which also helped in refining the control rules, summarized in table 15.1. From the table, one easily conceptualizes the linguistic descriptions of the fuzzy control rules; for example,

R_1: If *arrival* is *almost_none* and *queue* is *very_small*, then *extension* is *very_short*

. . .

R_{16}: If *arrival* is *too_many* and *queue* is *long*, then *extension* is *very_short*

Note that although the table summarizes sixteen rules, only eleven are necessary.

It was found experimentally that the min t-norm and the max-min compositional rule of inference could conveniently model the connective *and* and the decision-making logic, respectively. The union operator modeled by the max s-norm performed rule aggregation. The center-of-area method (Favilla et al. 1992) was used for defuzzification. Both aggregation and defuzzification schemes were determined experimentally.

15.1.3 State Machine

The state machine controls the predefined sequence of states through which the traffic controller cycles. One state is assigned to each phase of the traffic light (red, yellow, green) along with one default state for when

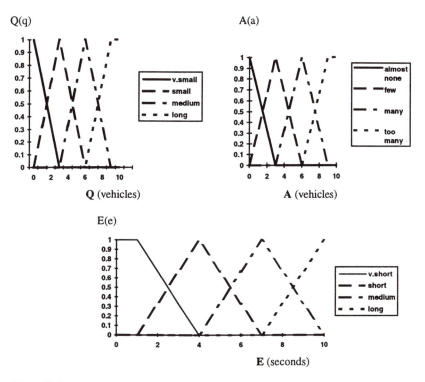

Figure 15.2
Fuzzy sets associated with *queue, arrival,* and *extension*

Table 15.1
Fuzzy control rules

Queue\Arrival	almost_none	few	many	too_many
very_small	very_short	short	medium	long
small	very_short	very_short	short	medium
medium	very_short	very_short	very_short	short
long	very_short	very_short	very_short	very_short

no incoming traffic is detected. This default state corresponds to the green time for a specific approach (usually the main one, where the probability of having incoming traffic is highest). Transition occurs within the pre-defined sequence if either the current state has reached its maximum time, the fuzzy controller decides no extension is needed, or no incoming traffic is being detected. A state can be skipped if no vehicles are queued in the corresponding approach.

15.1.4   Adaptation Methods

Traffic flow levels depend on time. During peak hours, for instance, they are much higher than during early morning hours. Adaptation methods aim to adjust the fuzzy controller to a broad range of traffic conditions.

The fuzzy controller acts in the feedback loop, whereas the adaptive module performs in monitoring mode to bring the fuzzy controller to its best operational conditions. Two different adaptive methods were derived (Favilla, Machion, and Gomide 1993), namely, statistical adaptive and fuzzy-adaptive.

In the statistical-adaptive method, at ten-second intervals, the sensors relay information about the number of vehicle arrivals in each lane of the intersection's approaches. The maximum value of these measurements is stored. This procedure is repeated eighteen consecutive times. The average (M) and standard deviation (σ) are computed based on this data set. Next, a new upper limit (ul) is set for the universes of the fuzzy sets A and Q according to the following simple procedure:

Procedure: Determine Upper Limit

Begin

$\quad$ $uladapt = (M + 3.\sigma)$

$\quad$ if $uladapt \leq 10$

$\quad$ then $new\_ul = uladapt$

$\quad$ else $new\_ul = current\_ul$

End

Note that this adjustment method is applied to the universes corresponding to the variables that appear in the antecedent part of the fuzzy control rules. The rationale behind this is the fact that what few, many or too many vehicles, as well as small, medium or long queues mean really depends on the traffic conditions. In other words, the meanings of these labels are context dependent. Therefore, the fuzzy controller becomes more sensitive to traffic flow variations. This may also be regarded as a statistical scheme for automatic scaling. However, since it has a time lag of three minutes (because it computes the upper limit for the next three minutes based on data acquired in the last three minutes), it may cause some inefficiency under certain traffic conditions. Thus, it should be used carefully.

The fuzzy-adaptive method consists of adjusting the universe of membership functions for the output variable by modifying the variable's upper limit. To accomplish this task, a fuzzy rule–based monitor (FLM) is employed. The FLM is structurally identical to the fuzzy controller but has input variables defined as follows:

1. Residual queue at the end of the green phase: defined on the universe of discourse **D**, with generic value of queue length, in number of vehicles, being $d \in \mathbf{D}$

2. Queue variation during the green phase: defined on the universe of discourse **V**, with elements $v \in \mathbf{V}$ representing number of vehicles

In this case, the output variable is upper limit variation (*Ul*) defined on the universe of discourse **U**, a time interval in seconds. This decision variable determines how much to compress or expand the membership functions associated with the linguistic variable E, but with upper limit values constrained to the interval [4,20] seconds. Figure 15.3 illustrates the compression-expansion effect, considering upper limits of (a) 10 and (b) 20, respectively. The rationale behind this scheme is analogous to that for the statistical-adaptive method. Based on the fuzzy-adaptive method, a specific value for *Ul* can be obtained for each phase, depending on the flow levels of the approaches considered in that phase. The procedure for computing the necessary adjustments is as follows:

Procedure: Determine Upper Limit Variation

Begin

    Set *i* of approaches ← green phase

    If beginning of the phase

        biggest queue ← biggest queue *j* (set *i*)

    If end of the phase

        determine upper limit variation (*biggest queue*)

        determine residual queue (*biggest queue*)

    Call procedure Fuzzy_Adaptive

        new_ul (set *i*) = *determined_ul* (set *i*)

End

Figure 15.4 shows the membership functions for the fuzzy supervisor, and table 15.2 summarizes the decision rules of procedure Fuzzy_Adaptive.

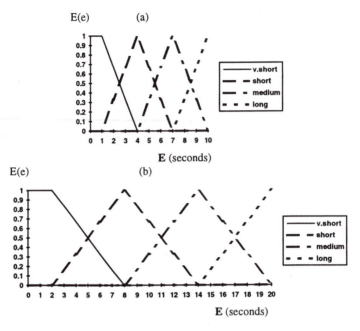

Figure 15.3
Compression-expansion effect of upper limit adjustment

The fuzzy monitor employs the same type of operators, decision-making logic, rule aggregation, and defuzzification method used in the fuzzy controller.

To verify the effectiveness of the fuzzy traffic controller, Machion (1992) performed a simulation study to compare fixed-time schemes with fuzzy control schemes. Figure 15.5 depicts the physical layout of the intersection used in Machion's study. Figure 15.6 shows the phase plans, which set the allowed turning movements at the intersection. Based on the traffic flow characteristics, the data set was organized into four groups to facilitate visualization of results and to make simulation experiments easy. Each group is called here a period, and when taken together, the periods are representative of the traffic flow that occurs over the course of a day. The simulation horizon was 3,600 seconds long. It was assumed that the sensing devices detect queues of up to ten vehicles per lane, as in actual intersections. However, for comparison purposes, simulation admitted measuring queues of any size. Figures 15.7 through 15.10 and tables 15.3 through 15.6 summarize the results.

The results clearly show that the fuzzy traffic controllers performed better than the fixed-time strategy, improving average delay time, average queue length, and maximum queue length significantly. Depending on

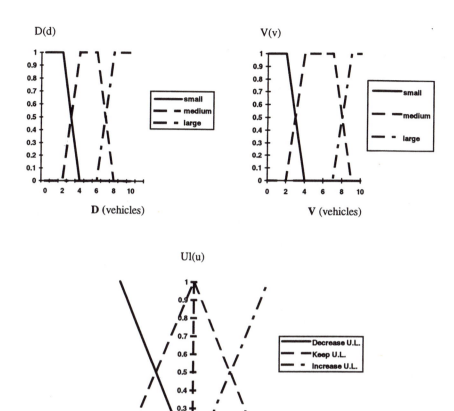

Figure 15.4
Fuzzy sets for residual queue, queue variation, and upper limit

Table 15.2
Decision rules for the fuzzy-adaptive procedure

Queue variation Residual queue	Small	Medium	Large
Small	decrease	decrease	decrease
Medium	increase	keep	decrease
Large	increase	increase	keep

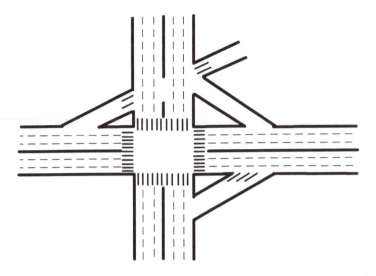

Figure 15.5
Intersection layout

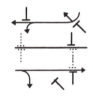

Phase A   Phase B   Phase C

Figure 15.6
Phase plans for the intersection

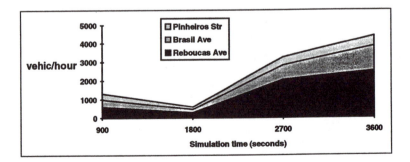

Figure 15.7
Traffic flow for period 1

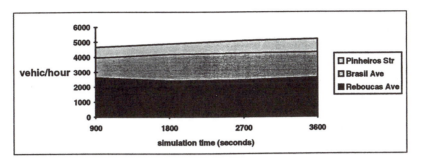

Figure 15.8
Traffic flow for period 2

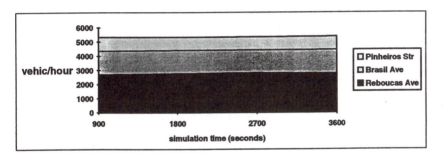

Figure 15.9
Traffic flow for period 3

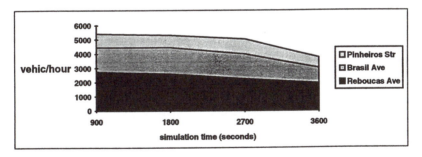

Figure 15.10
Traffic flow for period 4

Table 15.3
Results for period 1

Control measurement	Fixed time	Fuzzy without adaptation	Fuzzy with statistical-adaptive	Fuzzy with fuzzy-adaptive
Average delay (seconds)	16.02	8.94	8.33	9.75
Average queue (vehicles)	1.44	0.79	0.73	0.86
Maximum queue (vehicles)	15	14	14	10

Table 15.4
Results for period 2

Control measurement	Fixed time	Fuzzy without adaptation	Fuzzy with statistical-adaptive	Fuzzy with fuzzy-adaptive
Average delay (seconds)	55.99	11.28	12.38	11.94
Average queue (vehicles)	9.64	2.05	2.25	2.17
Maximum queue (vehicles)	51	16	23	17

Table 15.5
Results for period 3

Control measurement	Fixed time	Fuzzy without adaptation	Fuzzy with statistical-adaptive	Fuzzy with fuzzy-adaptive
Average delay (seconds)	30.51	20.96	21.00	13.82
Average queue (vehicles)	5.88	4.05	4.06	2.87
Maximum queue (vehicles)	32	34	34	20

Table 15.6
Results for period 4

Control measurement	Fixed time	Fuzzy without adaptation	Fuzzy with statistical-adaptive	Fuzzy with fuzzy-adaptive
Average delay (seconds)	30.16	10.55	10.77	12.79
Average queue (vehicles)	5.30	1.85	1.89	2.25
Maximum queue (vehicles)	33	18	18	14

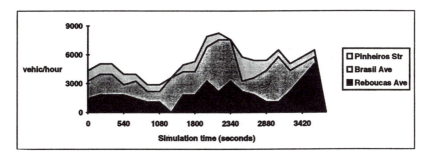

Figure 15.11
Variation in representative traffic flow densities

Table 15.7
Comparison of adaptation methods

Control measurement	Fuzzy without adaptation	Fuzzy with statistical-adaptive	Fuzzy with fuzzy-adaptive
Average delay (seconds)	31.29	31.45	20.15
Average queue (vehicles)	6.07	6.10	3.91
Maximum queue (vehicles)	72	72	58

the traffic flow levels, the adaptation strategies have been shown to be of value. To verify, in average terms, the performance of the adaptation methods, Machion simulated a period with a profile representative of the traffic flow variations that occur in a day (figure 15.11). Table 15.7 shows the results: The fuzzy controller with the fuzzy-adaptive method of making adjustments provided the best average performance.

15.2   Distributed Traffic Control

Fuzzy controllers have proven effective in controlling a single traffic inter-section, even when the intersection is somewhat complex. In certain instances, however, even if local controllers perform well, there is clearly no guarantee they will do so when intersections are coupled with traffic. In other words, the flow of vehicles in one direction depend on what is happening in neighbor intersections, as is the case in, for instance, down-town or high traffic-density areas. Therefore, more sophisticated decision structures must be built to develop efficient control policies, taking into account couplings and flow variations and considering the intersections' spatially distributed structure as well. Distributed system architectures offer an approach to handling such situations. In particular, the rationales for a distributed traffic control system include the following:

1. Adaptability: Logical, semantic, temporal, and spatial distribution allows a distributed system to provide alternative perspectives on emerging situations and potentially greater adaptive power.

2. Naturalness: Some problems, such as traffic control of a network of intersections, are better described as collections of separate agents, if is a better fit to a domain because elements are naturally distributed.

3. Specialization: Knowledge or action may be collected in specialized, bounded contexts, for purposes of control, extensibility, and comprehensibility.

4. Reliability: distributed systems may be more reliable than centralized systems because they provide redundancy, cross checking, and triangularization of results.

5. Autonomy: For protection and local control, parts of a system may be separated and isolated from one another which also improves reliability.

6. Resource limitations: Individual computational agents have bounded rationality, bounded resources for problem solving, and possibly bounded influence, necessitating cooperation and coordination to solve large problems.

A distributed system may potentially be cost effective, since it may involve a large number of simple computer systems of low unit cost. If communication costs are high, centralized intelligence with distributed sensing may be more expensive than distributed intelligence. System development and management, speed, and history are also relevant items within the distributed system framework.

In this section, we introduce a traffic control system built upon a distributed system whose architecture complies with the main attributes behind the rationales just described. The system comprises ideas of distributed artificial intelligence with an evolutive case-based mechanism; fuzzy sets and controllers describe the system state and provide local decisions.

15.2.1 Distributed Control System Architecture

The distributed control system architecture is composed of a network of cooperating, autonomous nodes (agents) capable of making their own control decisions and driving local devices. To cooperate, the nodes must have knowledge about system structure and state as well as global goals

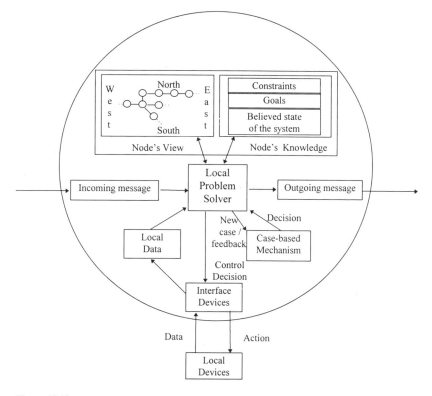

Figure 15.12
The structure of a node or agent

and constraints. In addition, nodes must share information through communication to achieve coordination.

Each node is assumed to have detailed knowledge about its neighborhood and topological knowledge about the overall network structure. Nodes can reason about the global problem and may cooperate to reach a solution once the application domain is known. The node knowledge, summarized in the upper portion of figure 15.12, provides the necessary information to cooperate with other nodes and to achieve system coordination. Nakamiti et al. (1994) provides a more detailed description of the architecture.

The local problem solver (LPS) is the node's basic processing unit. Its main task includes data acquisition and control at the local level, information exchange, and state updates. Ordinary situations are decided at the local level; that is, the node uses decision algorithms, rules or heuristics known to be effective in well-understood, typical situations. It may invoke a case-based mechanism (CBM) in handling unforeseen or more complex situations. The LPS's basic processing scheme, detailed later, can be condensed as in the following procedure:

Procedure: Local Problem Solver

Begin

   Receive_from_network_medium(information);

   Receive_from_local_devices(information);

   Update_system_view;

   Inform_neighbor_nodes;

   If not_able (Take_decision(view, rules_or_heuristics)) Then

     Send_CBM(situation);

     Receive_from_CBM(decision);

   Perform(decision);

   Send_neighbors(decision);

   Send_CBM(decision_performance)

End

An evolutive fuzzy case–based mechanism (Nakamiti and Gomide 1994) helps the LPS handle more complex situations in which the system is subject to dynamic disturbances. Clearly, it is not feasible to model all transient situations, but it is possible to have a memory of good past decisions that can be combined to provide, together with current information, an effective control decision. The basic CBM processing scheme, whose details are presented later in the traffic control context, is as follows:

Procedure: Case-Based Mechanism

Begin

   Receive_from_LPS(situation);

   Look_for_similar_past_cases(situation);

   Select_successful_cases;

   Combine_actions;

   Send_LPS(actions);

   Receive_from_LPS(performance);

   Decide_if_store(situation, actions, performance)

End

Additional details are provided in the application context below.

15.2.2   Distributed Traffic Control System

The distributed traffic control system consists of a number of local processors (nodes) located at each intersection. They communicate with neighbor processors through a communication link and decide the settings of the corresponding local traffic lights. They exchange messages, such as "Glicerio Street North is free" and "Traffic jam in the northern region," that encapsulate knowledge simply and allow the nodes to keep system state information updated. As in the case of fuzzy traffic controllers, sensors measure the traffic flow entering each intersection; these sensors provide local queue and arrival data to local processors, with the intention of optimizing traffic flow and reducing the average vehicle delay and average queue lengths to improve traffic quality. As usual in traffic control, the mechanism used to achieve performance requirements consists of changing the green time (and consequently the cycle length) according to vehicle arrivals and queue lengths.

15.2.3   Local Problem Solver

The local problem solver mainly controls the traffic flows at its designated intersection, it comparing the incoming traffic and queue lengths of related intersections in addition to its own. Local sensors and messages received from the network medium provide this information, based on which it decides whether to extend the current green time and informs its neighbors about its decision. This basic approach is used whenever traffic flows are similar to those used to settle the traffic lights according to Wohl and Martin (1967). In this case, the LPS makes fine adjustments to predefined settings.

As in the previous example, fuzzy sets are used to describe queues, vehicle arrivals, and green time extensions, as illustrated by figures 15.13

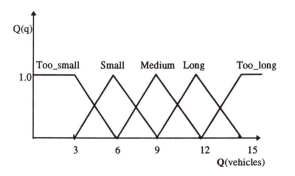

Figure 15.13
Fuzzy sets describing queue length

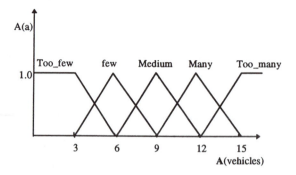

Figure 15.14
Fuzzy sets describing arrivals

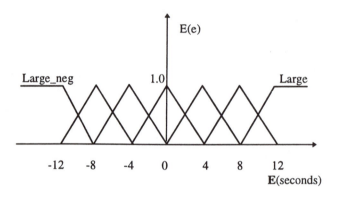

Figure 15.15
Fuzzy sets describing extensions

through 15.15. The LPS employs fuzzy control rules as a local decision procedure. Table 15.8 shows the green-time extensions for a *medium* local queue and *few* vehicle arrivals during the rush period. When the traffic flows depart from the expected ones, or when there is an unexpected situation (such as an accident, a game, or a parade), the strategy above may not be feasible. In these cases, the case-based mechanism provides the LPS with an appropriate decision as follows.

15.2.4 Evolutive Case-Based Mechanism

The LPS invokes the CBM whenever it has no effective solution for the current traffic situation, that is, when the traffic conditions differ substantially from the expected one. The CBM's main tasks include identification and retrieval of similar cases, combination of selected cases to generate a new decision, and case-based managing to store new cases and keep its size and content meaningful for later use. A forgetting scheme is also provided for discarding older or irrelevant cases.

Table 15.8
Control rules for the LPS

Lateral queue Next int. queue	Too_small	Small	Medium	Long	Too_long
Too_small	Medium	Small	No_ext	Small_neg	Medium_neg
Small	Medium	Small	No_ext	Small_neg	Medium_neg
Medium	Small	Small	No_ext	Small_neg	Medium_neg
Long	Small	No_ext	No_ext	Small_neg	Medium_neg
Too_long	No_ext	No_ext	Small_neg	Medium_neg	Large_neg

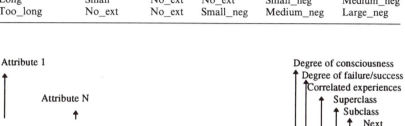

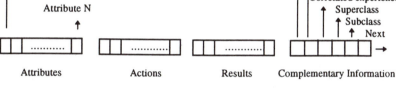

Figure 15.16
General structure of a case

To identify and retrieve the most similar cases, the CBM performs a tree search of the case base, which stores past experiences (memory of past successful cases) in structures like the one shown in figure 15.16. The attribute field identifies and describes the case characteristics, such as time, place, conditions of occurrence, and features. The action field contains the associated decisions (controls), whereas the result field stores the performance resulting from the decisions made. The complementary information field provides additional information about the case and for the tree search as well. Figure 15.17 shows a case example. Its attribute field describes

- local queue: 4 (*long*),
- lateral queue: 2 (*small*),
- next intersection queue: 3 (*medium*),
- previous intersection queue: 5 (*too_long*),
- next lateral queue: 2 (*small*),
- previous lateral queue: 1 (*too_small*),
- local waiting time: 5 (*too_long*),
- next intersection waiting time: 4 (*long*),
- previous intersection waiting time: 4 (*long*),

| 4 | 2 | 3 | 5 | 2 | 1 | 5 | 4 | 4 | 3 | 2 | | 2 | 2 | 1 | -2 | -1 | | 4 | 3 | 4 | 2 | 3 | | | 66 | | |

Attributes   Actions   Results   Complementary
                                                                                    Information

Figure 15.17
Case example

- lateral waiting time: 3 (*medium*),
- next lateral waiting time: 2 (*small*).

The case actions field specifies

- local green time extension: 2 (*medium*),
- next intersection extension: 2 (*medium*),
- previous intersection extension: 1 (*small*),
- next lateral extension: −2 (*medium_negative*),
- previous lateral extension: −1 (*small_negative*).

The results field describes

- local queue: 4 (*good*),
- next intersection queue: 3 (*regular*),
- previous intersection queue: 4 (*good*),
- lateral queue: 2 (*bad*),
- next lateral queue: 3 (*regular*).

The overall degree of failure/success for this case, considering weights of 1.5, 1.0, 1.0, 1.0, and 0.5 for the results part slots and a range of 5 for the result linguistic variable, is 0.66. This calculation will be made clearer later in the discussion of the case-based mechanism.

The values attached to each field slot can be conveniently coded to view each structure as a chromosome and each slot as a gene. Thus, a case base looks like the (partial) tree shown in figure 15.18. Retrieving the cases most similar to a current one involves a matching operation among the attributes identifying each case. For instance, consider the following example:

Current situation attributes:

| 4 | X | 1 | 5 | 2 | ? | 3 | 2 | X | 3 | 5 | 3 |

Stored case attribute:

| 5 | 1 | 1 | 3 | ? | 3 | X | 0 | X | 5 | 1 | 3 |

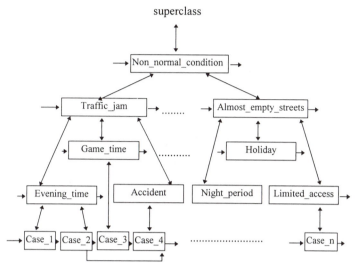

Figure 15.18
Partial tree structure of the case base

Clearly, very few positions match perfectly, although these cases may be very similar. Therefore, to perform a matching operation, the CBM must compute the degree of similarity between the current case and a stored case as follows:

$$\text{similarity degree} = \frac{\sum_{i=1}^{n} w_i s_i}{n},$$

where $s_i = 1$ if the ith attribute of the current case is X or the ith attribute of the stored case is X, or if the attribute has only one value, and otherwise

$$s_i = 1 - \frac{|(i\text{th attribute of\_current case}) - (i\text{th attribute of\_stored case})|}{\text{range of attribute alphabet}},$$

and where n is the number of attributes without '?', w_i is the weight of the ith attribute. The ith attribute of_case refers to the ith attribute of the case whose value is different from '?', and range stands for the number of linguistic terms of the linguistic variable minus one. For the example above, if we consider unit weights and a range of 5 for all attributes (for the attribute *Queue_length* of figure 15.13, the range would be 5 (*Too_long*) − 1 (*Too_small*) = 4), we would have a similarity degree of 0.78.

Obviously, instead of testing the entire the case base for similarity, we first select a specific subtree or a small group of subtrees that store the

most similar cases. This can be done following the same matching procedure to the group of attributes classifying the situation as in *Traffic_jam*
or *Almost_empty_streets*, for example. Thus, testing for similarity is performed only for cases belonging to the selected subtree(s) and eventually
for correlated cases as indicated by the respective pointers. As a result, we
have for each case in a subtree the corresponding degree of similarity and
the stored degree of failure/success as well. These values are a key to
selecting cases for combination.

After retrieving similar cases based on their attributes, the CBM combines the action fields of the two most successful cases through a genetic
algorithm. For this purpose, three possible strategies are

1. Take the actions considering their generic features for combination
(crossover); this is a good strategy for interrelated or fine-tuning actions.
For example, if cases i and j are selected, we have:

Action field of case i:

2	1	4	X	5	?	3	1	X	1	5	6

Action field of case j:

6	X	X	3	1	4	X	2	1	6	2	2

Resulting action field:

2	1	4	3	1	4	X	2	1	1	5	6

2. Take the actions individually, gene by gene; when both actions are the
same, we simply take their value, and for different actions we choose one
of them, avoiding Xs and ?s.

Action field of case i:

2	1	4	X	5	?	3	1	X	1	5	6

Action field of case j:

6	X	X	3	1	4	X	2	1	6	2	2

Resulting action field:

6	1	4	3	5	4	3	1	1	1	2	2

3. Take the actions individually, choosing any value, including the values
themselves, between those of the selected cases, as follows:

Action field of case i:

2	1	4	X	5	?	3	1	X	1	5	6

Action field of case j:

6	X	X	3	1	4	X	2	1	6	2	2

Resulting action field:

4	1	4	3	5	4	3	1	1	4	2	3

Whatever strategy the CBM chooses, the resulting action field will be sent to the local problem solver, which effectively applies the decisions. In the traffic control application, this field carries the green-time extensions for a set of traffic lights, as figure 15.17 illustrates.

The LPS, after applying the control decisions, measures the performance of the decisions by comparing the percentage differences of each result slot and assigning a linguistic value, then fills in the case results part. For example, if the local queue had 15 vehicles before and 11 vehicles after the decisions were implemented, the LPS will consider this a *good* result. It proceeds to compare the results and fill every slot in the corresponding field. Finally, it computes, using the same similarity degree equation, the degree of failure/success for the case. The degree of failure/success is used to find each case's chance for future use and to determine whether it should be stored.

The CBM provides a framework within which the current, best and most similar cases are combined to generate an adaptive control. Nevertheless, the CBM avoids significantly different controls (actions) because it only improves old cases using the current one to achieve adaptation. This may lead to local maxima. Some values or combinations of values may be missing as combinations proceed. To surpass this limitation, the values of the actions, or generic actions, may change at random with low probability (e.g., 0.04), depending on the combination strategy employed. This corresponds to the mutation operation in the context of CBM.

Case-based management takes place whenever new cases are generated. Old cases with a low degree of success are removed, whereas those with high degree of success are retained. This maintains the case base within a manageable size, an important issue in guaranteeing fast tree searches.

Overall, the distributed system tends to improve performance over time by adapting control decisions whenever significant traffic flow changes occur. This provides the necessary flexibility and adaptability traffic control systems demand.

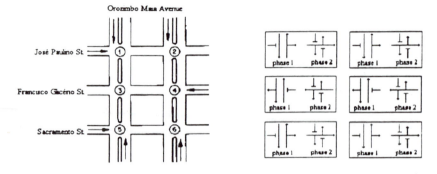

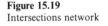

Figure 15.19
Intersections network

For comparison purposes, Nakatimi and Gomide (1996) simulated simultaneous, progressive, and distributed control systems (DCS). Recall that in a simultaneous system, all traffic lights along a street show the same color at the same time. In a progressive system, successive intersections have a common cycle length, but the timing of one traffic light in relation to the next is arranged to permit continuous movement of vehicles through the system, producing an effect commonly referred to as "green wave."

The simulation experiments considered a network of six intersections with eighteen traffic-lights, as sketched in figure 15.19. The cycle length and green time for the simultaneous and progressive systems correspond to actual observations. They were computed according to Wohl and Martin (1967) and, as usual, adjusted on site by traffic engineers to take into account each intersection's peculiarities. During the first 20,000 seconds of simulation the same arrival rates were obtained as were observed at the real intersections. Figure 15.20(a) shows the results for the three systems under consideration, computing average delays and queue lengths for the network as whole. The distributed control system presented better performance, demonstrating that its agents can cooperate and achieve coordination. In fact, its performance was about 24 percent better regarding average queue length (3.4 against 4.5 vehicles), and about 43 percent better for average delay (13.2 against 23.2 seconds).

In a second experiment, whose result is shown in figure 15.20(b), the vehicle arrival rates were increased by 30 percent after 3,000 seconds. The distributed control system proved capable of adapting to the traffic flow changes, with an average queue length of 8.9 vehicles and 38.4 seconds of average delay. The conventional approaches could not handle traffic flow changes properly, and the average queue length and delay increased con-

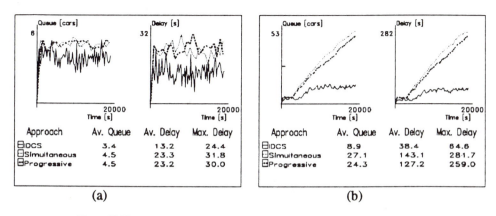

Figure 15.20
Simulation results: for (a) actual flows and (b) 30% above actual flow

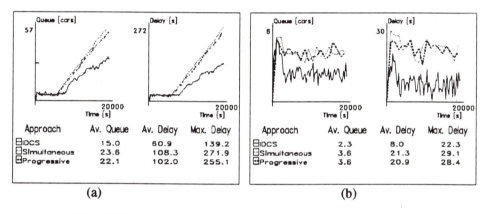

Figure 15.21
Simulation results: (a) 50% above and (b) 20% below rush flow

tinuously. In the simulation horizon, the DCS's average performance was about 63 percent better for average queue length and 70 percent better for average delays.

Figure 15.21(a) shows the simulation results when the rush hour arrival rates were increased by 50 percent after 3,000 seconds. The DCS adapted better, presenting in the simulation period an average queue length of 15.0 vehicles and 60.9 seconds of average delay, against 22.1 vehicles and 102.0 seconds for conventional approaches. Note, however, that even under DCS control, queues and delays still increased; this is because, for the specific traffic flows considered, the streets reached their physical capacities. In this circumstance, special measures must be taken to avoid overflow.

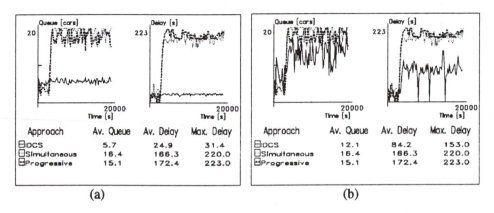

Approach	Av. Queue	Av. Delay	Max. Delay
⊟DCS	5.7	24.9	31.4
⊟Simultaneous	16.4	166.3	220.0
⊟Progressive	15.1	172.4	223.0

(a)

Approach	Av. Queue	Av. Delay	Max. Delay
⊟DCS	12.1	84.2	153.0
⊟Simultaneous	16.4	166.3	220.0
⊟Progressive	15.1	172.4	223.0

(b)

Figure 15.22
Results for a simulated accident at an intersection

When we consider arrival rates below the rush period levels, the DCS still shows better performance. Figure 15.21(b) depicts the simulation results for a traffic flow 20 percent below rush after 3,000 seconds. In this case, the DCS's average queue length was about 36 percent less, and the average delays about 62 percent less, than those of the conventional systems.

Next, Nakatimi and Gomide simulated a traffic accident at the 3,000th second. Figure 15.22(a) shows the average results for the whole system, and figure 15.22(b) shows the results for the intersection at which the accident occurred. Overall, the DCS performed about 62 percent and 85 percent better with respect to average queue length and delays, respectively, for the system as a whole. At the intersection, the difference was about 20 percent for average queue length (5.7 against 15.1 vehicles) and 50 percent for average delay (82.4 against 166.3 seconds).

In another experiment, traffic light settings were computed, following the directives of Wohl and Martin (1967), considering arrival rates 100 percent and 120 percent above rush flows. This scenario is similar to the traffic plan approach for enabling conventional settings to cope with changing flows during a time period. The arrival rates were increased by 20 percent after 10,000 seconds, as illustrated by figure 15.23(a). At the same time, the settings of the conventional systems were changed to match the 120 percent rate in an attempt to control the traffic load incurred. Again, the DCS performed better than conventional systems, as figure 12.23(b) shows.

In a final experiment, the arrival rates were assumed to be as profiled in figure 15.24(a). It was also assumed that, between 7,113 and 12,887 seconds, the conventional systems detected the variation in arrival rates

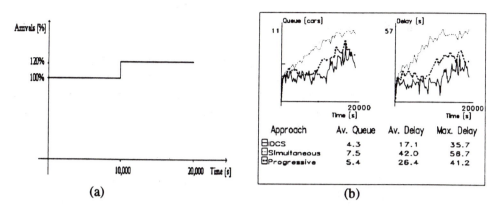

Figure 15.23
Simulation results updating conventional strategies after flow change

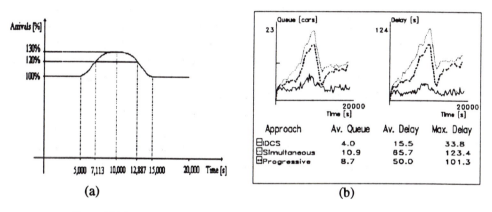

Figure 15.24
Simulation results for continuous flow variation

and were set to 120 percent of the base arrival rates; during the remaining period, they were set to 100 percent of the base rates. Figure 15.24(b) shows the results. The DCS's performance was once again superior when contrasted with that of simultaneous and progressive systems.

We may conclude that the distributed system presented herein does provide an effective approach for traffic control within a network of coupled intersections. It has greater adaptive power, as evidenced by emerging behavior in coping with time-varying flows of several intensities and unexpected situations. It better fits the domain problem, because a traffic network is naturally distributed, and because of its architecture, it provides isolation and autonomy for protection and local control once parts of the system are separated, as they may be when required.

15.3   Elevator Group Control

The problem we now discuss can be concisely stated as follows: A building has a number of elevators; provide an efficient schedule to assure smooth transportation in the building (figure 15.25). In fact, the smoothness requirement embraces several more specific items (Barney and Santos 1985; Gudwin et al. 1996):

1. Provide even service to every floor in the building

2. Minimize the time spent by passengers waiting for service

3. Minimize the time spent by passengers in moving from one floor to another

4. Serve as many passengers as possible at a given time

Because of the random nature of call time, call locations, and the destination of passengers, immense problems are encountered in attempting to satisfy all the above requirements. In general, the control strategy must be flexible enough to meet diverse conditions with respect to changes in passenger demands and different traffic patterns (up peak, down peak, special services, etc.) are concerned.

Several different policies have been proposed for elevator control, including the one known as the call allocation strategy (Barney and Santos 1985). The problem naturally lends itself to tasks suitable for representing and solving with the aid of fuzzy sets.

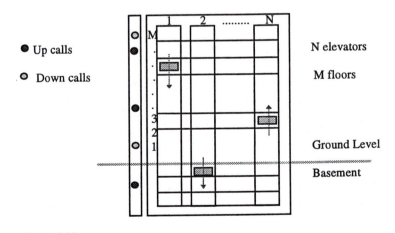

Figure 15.25
Elevator group control problem

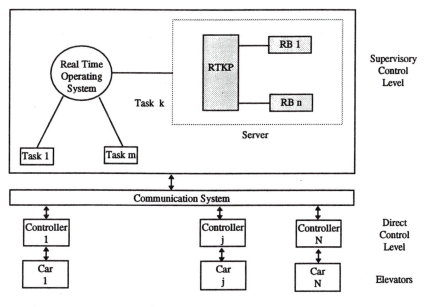

Figure 15.26
Elevator group control system: General structure

15.3.1   Elevator Group Control System

Figure 15.26 portrays the elevator group control system's general functional structure. In the two-level control hierarchy, the lower-level task is to command each elevator (car) to move up and down, to stop or start, and to open or close the door. Actually, this is a direct control level that performs the basic car movements. The higher level is a supervisory level to coordinate the movements of a group of cars to achieve a globally satisfactory performance. Coordinating decisions are based on system state (traffic conditions, hall calls, cabin calls, etc.) and are translated into commands for the direct control level.

The fuzzy group controller, a task of the overall control system, is encapsulated within a server to guarantee temporal isolation between itself and conventional real-time tasks. As figure 15.26 highlights, it is composed of a real-time knowledge processing module (RTKP) and several rule bases (RBs) to support focus-of-attention mechanisms (Gudwin et al. 1996).

15.3.2   Fuzzy Group Controller

As depicted in figure 15.27, the fuzzy group controller is divided into four basic parts. A preprocessor module transforms input information into the internal representation model used. A postprocessor module translates

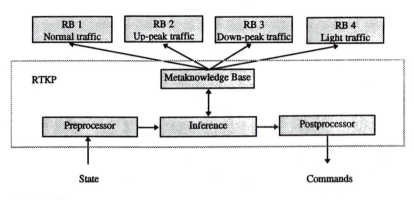

Figure 15.27
Fuzzy group controller structure

Table 15.9
Rules for normal traffic

A_time(i,j)\W_time(j)	Big_w	Medium_w	Small_w
Big_a	Medium	Low	Low
Medium_a	High	Medium	Low
Small_a	High	High	Medium

the internal representation model into output information (commands) required by the lower-level controllers. Between these two modules are an inference module for rule processing and a metaknowledge base for rule base activation.

Thus, the group controller comprises four separate rule bases, the selection of which is performed by metarules. A specific rule base of these metarules consists of two-valued predicates. The rules therein are classificatory; they identify the current pattern of traffic and select one among the remaining rule bases. Each rule base is responsible for handling a particular traffic pattern specific enough to merit a separate control protocol. We have formulated the rule bases for the four types of traffic: normal, up peak, down peak, and light traffic. Table 15.9 summarizes the detailed rules in these rule bases.

In table 15.9, $W\_time(j)$ and $A\_time(i, j)$ are linguistic variables denoting, respectively, the length of time the jth hall call is waiting for service and the length of time before the ith elevator expects to attend to the jth hall call, respectively. The rules in the table are processed whenever there are hall calls to be attended and the number of passengers within the ith elevator is less than its full service load. If the ith elevator is carrying passengers at capacity, the corresponding priority is set at *zero*.

Table 15.10
Rules for up-peak traffic

A_time(i,j)\W_time(j)	Big_w	Medium_w	Small_w
Big_a	Medium	Low↑ (Zero↓)	Low↑ (Zero↓)
Medium_a	High	Medium↑ (Zero↓)	Low↑ (Zero↓)
Small_a	High	High↑ (Zero↓)	Medium↑ (Zero↓)

Table 15.11
Rules for down-peak traffic

A_time(i,j)\W_time(j)	Big_w	Medium_w	Small_w
Big_a	Medium	Low↓ (Zero↑)	Low↓ (Zero↑)
Medium_a	High	Medium↓ (Zero↑)	Low↓ (Zero↑)
Small_a	High	High↓ (Zero↑)	Medium↓ (Zero↑)

Note that *priority* is a decision variable whose linguistic terms are *zero, low, medium* and *high*.

The rules in table 15.10 look similar to those in 15.9, but rule activation is very different. First, during up-peak conditions, rules are processed only when the number of passengers within elevators is below the maximum capacity. Otherwise, priorities are set to *zero*. Second, rules in the first row are processed independently of the elevator's direction of movement, whereas the decisions are set at *zero* in the shaded columns if hall calls are down calls, or as shown in the table if hall calls are up calls. In addition, a two-valued rule immediately activates to send elevators to the main floor whenever they becomes available.

Down-peak rules, shown in table 15.11, are processed under conditions similar to those for up-peak rules, but with the entries of the shaded columns reversed, and cars are not sent to the main floor simply because they are available. Rules for light traffic conditions are similar those for the normal traffic conditions, except that the number of elevators in service may change, for example, for energy saving and maintenance purposes.

The rules in the rule bases are viewed as Mamdani-type fuzzy conjunctions, and inference proceeds via max-min composition. Rule aggregation is performed after rule activation using the max aggregation operator. The center-of-area method is used for defuzzification. The mechanisms for encoding, decoding, and mapping are therefore

- for encoding: possibility computations,

- for mapping: fuzzy conjunction,

- for decoding: center of gravity using membership functions. ·

The fuzzy controller is also equipped with a mechanism of context dependency. Here, context adaptation adjusts universes in such a way that their intended label, e.g., *high* or *low*, depends on the traffic intensity. The scheme used is called the absolute limit context determination method, the simplest among a number of alternatives (Gudwin and Gomide 1994). Briefly, to define the context for a situation, samples are taken to generate an interval's upper and lower bounds.

15.3.3   Simulation Experiments

A series of simulation experiments was run in the following environment:

- Building and elevators:
  - Number of floors: 15 above main lobby and 3 below ground level
  - Height of each floor: 3.3 m
  - Number of elevators: 4
- Timings:
  - Time to open the door: 1 sec.
  - Time to close the door: 1 sec.
  - Time for each passenger to get in: 1 sec.
  - Time the door remains open: 1 sec.
  - Acceleration: 1 m/s^2
  - Deceleration: 1 m/s^2
  - Maximum load: 1600 kg.
  - Capacity: 20 persons

The main traffic patterns were considered; each experiment was carried out with and without context adaptation. In all experiments, the arrival request at the floors (hall calls) was modeled using a Poisson distribution. The various traffic scenarios are

1. Up peak: involves people entering the building in the morning between 7:45 and 8:45

2. Down peak: occurs between 17:15 and 18:45 and involves traffic exiting the building after working hours

3. All day: covers the entire day from 6:00 to 20:00, with the exception of up-peak and down-peak periods

Tables 15.12 and 15.13 summarize the main results of group control, expressed via average waiting time and maximum waiting time (both given in seconds). In all experiments the adaptation of context resulted in

Table 15.12
Average and maximum waiting time (no context adaptation)

	Average	Maximum
Down peak	25.00	184.0
Up peak	20.13	105.0
All day	14.95	209.0

Table 15.13
Average and maximum waiting time (context adaptation)

	Average	Maximum
Down peak	23.04	132.0
Up peak	17.00	92.0
All day	12.93	156.0

substantial gains in system performance. Figures 15.28 through 15.36 show the traffic profile, the waiting time histogram for up and down calls, and samples of the elevator movements.

15.4   Induction Motor Control

Most of the current research effort to develop speed control systems is devoted to alternating current (AC) machines, induction machines being of primary interest. Within the classes of AC machines, squirrel cage induction motors are particularly attractive because no brush or slip ring is needed, offering robust construction, operation and maintenance in industrial environments. However, induction motors are time-varying non-linear dynamic systems, and to be effective in high performance variable speed applications, more robust controllers than conventional PID are needed. Previous attempts have been reported as in, for example, Consoli et al. (1994), Mir, Zinger, and Elbuluck (1994), and Tang and Longya (1994).

The following discusses a fuzzy control scheme for variable speed, constant voltage/frequency operation (Caminhas, Tavares, and Costa 1995), based on a technique known as scalar control.

15.4.1   Speed Control of AC Machines

The main categories of AC machine speed control are scalar and the vector techniques. Both can achieve machine control by current or voltage feeding, the latter with constant voltage/frequency (v/f) ratio being the most popular for the scalar case. This scalar technique can be easily understood by inspecting figure 15.37, from which we see that

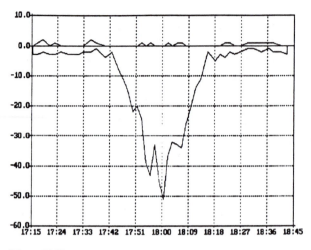

Figure 15.28
Profile for down-peak traffic

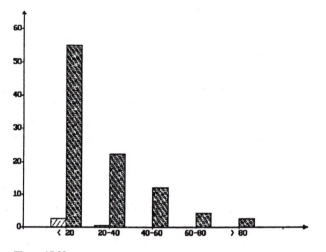

Figure 15.29
Histogram for down peak (with context adaptation)

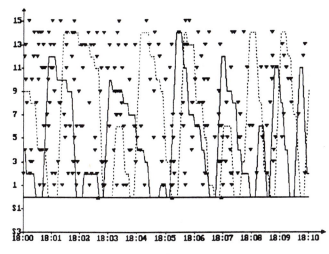

Figure 15.30
Sample trajectories for two elevators, down-peak traffic

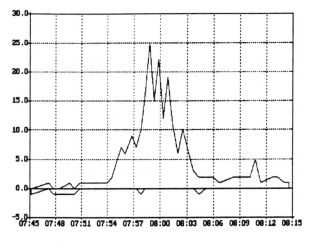

Figure 15.31
Profile for up-peak traffic

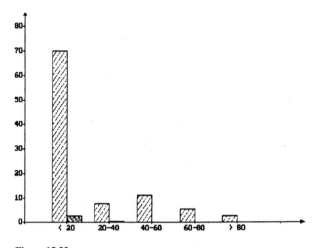

Figure 15.32
Histogram for up peak (with context adaptation)

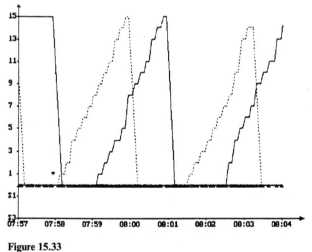

Figure 15.33
Sample trajectories for two elevators, up-peak traffic

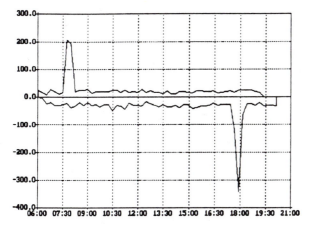

Figure 15.34
Profile for all-day traffic

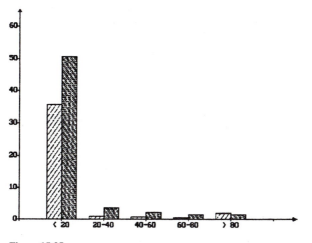

Figure 15.35
Histogram for all-day traffic (with context adaptation)

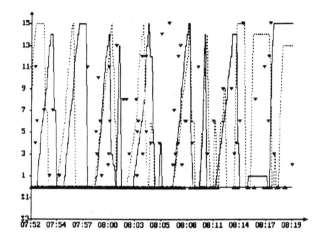

Figure 15.36
Sample trajectories for two elevators, all-day traffic

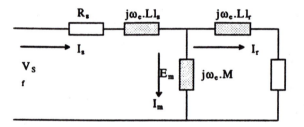

Figure 15.37
Equivalent circuit of an induction motor

$$E_m = j.M.\omega_e.I_m,$$

where M and I_m are the inductance and magnetizing current, respectively. Neglecting saturation, the magnetic flux is $M.I_m$, that is, proportional to the E_m/ω_e ratio. To achieve an efficient dynamic performance, the magnetic flux should be kept constant. Neglecting the stator impedance $(R_s + j\omega_e.Ll_s)$, the flux can be hold approximately constant if the v_s/f ratio is also kept constant. However, in low-speed conditions, the voltage drop in the stator resistance is significant when compared with E_m (Bose 1986; Murphy and Turnbull 1988), which means deterioration of the maximum machine torque due to the magnetic flux drop (see figure 15.38). Therefore, in developing speed control systems, this effect should be taken into account to improve dynamic efficiency.

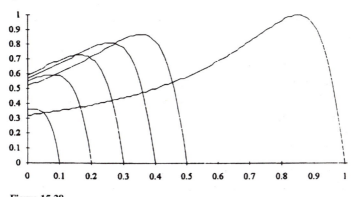

Figure 15.38
Torque speed curves at constant v_s/f

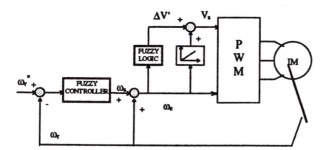

Figure 15.39
Block diagram of the drive system

15.4.2   Fuzzy Control Strategy

The machine drive system under consideration is composed of a squirrel cage induction motor, a pulse-width modulation (PWM) inverter, a mechanical load, and the control system, as shown in figure 15.39. The fuzzy controller provides the slip motor speed, ω_s, which is added to the electrical motor speed to get the stator frequency, ω_e. The voltage amplitude is obtained from ω_e and $\Delta V'$. The V_s and ω_e values are used to generate the PWM command signal:

$$Vs = K.\omega_e + \Delta V'.$$

The $\Delta V'$ term is the key to compensating the stator resistance effect which is modeled by the following fuzzy rules, where $|\omega_e|$ denotes the absolute value of ω_e:

- R_1:   If $|\omega_e|$ is *positive_big*, then $\Delta V'$ is *zero*.
- R_2:   If $|\omega_e|$ is *positive_medium*, then $\Delta V'$ is *positive_small*.

- R_3: If $|\omega_e|$ is *positive_small*, then $\Delta V'$ is *positive_medium*.
- R_4: If $|\omega_e|$ is *zero*, then $\Delta V'$ is *positive_big*.

Superior results are obtained with this scheme when compared to those obtained using the classical practice of adding, in the $K.\omega_e$ term, a constant $\Delta V'$ or a $\Delta V' = R_s.I_s$ value. Note that, in this scheme, I_s measurements are not needed.

15.4.3   Controller Design

The speed error and its variation are taken as controller inputs and the slip speed as the controller output. The speed error normalization is context dependent (Gudwin and Gomide 1994) and performed according to the variation in the reference speed. To normalize the variable for the universe of speed error variation, the dynamic speed equation is considered as a guide:

$$J.\frac{d\omega_r}{dt} = Tem - B.\omega_r - T_L.$$

For simplification purposes, the load and friction torque are negleted. Moreover, the electromagnetic torque *Tem* is assumed to be constant at 1.5 of its nominal value (T_N). Thus, the dynamic speed equation becomes

$$J.\Delta\omega_r = 1.5.T_N.\Delta t.$$

The motor parameters are $J = 0.016\ K.m^2$, $T_N = 8.5\ N.m$. For simulation purposes, the sampling period was taken as $\Delta t = 500\ \mu sec$. Hence, from the equation above, the resulting speed variations are $\Delta\omega_r = 0.4\ rad/sec$ and, for negative torque $\Delta\omega_r = -0.4\ rad/sec$, for each sampling period. The error variation is defined by

$$\Delta e(k) = e(k) - e(k-1) = [(\omega_r^*(k) - \omega_r(k)] - [\omega_r^*(k-1) - \omega_r(k-1)],$$

which, assuming a constant reference speed, can be rewritten as

$$\Delta e(k) = \omega_r(k-1) - \omega_r(k) = -\Delta\omega_r.$$

We therefore conclude that the universe for the speed error variation is $[-0.4, 0.4]$. The controller output must also take into account the slip speed limit to restrict the inverter maximum current. Here we assume a maximum slip speed of 30 rad/sec, which induces the universe $[-30, 30]$. The normalized universes of all variables handled by the fuzzy controller are assumed to be discretized into 25 points, not homogenously distributed.

The input and output linguistic variables have seven terms each, labeled *NB* (*negative big*), *NM* (*negative medium*), *NS* (*negative small*), *ZE* (*zero*), *PB* (*positive small*), *PM* (*positive medium*), and *PB* (*positive big*). The discrete membership function of each fuzzy set has the form

$$M_f(u) = \sum_{i=1}^{7} a_i / u_i,$$

where $a = [0.25, 0.5, 0.75, 1.0, 0.75, 0.5, 0.25]$.

The following table gives the control rules that compose the fuzzy controller:

$\Delta e \rightarrow$ $\downarrow e$	NB	NB	NP	ZE	PS	PM	PB
NB	*NB*	*NB*	*NB*	*NB*	*NM*	*NS*	*ZE*
NM	*NB*	*NB*	*NB*	*NM*	*NS*	*ZE*	*PS*
NS	*NB*	*NB*	*NM*	*NS*	*ZE*	*PS*	*PM*
ZE	*NB*	*N*	*NS*	*ZE*	*PS*	*PM*	*PB*
PS	*NM*	*NS*	*ZE*	*PS*	*PM*	*PB*	*PB*
PM	*NS*	*ZE*	*PS*	*PM*	*PB*	*PB*	*PB*
PB	*ZE*	*PS*	*PM*	*PB*	*PB*	*PB*	*PB*

The study adopted the max-min inference scheme and employed the center-of-area defuzification method. Simulation experiments were performed for a 2 HP induction motor. The integration step adopted was 50 μsec, and a sample period of 500 μsec was defined for control action purposes.

Figure 15.40 shows the steady-state curves for the torque/maximum torque ratio. As can be easily seen, the maximum torque remains constant within the whole operation range, which means that the maximum flux is kept constant, even when the motor is operating at low speed. Figure 15.41 shows the transient behavior for reference speeds within the 0.1–0.5 pu (per unit) range from the nominal speed, whereas figure 15.42 shows the transient behavior for a PI controller under the same operating conditions. The fuzzy controller is superior from the transient and stationary behavior point of view.

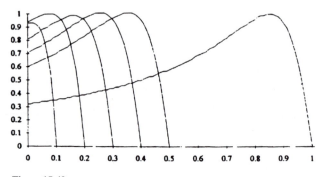

Figure 15.40
Torque speed curves with fuzzy control

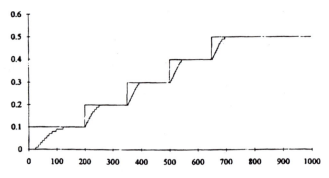

Figure 15.41
Speed response with fuzzy control

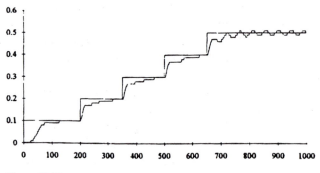

Figure 15.42
Speed response with PI control

15.5   Communication Network Planning

Modern communication networks offer a variety of services, for example, voice, data, video and image information transmission. For these purposes, the network must handle large volumes of information traffic and must comply with the evolution of information over time. Network planning is a major problem; since traffic demand is partly known, it involves requirements the can only be realistically described qualitatively, and constraints must be satisfied as well.

Communication networks are often planned in three main steps: external network planning, switch planning, and transmission planning. The first step aims to find satisfactory locations and the respective dimensioning of the central offices (CO) to provide an appropriate balance between equipment and access cost. The second step establishes inter-CO trunk capacities and dimensioning of switching equipment. The third step concerns routing plans, routing of fiber optic cables, and using cable galleries.

In particular, current approaches for transmission planning emphasize the concentration of large flows of information into small number of links to reduce cost. The goal is to have as many logical connections as possible in a physically small network. Intelligent routers placed at strategic locations make this goal feasible.

An essential step in transmission planning concerns grouping the COs into clusters based on their distances and demands. Solving the clustering problem is crucial for optimum network planning, and it is often solved by classical methods. This section focuses on this important clustering step.

15.5.1   Communication Network Model

The communication network is modeled by an undirected graph $G(N, A)$ consisting of a set N of nodes and a set A of arcs representing the COs and their links, respectively. Each pair of nodes $(i, j) \in N^2$ has an associated demand d_{ij}, and each arc between adjacent nodes has length l_{ij}, the distance between nodes i and j. The demand is only partially known and can be conveniently viewed as a fuzzy number. Thus, the problem of interest can be briefly stated as follows: Given $G(N, A)$, cluster the nodes of N into groups with a similar proximity index. The proximity index between nodes is defined by an $n \times n$ matrix P, $n = \mathrm{Card}(N)$, whose entries $[p_{ij}]$ are a function of the distance and the demand between nodes. Since the goal is to derive clusters with higher internal demand and closer nodes, we may set p_{ij} as follows:

$$p_{ij} = \alpha d_{ij} + (1 - \alpha)(1/l_{ij}), \quad i, j = 1, \ldots, n,$$

where $\alpha \in [0, 1]$ is a parameter chosen to weight the desired balance between demand and distance. The length l_{ij} for nodes other than adjacent nodes can be found by computing the shortest path from node i to node j. The Dijkstra shortest-path algorithm is among the most efficient (Johnsonbaugh 1993) for computing this distance. Since d_{ij} is a fuzzy number and l_{ij} is a real number, p_{ij} is also a fuzzy number.

15.5.2   Clustering Procedure

The clustering procedure has a self-organizing nature and assumes a topological structure among the cluster units. There are m cluster units arranged in a one- or two-dimensional array; the inputs are n-tuples of fuzzy numbers, $n > m$. A weight vector for a cluster unit serves as an exemplar of the input patterns associated with that cluster. The weight vector is also viewed as n-tuples of fuzzy numbers. During the self-organizing process, the cluster unit whose weight vector most closely matches the input pattern is chosen as the winner. The winner unit updates its weights by adjusting the modal values of the respective fuzzy numbers, assumed to have triangular membership functions. Thus, the inputs and weights are vectors whose components are triangular fuzzy numbers with modal values p_{ij} and w_{ik}, as depicted in figure 15.43. The triangle shape is assumed to be known and, of course, depends on the uncertainty description. Clearly, the input vectors are the columns $p_j = [p_{lj} \ldots p_{nj}]$, $j = 1, \ldots, n$, of matrix P. Similarly, the weight vectors are $w_k = [w_{1k} \ldots w_{nk}]$, $k = 1, \ldots, m$.

In communication network planning, clustering may be used for many purposes. One aims at organizing the overall network. For instance, we

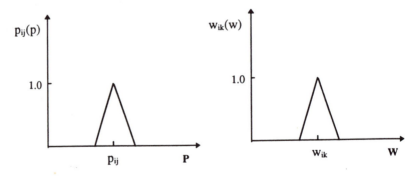

Figure 15.43
Input and weight vector components

may find clusters and assign the closest CO to the weight exemplar as a hub for that cluster. Alternatively, we can assume each hub as given (often called *seeds* in the communication planning jargon), because in the most practical instances these are in existing buildings. Thus clustering must be pursued under the constraint that a set $H = \{h_k / h_k \in N, k = 1, \ldots, m\}$ of hubs is given. Operations research (Hillier and Lieberman 1990), heuristics (Klincewicz 1991), and neural networks (Caminhas and Tavares 1994) are examples of techniques that can be used. Here we provide a clustering procedure for the constrained case only, since the unconstrained case becomes clear afterwards. The essence of the procedure relies on using the possibility measure as a similarity index among the cluster units. The procedure may be viewed as a learning scheme for a self-organizing neurofuzzy network that is supervised for the constrained case and unsupervised for the unconstrained case. The details of the clustering procedure are as follows. (See also chapter 11.)

Clustering Procedure

Step 1: Initialize modal values w_{kj} such that $w_{kj} = p_{kj}$, $h_k \in H$, $k = 1, \ldots, m$, $j = 1, \ldots, n$.

Initialize the triangle parameters and the learning rate $\lambda \in (0, 1]$.

Step 2: Choose an $L \in N$.

Step 3: If $(L = h_k, k \in \{1, \ldots, m\}$ and $h_k \in H)$ then go to Step 5,

Step 4: Call Competition

Call Adjustment

$L = C$

Go to Step 5

Step 5: Repeat

Call Competition

Call Adjustment

Until $C = L$

Step 6: If (*Convergence = False*) then update λ and go to Step 2

Otherwise Stop.

Competition: For each weight vector $w_k, k = 1, \ldots, m$, compute

$$M(k) = \frac{1}{n} \sum_{i=1}^{n} \left\{ \max_{w \in W} [\min(p_{ij}(p), w_{ik}(w))] \right\}.$$

Find index C such that $M(C)$ is a maximum.

Return.

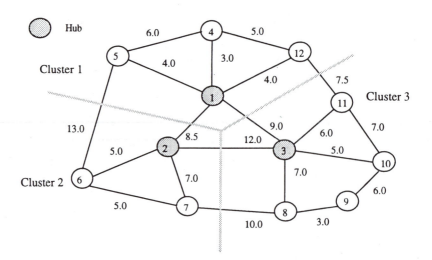

Figure 15.44
Example network with 12 nodes

Adjustment: Update the vector w_c as follows:

$$w_{cj}(\text{new}) = [w_{cj}(\text{old}) - \lambda p_{Lj}(\text{old})].$$

Return.

After the procedure converges, we may present the input data to find the cluster to which it belongs with a certain degree of matching. Note that the competition step involves possibility measure calculations averaged over the respective data set and a winner-take-all scheme.

The clustering procedure was used to cluster the communication networks depicted in figures 15.44 and 15.45. Tables 15.14 to 15.17 collect the clustering results. The solutions provided by the fuzzy clustering method are more realistic from the practical point of view, since planners usually have low confidence when crisp solutions are provided, as happens with the classical methods. The detailed demand data can be found in Figueiredo et al. 1996).

15.6   Neurocomputation in Fault Diagnosis of Dynamic Systems

Analytical redunduncy is a technique for fault detection and diagnosis of dynamic systems (FDD), which has been approached using several methodologies. These include parity space (Himmelblau 1978; Patton, Frank, and Clark 1989), state estimation, parameter estimation (Willsky 1990), knowledge-based systems and pattern recognition (Frank 1990).

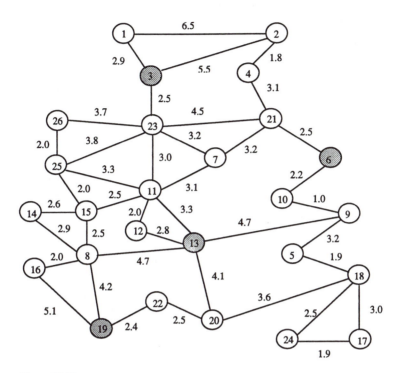

Figure 15.45
Example network with 26 nodes

Table 15.14
Clustering results with no seeds (12 nodes)

No seeds			Clusters (nodes)		
Experiment	Hubs	α	1	2	3
1	2, 1, 9	1	2, 3, 6	1, 4, 5, 12	7, 8, 9, 10, 11
2	1, 2, 11	0.75	1, 4, 7, 12	2, 3, 5, 6	8, 9, 10, 11
3	1, 2, 11	0.5	1, 4, 5, 12	2, 3, 9, 10	6, 7, 8, 11
4	1, 2, 11	0.25	1, 4, 5, 12	2, 3, 9, 10	6, 7, 8, 11
5	1, 2, 10	0	1, 3, 9, 11	2, 4, 5, 12	6, 7, 8, 10

Table 15.15
Clustering results with weights initialized with seeds (12 nodes)

Seeds: nodes 1, 2, 3			Clusters (nodes)		
Experiment	Hubs	α	1	2	3
1	1, 2, 3	1	1, 5, 12	2, 4, 6	3, 7, 8, 9, 10, 11
2	1, 2, 3	0.75	1, 4, 5, 12	2, 6, 7, 11	3, 8, 9, 10
3	1, 2, 3	0.5	1, 4, 5, 7	2, 6, 8, 10	3, 9, 11, 12
4	1, 2, 3	0.25	1, 4, 5, 6	2, 7, 8, 9	3, 10, 11, 12
5	1, 2, 3	0	1, 5, 7, 10	2, 4, 6, 8	3, 9, 11, 12

Table 15.16
Clustering results with no seeds (26 nodes)

No seeds			Clusters (nodes)			
Experiment	α	Hubs	1	2	3	4
1	1	8, 4, 11, 23	2, 6, 8, 13, 15, 16, 17, 19, 20, 22, 24, 25, 26	1, 3, 4, 5, 12, 14	9, 11, 18	7, 10, 21, 23
2	0.75	12, 23, 6, 9	8, 12, 16, 17, 18, 19	1, 20, 21, 22, 23, 24, 25	2, 3, 4, 5, 6, 15, 26	7, 9, 10, 11, 13, 14
3	0.5	9, 13, 18, 23	5, 6, 7, 8, 9, 25	10, 11, 12, 13, 15, 16	17, 18, 19, 20, 21, 22	1, 2, 3, 4, 14, 23, 26
4	0.25	21, 22, 26, 6	1, 11, 12, 13, 14, 21	7, 8, 9, 10, 15, 22	16, 17, 18, 19, 20, 23, 26	2, 3, 4, 5, 6, 24, 25
5	0	26, 11, 19, 25	3, 4, 5, 12, 20, 26	6, 7, 8, 9, 10, 11	13, 14, 15, 16, 17, 18, 19	1, 2, 21, 22, 23, 24, 25

Table 15.17
Clustering results with weights initialized with seeds (26 nodes)

Seeds: nodes 3, 6, 13, 19			Clusters (nodes)			
Experiment	α	Hubs	1	2	3	4
1	1	3, 6, 13, 19	1, 2, 3, 4, 12	5, 6, 8, 14, 15, 16, 17, 20, 22, 24, 25, 26	9, 11, 13, 18	7, 10, 19, 21, 23
2	0.75	3, 6, 13, 19	3, 4, 5, 11, 12, 16, 20	6, 7, 9, 10, 22	8, 13, 14, 15, 17, 18, 21	1, 2, 19, 23, 24, 25, 26
3	0.5	3, 6, 13, 19	3, 4, 5, 8, 9, 11	6, 7, 10, 12, 20, 23	13, 14, 15, 16, 17, 18, 21	1, 2, 19, 22, 24, 25, 26
4	0.25	3, 6, 13, 19	3, 9, 17, 18, 20, 26	1, 6, 8, 11, 21, 22	13, 14, 15, 16, 23, 24, 25	2, 4, 5, 7, 10, 12, 19
5	0	3, 6, 13, 19	1, 3, 4, 5, 9, 10	6, 8, 11, 12, 20, 22	13, 14, 15, 16, 17, 18, 21	2, 7, 19, 23, 24, 25, 26

Most of these techniques and those related to them assume that a mathematical model of the system is available. Probabilistic, heuristic, neural, and fuzzy set–based techniques, for example, can tackle pattern classification and recognition problems.

Here we introduce a system constructed using fuzzy set models of neurons and networks presented in chapter 11. The neurofuzzy network has a feedforward structure whose connections and weights are determined by training. Net topology encodes a rule base encompassing rules with uncertainty and certainty factors. Actually, these factors are viewed as network weights. Weights are updated by a competitive learning procedure. The processing time involved is not sensitive to the input space partition and is fast, features which are desirable in real-time applications. Figure 15.46 summarizes the system's structure.

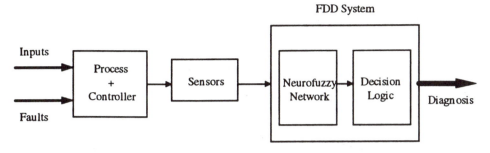

Figure 15.46
Structure of the FDD system

15.6.1 Neurofuzzy Network Structure and Learning

Consider a set of training patterns $X_p = (x_{p1}, \ldots x_{pi}, \ldots x_{pNe})$, $p = 1, 2, \ldots, m$, where $x_{p1} \ldots x_{pNe}$ are the Ne coordinates of X_p. In FDD problems, these coordinates can be any input, output, state, or combination of these variables considered essential for fault detection and diagnosis. Assume the training patterns are grouped into Nc classes, $C_1, \ldots, C_p, \ldots, C_{Nc}$. Caminhas, Tavares, and Gomide (1995) developed a neurofuzzy network for pattern classification problems based on AND and OR logic neuron models. This section introduces a generalized model of the neurofuzzy networks and the associated learning procedure (Caminhas, Tavares, and Gomide 1996b). Caminhas, Tavares, and Gomide (1996a) have shown that this model has several advantages when compared with learning vector quantization approaches (Kohonen 1990). Figure 15.47 shows the generalized neurofuzzy network structure considering Ne inputs and Nc output classes. Note that this net structure encodes a set of rules of the type "If $(x_{p1}$ is A_{k1} with uncertainty $w_{j1})$ ... and $(x_{pi}$ is A_{ki} with uncertainty $w_{ji})$ ... and $(x_{pNe}$ is A_{kNe} with uncertainty $w_{jNe})$, then the pattern is class C_l with certainty v_j," where $x_{p1} \ldots x_{pNe}$ are the Ne input coordinates; Z_l is the output level for class l; Ns is the number of fuzzy sets assembling the input partition; v_j is the certainty degree of the jth encoded rule; w_{ji} is the uncertainty degree of the ith input of rule j. The AND and OR are logic neurons, and as we learned in chapter 11,

$$y_j = \mathop{T}_{i=1}^{Ne} (w_{ji} \text{ s } x_{ji}) \quad \text{and} \quad z_l = \mathop{S}_{j=1}^{n_l} (v_j \text{ t } y_j)$$

where n_l is the number of AND neurons connected to the lth OR neuron.

15.6.2 Learning Algorithm

Briefly, the learning algorithm is based on a supervised, competitive strategy with the following key steps: determination of the membership

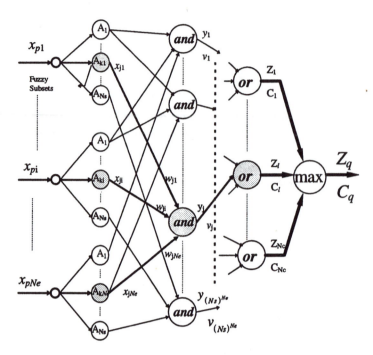

Figure 15.47
Neurofuzzy network structure

functions; generation of network connections and weight initialization; determination of active AND and OR neurons; fuzzification; determination of the winning class; and weight updating. These steps are detailed below.

Step 1: Membership functions generation. For simplicity, we assume triangular membership functions. Clearly, any other function could be used, but they must overlap as shown in figure 15.48, where $\Delta_i = (X_{i\max} - X_{i\min})/(N_s - 1)$. The parameters $X_{i\min}$ and $X_{i\max}$, for $i = 1, 2, \ldots, Ne$, are found through

$$X_{i\min} = X_{i\min}^{C_{q1}} + \frac{X_{i\min}^{C_{q2}} - X_{i\min}^{C_{q1}}}{N_s}, \quad X_{i\max} = X_{i\max}^{C_{q3}} + \frac{X_{i\max}^{C_{q3}} - X_{i\max}^{C_{q4}}}{N_s},$$

where $X_{i\min}^{C_{q1}} \le X_{i\min}^{C_{q2}}$ are the least values of the coordinate $x_{pi'}$, $\forall p = 1, 2, \ldots, m$, for the two classes C_{q1} and C_{q2}, and $X_{i\max}^{C_{q3}} \ge X_{i\max}^{C_{q4}}$ are the maximum values of the coordinate $x_{pi'}$, $\forall p = 1, 2, \ldots, m$, for the two classes C_{q3} and C_{q4}.

Step 2: Generation of network connections; weights w_{ji} and v_{ij} initialization. This step establishes the connections between the AND and OR

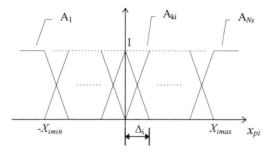

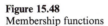

Figure 15.48
Membership functions

neurons and can be viewed as an automatic rule generation procedure. The scheme is an adaptation of Nozaki, Ishibuchi, and Tanaka 1994 and Wang and Mendel (1992).

For $j = 1, \ldots, (Ns)^{Ne}$ do:

(a) connection generation:

compute β_l for $l = 1, 2, \ldots, Nc$, $\beta_l = \sum_{X_p \in C_l} T_{i=1}^{Ne} x_{ji}$, where $x_{ji} = A_{ji}(x_{pi})$, and

Determine class $q(C_q)$ such that $\beta_q \leftarrow \max_l \{\beta_1, \ldots, \beta_l, \ldots, \beta_m\}$.

Thus, the jth AND neuron is connected to the qth OR neuron. Note that, by construction, any index j is such that $j = v'_{Ne} + \sum_{i=2}^{Ne} (v'_{Ne-i+1} - 1) . N_s^{i-1}$ where v' is a vector containing the indexes of the fuzzy sets for the input coordinates.

(b) Initialize weights w_{ji} and $v_{j'}$, setting:

$$ w_{ji} = 0, \ \forall i = 1, 2, \ldots Ne, \quad \text{and} \quad v_j = \frac{\beta_q - \sum_{l \neq q} \beta_l / (Nc - 1)}{\sum \beta_l}. $$

Weight v_j could be set at 0. However, it has been found experimentally that the scheme above peforms better.

Step 3: Determination of active AND and OR neurons. Given membership functions as shown in figure 15.48, clearly at most two of them have nonnull values for any given input pattern. The membership functions $\mu_{Aji}(x_{pi})$ with nonnull values are called active sets. Since the number of AND neurons depends on the input partition, from the total of $(Ns)^{Ne}$ at most 2^{Ne} can be active. They are found as follows:

For a given pattern p and for $i = 1, \ldots, Ne$ compute:

$$
u_i^1 = \begin{cases} 1 + INT\left(\dfrac{x_{pi} - X_{i\min}}{\Delta_i}\right), & \text{if } X_{i\min} < x_{pi} < X_{i\max} \\ 1, & \text{if } x_{pi} \leq X_{i\min} \\ Ns, & \text{if } x_{pi} \geq X_{i\max}, \end{cases}
$$

$$
u_i^2 = \begin{cases} u_i^1 + 1, & \text{if } X_{i\min} < x_{pi} < X_{i\max} \\ u_i^1, & \text{otherwise,} \end{cases}
$$

where $INT(*)$ denotes the integer part of $(*)$, whereas u_i^1 and u_i^2 are the indexes of the two active fuzzy sets for the ith. If Pa is the number of coordinates whose values fall in the interval $X_{i\min} < x_{pi} < X_{i\max}$, then the number of active AND neurons is $Na = 2^{Pa} \leq 2^{Ne}$. Also, let $\mathbb{L}^*$ be the set of active AND neurons. This set, together with the connections defined in step 2, determine the set $\mathbb{Q}^*$ of active OR neurons.

Step 4: Fuzzification. Membership degrees are then computed for a given input pattern.

Step 5: Determination of the winning class.

Compute $Z_l \; \forall \, l \in \mathbb{Q}^*$:

$$
Z_l = \overset{n_l}{\underset{j=1}{S}} \, (v_j \mathbin{\text{t}} y_j), \; \forall j \in \mathbb{L}^*, \quad \text{and} \quad y_j = \overset{Ne}{\underset{i=1}{T}} \, (w_{ji} \mathbin{\text{s}} x_{ji}).
$$

Let q be the index of the winning class (C_q). Thus $Z_q = \max\{Z_l\}$, $\forall l \in \mathbb{Q}^*$.

Step 6: Weights updating.

Let C_q be the winning class found in step 5. If C_q is the correct class, then set

$$
v_J(k+1) = v_J(k) + \alpha_1 \cdot [1 - v_J(k)] \quad \text{and}
$$
$$
w_{JI}(k+1) = w_{JI}(k) + \alpha_2 \cdot [1 - w_{JI}(k)],
$$

where I and J are found as follows:

$$
\Omega_J = \max_j \, \{[v_j \mathbin{\text{t}} y_j] \; \forall j \in \mathbb{L}^* \text{ and } X_p \in C_q\},
$$

$$
\Omega_I = \min_i \, \{[x_{ji} \mathbin{\text{s}} w_{ji}], \; \forall i = 1, \ldots Ne \text{ and } j = J\};
$$

otherwise set

$$
v_J(k+1) = v_J(k) - \alpha_3 \cdot v_J(k),
$$

$$
w_{JI}(k+1) = w_{JI}(k) - \alpha_4 \cdot w_{JI}(k),
$$

where α_1, α_2, α_3, α_4 are the learning rates with $0 \leq \alpha_1 \ll \alpha_3 < 1$ and $0 < \alpha_2 \ll \alpha_4 < 1$.

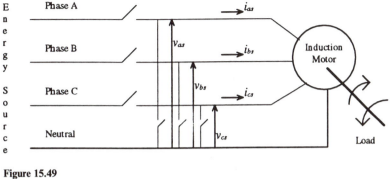

Figure 15.49
Dynamic system

15.6.3   Fault Detection and Diagnosis

Consider now fault detection and diagnoses of the dynamic system composed of an energy source, an induction motor, and a mechanical load, as shown in figure 15.49. This is a nonlinear, time-varying system of seventh order (Bose 1986). Three neurofuzzy networks are used, one for each phase: A, B, and C (figure 15.50). The network inputs are the current, current variation, voltage and voltage variation, respectively. Three outputs are associated with each network, representing three classes of faults: C_1 (normal operation), C_2 (phase-neutral short circuit), and C_3 (open phase/no energy source). The outputs of networks A, B, and C correspond to the indexes of the winning classes, that is 1, 2, or 3. Based on these characteristics, a decision logic can easily be constructed to detect the fault and decode it into the corresponding diagnosis. Table 15.18 summarizes the decision logic.

15.6.4   Simulation Results

Simulation experiments were run for a number of randomly generated faults. Figure 15.51 shows the results after the following events: phase A, B, and C open; normal operation; short-circuit between phases B, C, and neutral; normal operation; and phases A, C open. The figure also shows details emphasizing the time instant before and after a short circuit occurred between phases B, C, and neutral. It takes about 4 ms to detect and diagnose a fault, corresponding to two sample periods.

The neurofuzzy nets were trained under noiseless current and voltage measurements and a constant load. However, during network testing, measurements were considered noisy and load time varying. As figure 15.51 shows, the system is robust against measurement noise and load variations. These features, and the fast response time as well, indicate the

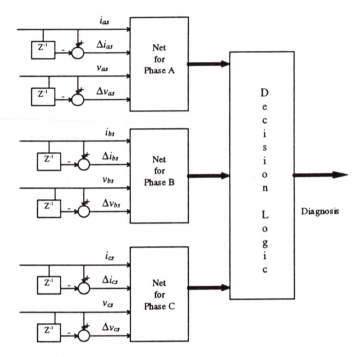

Figure 15.50
System structure for the FDD example

Table 15.18
Decision logic

Fault index	Output of net A	Output of net B	Output of net C	Fault diagnosis
01	1	1	1	normal operation
02	2	1	1	short-circuit between phase A and neutral
03	1	2	1	short-circuit between phase B and neutral
04	1	1	2	short-circuit between phase C and neutral
05	2	2	1	short-circuit between phases A, B and neutral
06	2	1	2	short-circuit between phases A, C and neutral
07	1	2	2	short-circuit between phases B, C and neutral
08	2	2	2	short-circuit between phases A, B, C and neutral
09	3	1	1	phase A open
10	1	3	1	phase B open
11	1	1	3	phase C open
12	3	3	1	phases A, B open
13	3	1	3	phases A, B, C open
14	1	3	3	phases B, C open
15	3	3	3	phases A, B, C open

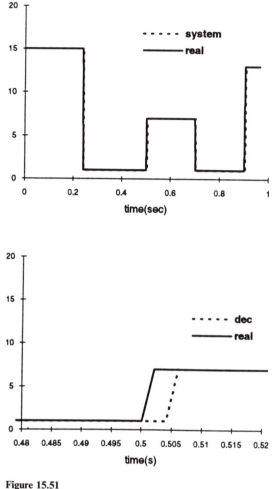

Figure 15.51
Simulation results

feasibility of the approach in real-time situations. The triangular norms used were the max s-norm and the min t-norm, and learning rates were set as $\alpha_1 = \alpha_3 = 0.001$ and $\alpha_2 = \alpha_4 = 0.1$.

15.7   Multicommodity Transportation Planning in Railways

Multicommodity network flow problems arise when several items (commodities) share arcs in a capacitated network, a type of network typically found in communication systems, urban traffic systems, multiproduct production distribution systems, railway systems, and many others as well. In particular, the efficient allocation of vehicles to conform with transportation demand as well as the distribution of empty vehicles from locations where they have been unloaded to locations where they will be reloaded is a problem common to all modes of freight transportation. In the case of railways, decision making must consider allocation of loaded wagons and distribution of empty wagons to minimize transportation costs, subject to practical and physical constraints such as the total number of wagons, track capacities, wagon types, and balancing. A convenient way to model a railway network is to look at it as a graph where each node represents a station and each arc a line section. In addition, if we assume that a linear objective function models the transportation cost, the decision model can be put in the form of the following linear programming model (Mendes, Yamakami, and Gomide 1996):

$$\min \sum_k \left(\sum_j (C_{jk} . X_{jk}) + C'_k . X'_k \right)$$

$$\hat{A}_j . X_j = r_j; \quad j = 1, \ldots, J$$

$$\sum_{j \in J^k} \tilde{A}_j (X_{jk} + X'_k) = 0; \quad k = 1, \ldots, K$$

$$\sum_j E_j . X_j^i \le b_i; \quad i = 1, \ldots, I$$

$$\sum_{j \in J^k} (T_j . \delta_j . X_{jk}) + T' . X'_k \le f_k$$

$$X_j = \sum_k X_{jk}; \quad j = 1, \ldots, J$$

$$X_{jk} \ge 0, \ X'_k \ge 0; \quad \forall_j \ \forall_k$$

where

j: index the products;

k: index the wagon types;

i: index the line sections;

J: total number of products;

K: total number of wagon types;

I: total number of arcs in the graph;

C_{jk}: vector of unit allocation costs C_{jk}^i for wagon k transporting product j;

C'_k: vector of unit distribution costs $C_{jk}^{\prime i}$ of an unloaded wagon k in line section i to get product j;

X_{jk}:   vector of flows X_{jk}^i of the product j tranported by wagon k;

X_k':   vector of flows $X_{jk}^{\prime i}$ for empty wagon k to get product j;

J^k: set of products transported by wagon of type k;

b_i: capacity of the line section i;

T_j: travel time to transport product j from the supply to the demand node;

T':   vector of travel times for empty wagons;

f_k: total number of wagons of type k;

δ_j: vector indicating which line sections are use to transport product j;

E_j: weight of a wagon loaded with product j;

r_j: total amount of product j to be transported.

The matrices $\hat{A}_j$ and $\tilde{A}_j$ denote the incidence matrices of the original network graph with an extra arc added from origin to destination for each product j (the dotted arcs in figure 15.52). This addition puts the model in the standard form. Thus, added arcs play the role of slack variables. For instance. let us consider the simple network shown in figure 15.52, where three products are to be transported and two types of wagon are available. Tables 15.19, 15.20 and 15.21 report the remaining data. The linear programming model becomes:

$$\min \sum_{k=1}^{2} \left(\sum_{k=1}^{3} (C_{jk} X_{jk}) + C_k' X_k' \right)$$

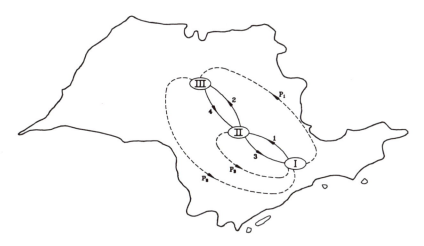

Figure 15.52
Example railway network

Table 15.19
Line sections data

Arcs	Origin	Destination	Distance (km)	Maximum capacity (t)	Empty-time (day)
1	I	II	136	3000	1
2	II	III	140	3000	1
3	II	I	136	1000	1
4	III	II	140	3000	1

Table 15.20
Products data

Product	Origin	Destination	Quantity (t)	Cost ($)	Time (day)	Capacity (t)	Wagon type
P1	I	III	500	4	3	50	1
P2	III	I	1000	4	3	50	1 and 2
P3	II	I	1000	3	3	50	2

$$\hat{A}_j . X_j = r_j; \quad j = 1, \ldots, 3$$

$$\sum_{j \in J^k} \tilde{A}_j (X_{jk} + X'_k) = 0; \quad k = 1, \ldots, 2$$

$$\sum_j E_j . X_j^j \le b_i; \quad i = 1, \ldots, 4$$

$$\sum_{j \in J^k} (T_j . \delta_j . X_{jk}) + T' . X'_k \le f_k$$

$$X_j = \sum_{k=1}^{2} X_{jk}; \quad j = 1, \ldots, 3$$

$$X_{jk} \ge 0, \ X'_k \ge 0; \quad \forall_j \ \forall_k.$$

$$\hat{A}_1 = \begin{bmatrix} 1 & 0 & -1 & 0 & 1 \\ -1 & 1 & 1 & -1 & 0 \\ 0 & -1 & 0 & 1 & -1 \end{bmatrix} \quad \hat{A}_2 = \begin{bmatrix} 1 & 0 & -1 & 0 & -1 \\ -1 & 1 & 1 & -1 & 0 \\ 0 & -1 & 0 & 1 & 1 \end{bmatrix}$$

$$\hat{A}_3 = \begin{bmatrix} 1 & 0 & -1 & 0 & -1 \\ -1 & 1 & 1 & -1 & 1 \\ 0 & -1 & 0 & 1 & 0 \end{bmatrix} \quad \tilde{A}_j = \begin{bmatrix} 1 & 0 & -1 & 0 & 0 \\ -1 & 1 & 1 & -1 & 0 \\ 0 & -1 & 0 & 1 & 0 \end{bmatrix}$$

Note that, for notation simplicity, the cost vectors are assumed to be row vectors, and the decision variables X_{jk} and X'_k should be viewed as augmented variables (including slack and auxiliary variables) to conform with

Table 15.21
Wagon characteristics

Type	Maximum (units)	Empty car weight (t)
1	50	25
2	30	25

matrices' dimensions. This causes no problem once we define appropriate coefficients for the added variables in the objective function.

The linear programming model above is useful only when the decision maker is absolutely confident about the unit costs, transportation demand, travel times and amount of each type of wagon, a rare situation in practice. Most often, decision makers have good approximate values for those parameters and are able to admit flexibility in certain constraints. In addition, usually they prefer a range of solutions fitting their approximate intuition about the problem to a unique solution, as is provided by conventional decision models. One way to handle this is to develop a fuzzy linear multicommodity programming model to include imprecision in costs, coefficients and constraints. Thus, we bring in Zimmermann's method to handle the cost function and combine it with Delgado, Verdegay, and Vila's method to take fuzzy coefficients and constraints into account. As discussed in chapter 14, the following fuzzy linear programming model results:

$$\min_{f} \sum_{k=1}^{2} \left(\sum_{j=1}^{3} (C_{fjk} X_{jk}) + C'_{fk} X'_{k} \right)$$

$$\hat{A}_j . X_j = r_j; \quad j = 1, \ldots, 3$$

$$\sum_{j \in J^k} \tilde{A}_j (X_{jk} + X'_k) = 0; \quad k = 1, \ldots, 2$$

$$\sum_{j} E_j . X_j^i \leq_f b_{fi}; \quad i = 1, \ldots, 4$$

$$\sum_{j \in J^k} (T_{fj} . \delta_j . X_{jk}) + T' . X'_k \leq_f f_{fk}$$

$$X_j = \sum_{k=1}^{2} X_{jk}; \quad j = 1, \ldots, 3$$

$$X_{jk} \geq 0, \, X'_k \geq 0; \quad \forall_j \, \forall_k.$$

The following auxiliary optimization problems are induced. The first is a version of the original linear programming model for verifying the effect of relaxing cost bounds to consider cost tolerance; the second and third are generated from the fuzzy linear programming model to include fuzziness in the indicated coefficients and the constraints. Thus we have:

$\min \theta$

$$\sum_{k=1}^{2} \left(\sum_{j=1}^{3} (C_{jk}^o X_{jk}) + C_k'^o X_k' \right) \leq r_0^o + \theta t_0^o$$

$$\hat{A}_j^o . X_j = r_j^o + \theta t_j^o \quad j = 1, \ldots, 3$$

$$\sum_{j \in J^k} \tilde{A}_j^o (X_{jk} + X_k') = 0; \quad k = 1, \ldots, 2$$

$$\sum_j E_j^o . X_j^i \leq b_i^o + \theta t_i'^o; \quad i = 1, \ldots, 4$$

$$\sum_{j \in J^k} (T_j^o . \delta_j^o . X_{jk}) + T'^o . X_k' \leq f_k^o + \theta t_k''^o$$

$$X_j = \sum_{k=1}^{2} X_{jk}; \quad j = 1, \ldots, 3$$

$$X_{jk} \geq 0, X_k' \geq 0; \quad \forall_j \, \forall_k$$

$$\theta \in [0, 1]$$

$\min \theta$

$$\sum_{k=1}^{2} \left(\sum_{j=1}^{3} (\underline{C}_{jk} X_{jk}) + \bar{C}_k' X_k' \right) \leq \bar{r}_o + \theta \bar{t}_o$$

$$\hat{A}_j . X_j = \underline{r}_j + \theta \underline{t}_j \quad j = 1, \ldots, 3$$

$$\sum_{j \in J^k} \tilde{A}_j (X_{jk} + X_k') = 0; \quad k = 1, \ldots, 2$$

$$\sum_j E_j . X_j^i \leq \underline{b}_i + \theta \underline{t}_i'; \quad i = 1, \ldots, 4$$

$$\sum_{j \in J^k} (\bar{T}_j . \delta_j . X_{jk}) + T' . X_k' \leq \underline{f}_k + \theta \underline{t}_k''$$

$$X_j = \sum_{k=1}^{2} X_{jk}; \quad j = 1, \ldots, 3$$

$X_{jk} \geq 0, X'_k \geq 0; \quad \forall_j \forall_k$

$\theta \in [0, 1]$

$\min \theta$

$$\sum_{k=1}^{2} \left(\sum_{j=1}^{3} \bar{C}_{jk} X_{jk} \right) + \underline{C}'_k X'_k) \leq \underline{r}_o + \theta \underline{t}_o$$

$$\hat{A}_j . X_j = \bar{r}_j + \theta \bar{t}_j \quad j = 1, \ldots, 3$$

$$\sum_{j \in J^k} \tilde{A}_j (X_{jk} + X'_k) = 0; \quad k = 1, \ldots, 2$$

$$\sum_j E_j . X^i_j \leq \bar{b}_i + \theta \bar{t}'_i; \quad i = 1, \ldots, 4$$

$$\sum_{j \in J^k} (\underline{T}_j \delta_j X_{jk}) + T' X'_k \leq \bar{f}_k + \theta \bar{t}''_k$$

$$X_j = \sum_{k=1}^{2} X_{jk}; \quad j = 1, \ldots, 3$$

$X_{jk} \geq 0, X'_k \geq 0; \quad \forall_j \forall_k$

$\theta \in [0, 1]$.

Since these are conventional linear programming models, they can be solved using standard methods to provide the following solutions, respectively:

$X_{11} = (441.50, 441.50, 0, 0)$

$X_{21} = (0, 0, 291.5, 291.5)$

$X_{22} = (0, 0, 0, 0)$

$X_{32} = (0, 0, 375, 0)$

$X'_1 = (0, 0, 75, 75)$

$X'_2 = (187.5, 0, 0, 0)$ cost: 4,394.5

$X_{11} = (281.25, 281.25, 0, 0)$

$X_{21} = (0, 0, 281.25, 281.25)$

$X_{22} = (0, 0, 0, 0)$

$X_{32} = (0, 0, 270, 0)$

$X'_1 = (0, 0, 0, 0)$

$X'_2 = (135, 0, 0, 0)$ cost: 4,027.5

$$X_{11} = (550, 550, 0, 0)$$
$$X_{21} = (0, 0, 320.85, 320.85)$$
$$X_{22} = (0, 0, 412.5, 412.5)$$
$$X_{32} = (0, 0, 0, 0)$$
$$X_1' = (0, 0, 114.57, 114.57)$$
$$X_2' = (206.25, 206.25, 0, 0) \qquad \text{cost: } 4{,}491.7$$

In all cases, $\theta = 0$, which means that the decision maker can evaluate and compare the effects of problem uncertainty with the expected result provide by the "precise" case and select the most appropriate solution for the transportation planning problem.

References

Al-Khalili, A. 1985. Urban traffic control—A general approach. *IEEE Trans. on Systems, Man, and Cybernetics* SMC-15 (2):260–71.

Barney, G., and S. Santos. 1985. *Elevator Traffic Analysis, Design and Control*, IEE Control Engineering Series 2. London: Peter Peregrinus.

Bose, B. 1986. Power Electronic A.C. Drives. Englewood Cliffs, NJ: Prentice Hall.

Caminhas, W., and H. Tavares. 1994. Telecommunication network planning: central offices clustering using neural nets (in Portuguese). *Proc. First Brazilian Congress on Neural Networks*, Itajubá, Brazil, 364–68.

Caminhas, W., H. Tavares, and P. Costa Jr. 1995. Fuzzy control of induction motor drives. *Proc. Sixth IFSA World Congress*, São Paulo, Brazil, vol. 1, 565–68.

Caminhas, W., H. Tavares, and F. Gomide. 1995. Competitive learning of fuzzy, logical neural networks. *Proc. Sixth International Fuzzy Systems Association World Congress*, São Paulo, Brazil, vol. 2, 639–42.

Caminhas, W., H. Tavares, and F. Gomide. 1996a. *A Self-Organizing Clustering Procedure for Communication Network Planning*. Internal Report DCA/FEEC/UNICAMP, Campinas, Brasil.

Caminhas, W., H. Tavares, and F. Gomide. 1996b. A neurofuzzy approach for fault diagnosis of dynamic systems. *Proc. IEEE International Conference on Fuzzy Systems*, New Orleans, LA, 2032–37.

Consoli, A., E. Gerruto, A. Raciti, and A. Testa. 1994 Adaptive vector control of induction motor drives based on a neuro-fuzzy approach. *Proc. IEEE Power Electronic Specialist Conference*, Taipei, Taiwan, 225–32.

Favilla, J., A. Machion, F. Gomide, and R. Gudwin. 1992. Adaptive fuzzy logic–based urban traffic control. *Proc. Third Annual Symposium of I.A.K.E.*, Washington, DC, 570–77.

Favilla, J., A. Machion, and F. Gomide. 1993. Fuzzy traffic control: adaptive strategies. *Proc. IEEE International Conference on Fuzzy Systems*, San Francisco, CA, 506–11.

Figueiredo, M., W. Caminhas, H. Tavares, and F. Gomide. 1996. *Self-Organizing Fuzzy Clustering In Communication Network Planning*. RT-DCA Internal Report, Unicamp, Campinas, São Paulo, Brazil.

Frank, P. 1990. Fault diagnosis in dynamic systems using analytical and knowledge-based redundancy—A survey and some new results. *Automatica.* 26:459–74.

Gudwin, R., and F. Gomide. 1994. Context adaptation in fuzzy processing. *Proc. Brazil-Japan Joint Symposium on Fuzzy Systems*, Campinas, Brazil, 15–20.

Gudwin, R., F. Gomide, M. Andrade Netto, and M. Magalhães. 1996. Knowledge processing in control systems. *IEEE Trans. on Knowledge and Data Engineering* 8 (1):106–19.

Hillier, F., and G. Lieberman. 1990. *Introduction to Operations Research.* New York: McGraw-Hill.

Himmelblau, D. 1978. *Fault Detection and Diagnosis in Chemical and Petrochemical.* Amsterdam: Elsevier.

Johnsonbaugh, R. 1993. *Discrete Mathematics,* New York: Macmillan.

Klincewicz, J. 1991. Heuristics for the p-hub loction problem. *European Journal of Operations Research* 53:25–37.

Kohonen, T. 1990. Improved versions of learning vector quantization. *International Joint Conference on Neural Networks,* San Diego, CA, 545–50.

Machion, A. 1992. Fuzzy traffic control (in Portuguese). Master's thesis, Unicamp, FEEC/DCA, Campinas, São Paulo, Brazil.

Mendes, R., A. Yamakami, and F. Gomide. 1996. *Fuzzy Multicommodity Transportation Problem: Modeling and Application.* Internal Report RT-DT-DCA, Unicamp, Campinas, São Paulo, Brazil.

Mir, S., D. Zinger, and M. Elbuluk. 1994. Fuzzy controller for inverter fed induction machines. *IEEE Transactions on Industry Applications* 30 (1):78–84.

Murphy, J., and F. Turnubull. 1988. *Power Electronic Control of AC Motors.* New York: Pergamon Press.

Nakamiti G., R. Freitas, J. Prado, and F. Gomide. 1994. Fuzzy distributed artificial intelligence systems. *Proc. IEEE International Conference on Fuzzy Systems,* Orlando, FL, 462–67.

Nakamiti G., and F. Gomide. 1996. Incorporating fuzzy sets in intelligent distributed systems. *Proc. Biennial Conference of the North American Fuzzy Information Processing Society,* Berkeley, CA, 433–37.

Nakamiti G., and F. Gomide. 1994. An evolutive fuzzy mechanism based on past experiences. *Proc. Second European Congress on Intelligent Techniques and Soft Computing,* Aachen, Germany, 1211–17.

Nozaki, K., H. Ishibuchi, and H. Tanaka. 1994. Trainable fuzzy classification systems based on fuzzy if-then rules. *Proc. IEEE International Conference on Fuzzy Systems,* Orlando, FL, 498–502.

Pappis, C., and E. Mamdani. 1977. A fuzzy logic controller for a traffic intersection. *IEEE Trans. on Systems, Man, and Cybernetics* vol. SMC-7, 707–17.

Patton, R., P. Frank, and R. Clark. 1989. *Fault Diagnosis in Dynamic Systems: Theory and Applications.* New York: Prentice-Hall.

Robertson, D. 1979. Traffic models and optimum strategies of control: A review. *Proc. International Symposium on Traffic Control Systems,* University of California, vol. 1, pp. 262–88.

Tang, Y., and X. Longya. 1994. Adaptive fuzzy control of a variable speed power generating system with doubly excited reluctance machine. *Proc. IEEE Power Electronic Specialist Conference,* Taipei, Taiwan, 377–84.

Wang, L., and J. Mendel. 1992. Generating fuzzy rules by learning from examples. *IEEE Trans. on Systems, Man, and Cybernetics* 22 (6):1414–27.

Webster, F. 1958. Traffic Signal Settings. Road Research Technical Paper 39, Road Research Laboratory, London.

Willsky, A. 1990. A survey of design methods for failure detection in dynamic systems. *Automatica* 12:601–11.

Wohl, M., and B. Martin. 1967. *Traffic System Analysis for Engineers and Planners.* New York: McGraw-Hill.

Wu, T. 1992. Fiber network service survivability. Norwood, Mass.: Artech House.

Index